Reviews for
The Making of the British Army

'A romantic history of the British Army that stirs the blood.'
Charles Moore, *Daily Telegraph*

'An important book, because it shows how history has not just
shaped the Army, its traditions and its ethos, but also how it has
formed British strategy, for better and for worse.'
Antony Beevor, *The Times*

'Lucid, absorbing . . . Mallinson combines a professional's feel for
his subject with a populist touch.'
Christopher Sylvester, *Daily Express*

'Fascinating . . . clear and concise . . . important. It is hard to see
this book being bettered in the near future.'
Simon Heffer, *Daily Telegraph*

'A compelling history of the British Army.'
Emmanuelle Smith, *FT*

'Thought-provoking and endlessly entertaining.'
Trevor Royal, *The Herald (Scotland)*

'Whether he is unpicking the close stitching of a battle or
lyricising the hero of the book, the common soldier, his touch
and judgment are compelling.'
David Edelsten, *The Field*

'Mallinson is surely right to stress the one enduring quality of the
British Army: "operational resilience".'
Saul David, *Spectator*

'An admirable introduction to an always controversial subject.'
M R D Foot, *Literary Review*

TRANSWORLD PUBLISHERS
61–63 Uxbridge Road, London W5 5SA
A Random House Group Company
www.rbooks.co.uk

THE MAKING OF THE BRITISH ARMY
A BANTAM BOOK: 9780553815405

First published in Great Britain
in 2009 by Bantam Press
an imprint of Transworld Publishers
Bantam edition published 2011

This book is a work of non-fiction.

A CIP catalogue record for this book
is available from the British Library.

Addresse   anies outside the UK
can be found at: www.randomhouse.co.uk
The Random House Group Ltd Reg. No. 954009

The Random House Group Limited supports The Forest Stewardship Council
(FSC), the leading international forest certification organisation. All our titles that
are printed on Greenpeace approved FSC certified paper carry the FSC logo. Our
paper procurement policy can be found at
www.rbooks.co.uk/environment

Typeset in 11/13pt Minion by Falcon Oast Graphic Art Ltd.
Printed in the UK by CPI Cox & Wyman, Reading, RG1 8EX.

2 4 6 8 10 9 7 5 3 1

The Making of the British Army

Allan Mallinson

BANTAM BOOKS

LONDON • TORONTO • SYDNEY • AUCKLAND • JOHANNESBURG

To
SUE
a soldier's wife who followed the drum
and
from German lodgings or Roman villa
made a home

And in gratitude to
Sir John Keegan,
who taught a great many of us their history

Contents

Maps

Illustrations

Introduction

All civilizations owe their origins to the warrior; their cultures nurture the warriors who defend them, and the differences between them will make those of one very different in externals from those of another.

Sir John Keegan, *A History of Warfare*

I came to write this book first and foremost as a soldier of thirty-five years. How and why 'my' army had become what it was – extraordinarily capable in spite of its small size – occupied me more and more. A soldier lives daily with his heritage – the uniforms, the pictures on the walls, the names of things, how people talk, what they do, and how they do it; never more so than on operations. And when you live with your history day in, day out, for thirty-five years you begin to see it in a different way from what is sometimes written in the history books, for certain things gain in significance, while others become mere 'noise'. So I began 'jobbing back' into history to try to understand what it was in our past that made us tick today. And when I made a link in the past I found that it immediately demanded I reach back even further, for there was never the equivalent of a military 'big bang' when suddenly the elements of the army were created. Indeed, even in the mists of the Dark Ages there are things that resonate in the modern army. Certainly I cannot believe that, say, Alfred's victory over the Danes at Ethandune on the downs overlooking the present-day battle-training area of Salisbury Plain is entirely unconnected with the spirit of that training.

So where to begin the story of the making of the British army? The Anglo-Saxon *fyrd* was a levy of free farmers assembled for a definite and short period, for the crops had to be sown and the harvest gathered. The Vikings with whom they battled were no better trained, however. The Normans were altogether better organized, but a single Roman legion would have routed William at Hastings, as it would any of the medieval peasant armies that fought in the Wars of the Roses, or the innumerable clan wars north of the border. Henry V's magnificent longbowmen would have found the Roman *testudo* hard to crack. And even gunpowder would not have dismayed a legion too much before the end of the sixteenth century, for muskets and artillery were then crude and cumbersome. The point was that Roman soldiers were professionals: the legions trained fulltime. For twelve hundred years and more, neither England nor Scotland had a standing army; and without a standing army there could be no continuity. So a study of *the* British army is therefore best begun at the point from which there is unbroken continuity: the seventeenth century, at the restoration of Charles II to the thrones of England and Scotland, when events at last forced a standing army on the nation.

The next question is what to leave out, which almost every soldier I have spoken to about the book has asked. In a way everything matters, and the army has seen so much action. Conversely, however, it has never been monolithic: what happened in one part of the army, in one regiment perhaps, dramatic though it may have been, did not always change the overall picture. What matters are those people and events in the past that have made the army what it is today – an army emerging from a long and bruising campaign in Iraq, and fighting another in Afghanistan. I am not writing *the* history of the army; I am trying to explain its present in terms of its past.

There are of course different ways of doing this, but I felt that a continuous narrative would give the reader the best sense of the sweep of time, and therefore of the extraordinary way in which the army has advanced and retreated in size and efficiency over the years. It also puts into context some of the heroic actions that have imprinted themselves indelibly on the army's collective mind. And since any history of the British army is a part of the nation's story, it seems best to describe it from the perspective of Britain's history, which is not to say solely from the perspective of the history of these islands: try writing the history of the world without the British army. Of the 192 member states of the United Nations, the army has fought in or with well over half of these states or their predecessor polities.

Today only the United States and Britain (and possibly France) are capable of mounting independent operations of any scale overseas. This is not solely because of their armies' capabilities: overseas campaigns are joint operations, involving army, navy and air force. Nor is it just a matter of size: there are many larger armed forces than Britain's. The Indian army alone has five times the number of troops in Britain's three services put together (and as many again in the reserve). Even the Japanese army, or 'Ground Self-Defense Force' as constitutionally it has to be called, has some 40,000 more troops than the British. It is the range and balance of capabilities within a nation's armed forces that determine whether or not they are able to mount an independent campaign far from home. A force must be able to acquire intelligence on the enemy; manœuvre against him; bring fire support to bear (from land, sea and air); protect against the enemy's own fire; and sustain itself. And when all these capabilities are lined up, a further capability – to command them – becomes key, for as Montgomery's American counterpart in Normandy, Omar Bradley, said, 'Congress can make a general, but only communications can make him a commander.'

But 'Congress and communications' alone do not make a good general. 'Generalship' is a separate ingredient. When I was military attaché in Rome, an impressive Alpini brigadier once asked me who were the ten greatest British generals. I thought for a moment, then answered that there wouldn't be much debate about the top five, but for the rest . . . He stopped me, and with a sigh said: 'The point is, we don't have a single one. How do you think that makes an Italian officer feel?'

Generalship does indeed breed generalship. Sir David Richards, who in October 2010 became the new chief of the defence staff, led NATO forces in Afghanistan in 2006 during the Taleban offensive which looked as if it might succeed; giving evidence to the House of Commons defence committee afterwards, he described how he would retire to his office and 'have a conversation with Slim and Templer'. Field Marshals Slim and Templer had died in 1970 and 1979 respectively. They had both, however, been singularly successful fighting off the back foot east of Suez, and by extension had things to say about taking the fight to the enemy in Afghanistan today. When good generalship is a tradition, it becomes sustaining. Indeed, the same is true all the way down – to individual regiments, to the serjeants' mess, and to the most junior soldier. When asked why the Rifle Brigade had fought so well in the defence of Calais in 1940, their commanding officer replied simply, 'The regiment had always fought well, and we were among friends.'

The army's experience of mastering these various capabilities, the 'functions in combat' (manœuvre, fire, communications, logistics, etc.), and generalship is an important part of this book. But *The Making of the British Army* is first and foremost a human story, for the old adage has much truth: navies and air forces are about manning equipment, whereas armies are about equipping men. And this is why military operations, as opposed to air or naval, are less 'scientific', for the human factor is an elusive one. Other armies have fielded technically more able men than the British, men more ideologically driven, more combative (and certainly more brutal): for example, Max Hastings in his 1984 book *Overlord* argues that man-for-man the Germans were by far the best of the six armies in Normandy forty years earlier. But no army has so consistently fielded such all-round good soldiers – *regular* soldiers, certainly – as the British. Raymond Seitz, the former US ambassador in London, an anglophile but not an over-sentimental one, observes in his memoir *Over Here*: 'I know nothing kinder than an English nurse nor braver than a British soldier.'

There is another element in the making of the British army that is common to all armies, and that is the unique nature of war itself, and consequently of soldiering. The centurion in St Matthew's Gospel who says, 'For I am a man under authority, having soldiers under me: and I say to this man, Go, and he goeth; and to another, Come, and he cometh,' is saying that he is a man apart. Shakespeare's vision of the soldier in *As You Like It* is also of a man apart – 'Full of strange oaths ... Jealous in honour, sudden and quick in quarrel, seeking the bubble reputation even in the cannon's mouth.'

The relationship between the warrior and society, between war and civilization, is intimately and personally explored in a little book called *Fusilier* by the late Reverend

Professor John McManners, sometime Regius Professor of Ecclesiastical History at Oxford, fellow and chaplain of All Souls, and alumnus of the theological college in which I was studying before I decided to join the army. In 1945 Jack McManners was a temporary major in the Royal Northumberland Fusiliers, one of the toughest infantry regiments the army has ever mustered. He had joined straight from Oxford in 1939 with a first in history, and was soon in action with the grenade and the bayonet in North Africa. Indeed, he was in action more or less continuously until the end of the war. In 2002 he turned his lifetime's 'recollections and reflections' on the war into *Fusilier*, in which he writes:

> People tell academics and clergy to look at what the 'real world' is like. By this they mean dictating letters, selling and buying shares, instituting manufacturing processes, tapping information into computers. But behind their world is the real world they have forgotten: the battlefield. Here is the ultimate reason of the social order written in letters of lead and shards of steel.

The Making of the British Army is about the battlefield, the place where, ultimately, the peace of 'the social order' is decided: and it is indeed written in letters of lead and shards of steel.

1

The First Dry Rattle of New-drawn Steel

Edgehill, 23 October 1642

ROBERT BERTIE, EARL OF LINDSEY, LAY IN AGONY ON A PILE OF straw while his son, Lord Willoughby de Eresby, tried to staunch the flow of blood. Veteran of many a battle on the Continent, the earl was now a year short of sixty, bald and gaunt, but keen-eyed still. The musket ball was lodged deep in his thigh. Seeing his father fall Lord Willoughby had rushed to his side, only to be taken prisoner with him. It was evening, pitch dark and bitter cold. Outside the dimly lit barn which served as a Parliamentarian dressing station, 4,000 men lay dead or wounded on the gentle Warwickshire hillside near the little village of Radway: 'The field was covered with the dead,' wrote one who survived; 'yet no one could tell to what party they belonged.' The cold was a blessing, some were saying: it would make the blood congeal, save them from bleeding to death. But Lord Willoughby could do nothing to stem the haemorrhage, and he doubly despaired that it should have come to this: the noble earl of Lindsey, who had begun the morning as the King's general-in-chief, felled in the mêlée by a common musket!

There had not been a pitched battle on English soil for 130 years.* There had not been much of a battle anywhere for an English army in all that time. There *was* no English army. When it came to pushing the Scots back across the border, or putting down the Irish, as occasionally it did, the King would drum up a scratch force, engage officers who had gained a bit of experience with one of the continental armies, hire foreign mercenaries (Italian cavalry had fought against the Scots at Flodden in 1513) – and, when the job was done, quickly pay them off again. Standing armies were expensive. When war with France or Spain threatened, it was the navy to which the nation looked for the safeguard of the realm. Britain was the 'sceptred isle', and doubly blessed by her geography: only Denmark and the Kingdom of Naples had so favourably short a land border with their nearest neighbour as that between England and Scotland. Most of the inhabitants of the British Isles never saw a musket, let alone carried one.

And so in the opening moves of the Civil War, King Charles I had mustered a scratch army, derisively called Cavaliers, to do battle with Parliament's scratch army, derisively called Roundheads, on a bright October morning in the green and pleasant English countryside between Stratford-upon-Avon and Banbury. The cores of both armies were the 'trained bands', the county militias under the lords lieutenant. But trained they scarcely were – certainly not *well*

---

* A pitched battle is an encounter of choice, of major forces, unlike a chance meeting such as a skirmish of advance guards and outposts, or where one side is forced to fight, such as a siege. There had been a heavy skirmish at Newburn in Northumberland in 1640 between Scots and English troops in what was known as the 'Second Bishops' War', a smaller-scale affair at Powick Bridge near Worcester the month before Edgehill, and a nasty but short skirmish at Solway Moss near Carlisle in 1542; but Edgehill was the first true pitched battle on English soil since Flodden in 1513.

trained – except for some of the London bands, for the half-century since the Armada had been years of military decline. 'Arms were the great deficiency,' wrote one Royalist eye witness at Edgehill, 'and the men stood up in the same garments in which they left their native fields.' They stood, indeed – both sides – in the ancient line of battle, as the Greeks and Romans had, the Royalists at the top of the grassy slopes of Edgehill above the Vale of the Red Horse, many 'with scythes, pitchforks, and even sickles in their hands, and literally like reapers descended to the harvest of death'. Without so much as the customary sash to show their allegiance, it was little wonder that when they fell 'no one could tell to what party they belonged'.

The cavalry were not much better found, although the Royalist horse, whose *élan* became synonymous with the very word 'Cavalier', were superior to Parliament's. They were led by Prince Rupert of the Rhine, King Charles's nephew, dashing and ardent, who at only twenty-three had seen more recent service, in the Netherlands and Germany, than any officer in the field that day. Of artillery – 'with which war is made', as Napoleon Bonaparte would famously pronounce a century and a half later – there was pathetically little: just forty-odd guns between the two sides, neither manœuvrable nor able to throw a great weight of shot. It wasn't that the country lacked the industrial base and technological know-how: since Tudor times there had been fifty iron foundries in the Kent and Sussex Weald capable of producing cannon as good as any in Europe, and of late the output of the gunpowder mills had been increasing in both quantity and quality. It was the skill to use the means of modern, continental warfare that was lacking. In Henry VIII's day every able-bodied nobleman had been blooded; now, not one in five had seen a battlefield. England, declared the fury in the Court's Twelfth Night masque of 1640, the last of Charles I's reign, was 'overgrown with peace'.

And so 28,000 men would do battle at Edgehill, with their

officers scarcely knowing what they were about. Few on either side had any illusions about their situation, however. Sir Edmund Verney, the royal standard-bearer, confided to his son in a letter the night before: 'Our men are very raw, our victuals scarce and provisions for horses worse. I daresay there was never so raw, so unskillful and so unwilling an army brought to fight.' He did not live to receive a reply. The same was true of the Parliamentarian army, although a certain religious zeal enlivened the ranks, like the rum ration of later wars.

Matters were made worse on the Royalist side by disputes among the senior officers. As King Charles's serjeant-major-general,* Jacob Astley, began his duty of forming up the infantry, a row broke out over what that formation should be. Astley, who had seen service with both the Dutch and Swedish armies, favoured the Swedish model of three ranks.† But his general-in-chief, Robert Bertie, the earl of Lindsey, favoured the Dutch model in which the infantry stood five ranks deep at least – a formation that was not able to cover as much ground but which was more solid and easier to control, especially with inexperienced troops. And Lindsey wanted also to keep the cavalry in close support, for the Parliamentary commander, the earl of Essex, had fought alongside the Dutch too; and Lindsey fancied therefore that he knew how Essex would fight.

Prince Rupert disagreed. Serjeant-Major-General Astley had once been his tutor, and so he, too, favoured the Swedish model – not least in using the cavalry independently of the infantry. As the lieutenant-general Rupert was not just in command of the cavalry, he was second-in-command of the army; and since King Charles himself was at its head, he

---

* 'Serjeant-major-general' later became simply 'major-general'. 'Serjeant' is now usually spelt with a *g*, although some regiments, notably The Rifles, retain the *j* spelling.

† In a 'rank', sometimes loosely called a line, soldiers stand side by side; in a 'file' they stand one in front of the other.

would answer only to the King. When Charles deferred to his nephew, Lindsey resigned his empty command and took his place instead at the head of the regiment he had raised in his native Lincolnshire. The earl of Forth, whose service had been with the great Swedish soldier-king Gustavus Adolphus, assumed the appointment, and the 'Swedish' troop dispositions were made.

There are better ways to begin a battle than with squabbling among senior officers and making infantrymen change their dispositions and then change back again. But to fight a battle without a common understanding of tactics is asking for trouble, then as now. It was what happened when armies were brought together only on the eve of battle, and when officers received their training – some of it by no means up to date – in very different schools. These things were only avoided by having a professional, standing army. Did Charles wish for such an army, now, as he faced the earl of Essex's men? Perhaps. But what if the standing army had sided with Parliament instead? After all, Britain was an island, the Scots were manageable and the Irish, for all their intractability, did not threaten the peace of England. Best leave professional armies to the continental powers, for see how they had fuelled a war of religion across Europe for the past twenty-five years!*

It took time to draw up 15,000 men in line, all but a couple of thousand of them on foot, especially the semi-feudal companies of countrymen with scythes, pitchforks and sickles. The trained bands, if not drilled as well as once they had been, were at least uniformly equipped with the matchlock musket and the pike – in a 3:1 ratio of pikemen to 'the shot'. In some of the more poorly drilled county militias the proportion of pikemen was greater, for the

---

* The Thirty Years War had ravaged the Continent – especially the north – since 1618, and would continue for another six years after Edgehill.

A few of the dozens of 'words of command' needed to get the musketeers of both Royalist and Parliamentary armies to handle loose gunpowder safely and to fire volleys.

matchlock was an unwieldy weapon, the barrel 4 feet long, so heavy, and so violent in its recoil, that it had to be fired from a rest driven into the ground. Inaccurate even over its short range, it was a crude device – in essence a steel pipe sealed at one end, with a thin bore-hole through to a 'pan' in which an initiating charge of powder was sprinkled and then sparked by a smouldering twist of rope (the slow match, hence 'matchlock') clamped in a trigger-operated lever. This fired the main charge – powder which had been poured down the barrel, with a lead or iron ball dropped in after it, all tamped tight by a ramrod. It was prone to misfire, for in rain the powder got damp and the match could go out. But with loose powder and glowing matches in close proximity, the risk of premature – and catastrophic – explosion was an even greater concern, and loading therefore proceeded at the pace of the slowest musketeer, to words of command more akin to a health and safety notice than to battlefield orders:

Take up your Match;
Handle your Musket;
Order your Musket;
Give Rest to your Musket;
Open your Pan;
Clear your Pan;
Prime your Pan;
Shut your Pan;
Cast off your loose Powder;
Blow off your Powder;
Cast about your Musket;
Trail your Rest;
Open your charge;
Charge with powder;
Charge with bullet;
Draw forth your Scouring Stick;

Shorten your Scouring Stick;
Ram Home;
Withdraw your Scouring Stick;
Shorten your Scouring Stick;
Return your Scouring Stick;
Recover your Musket;
Poise your Musket;
Give rest to your Musket;
Draw forth your Match;
Blow your coal;
Cock your Match;
Try your Match;
Guard and Blow;
Open your pan;
Present;
Give Fire!

With such deliberate drill the rate of fire was glacial, even with alternate ranks firing and reloading. It was fatal, not just ineffective, to discharge at too great a range, for if the fire fell short the enemy's musketeers and pikemen would be able to close with them before another volley (fire by the entire line) could be got off. And if it were cavalry advancing against the line there was scarcely time to get off a volley at all before the pikemen needed to take post in front. At Edgehill the pike they carried was 16 feet long, making the line even more unwieldy to manœuvre, for the pikemen had to wedge the butts in the ground and brace themselves to make a solid wall of steel against cavalry or the enemy's pikes. Little wonder, then, that even in the best-trained bands there were three of them to every musket.

At Rupert's urging, King Charles placed his cavalry – the 'horse' proper as well as the dragoons (who fought dis-mounted with sword and musket rather than from the saddle)* – on either end of the line to prevent his flanks from being turned, and to allow freedom of movement when the moment came to charge. And Rupert, in command of the stronger right wing of the cavalry, would be looking for just that opportunity, for in many a battle on the Continent he had seen the enemy's line scattered by a well-timed charge.

Opposite Rupert's wing, three-quarters of a mile or so down the hill and beyond a hedge, the left wing of the Parliamentary cavalry was well supported by musketeers and cannon. Indeed, Parliament's line, comprising three 'tertias' (brigades) of infantry, outnumbered the Royalists by 3,000 musketeers and pikemen; but this margin was less than the earl of Essex had hoped for (there were many stragglers behind him still after his rapid march from Worcester), and he therefore moved two cavalry regiments from his right to behind the infantry, leaving just one regiment of horse supported by dragoons and musketeers on that flank.

But this, to Essex's mind, did not matter, for Parliament was not going to attack first. After all, Charles had the advantage of the slope, and Rupert had a reputation for dash. Essex was not going to risk his infantry to the shock action of a cavalry charge as they advanced uphill. And so the morning passed with little but mutual jeering and a desultory and ineffective exchange of artillery. The battle-field was still a quietish sort of place until the lines came to close quarters; a man might say his prayers or play a game of cards until the moment came.

In the early afternoon Astley knelt down and in the hearing of all prayed: 'Oh Lord, Thou knowest how busy I must be this day. If I forget Thee, do not Thou forget me.' Then he

---

* A dragoon's horse was meant simply to be the means of moving quickly from one battle position to another. Dragoons would then have to be detailed as horse-holders, usually one dragoon to three horses, making them somewhat uneconomical.

rose, and with a 'Forward, boys!' led the Royalist line in a steady march down the hill. Half a mile on they halted and the cannon on both sides opened fire; but the smoke and noise was greater than the harm, and the guns soon fell silent again.

Essex, though dismayed by the passing of the day to no effect, and not least by Astley's half-advance, was not going to be tempted into attacking. But neither was he going to wait idly on Charles's whim. He decided to send dragoons to probe the Royalist right, following them up with horse and a few of the supporting musketeers from his left flank. It was about three o'clock, the sun was already low in the sky, and the Royalist right had little difficulty seeing them off. It was not exactly the opportunity Rupert had hoped for, but at this hour it was his best chance. He gave the order, and both wings of his cavalry began to advance, the plumed host surging forward at first in an amiable trot, for all the world like gentlemen taking their sport.

There was a tactic much favoured by Spanish cavalry, the *caracole*, in which successive lines of horse would canter elegantly up to the enemy line, wheel to the left and discharge their pistols. But Rupert was having none of this: he would have his men go at a gallop, firing as they collided with the enemy horse. Then, seizing sword from scabbard, they would forge a path through the mass of horse by sheer momentum. It was how Gustavus Adolphus's Swedes had borne down on so many of their German opponents. And there was only one way to deal with it – a counter-charge.*

Essex's cavalry, trained (in so far as they were trained) after the Dutch model, awaited the attack with pistol and carbine rather than preparing for a counter-charge. But

---

* Gustavus Adolphus was reckoned to be the finest soldier of his age, 'the first modern general', but it was charging thus at the battle of Lutzen that he had been killed ten years before Edgehill.

Rupert's cavalry, confident men on powerful horses, with the hill giving impetus to their advance, were a terrifying sight to men who might ride to church of a Sunday, or drive a plough in spring and autumn, but had never heard the thunder of so many unfriendly hooves. They fired an ineffective volley, turned and fled the field.

Rupert's men spurred after them, quickly overrunning the cannon and muskets on both flanks of the Parliamentary line. Without cavalry to cover them, the line might indeed have been rolled up from end to end, but the Cavaliers, high on the thrill of the chase, instead galloped on in pursuit of the fleeing Roundheads until, some miles on, they came upon their baggage train, where in time-honoured fashion they fell out to loot. A century and a half later the duke of Wellington would still be complaining about the cavalry's 'habit of galloping at everything'.

Seeing the collapse of the Parliamentarian flanks, the Royalist infantry now advanced boldly. But in the centre of the Parliamentarian line two brigades had stood firm, and with no Royalist cavalry in sight to oppose them, Essex counter-attacked with the two regiments of cavalry he had posted behind these stalwarts in the centre.

The situation suddenly looked dangerous for the King's side, for there was no mounted reserve, Charles having allowed his Life Guard to join Rupert's charge. But 'the foot soldiers stood their ground with great courage,' as one chronicler wrote, 'and though many of the King's soldiers were unarmed and had only cudgels, they kept their ranks, and took up the arms which their slaughtered neighbours left to them'.*

In the ensuing 'push of pikes', a cosy term to describe

---

* Edward Hyde, Earl of Clarendon, *History of the Rebellion and Civil Wars in England* (begun after 1649, published in successive volumes after the Restoration).

brutish hand-to-hand fighting, the Parliamentarians were just too strong, and at length the Royalist centre gave way. Indeed for a time it looked as if Charles would have to concede, but both sides had been badly shaken by their crude initiation to battle, and were rapidly exhausted by the close fighting. At the last minute some of Rupert's men came cantering back to put heart into the Royalists, and the earl of Essex prudently broke off battle. Neither side had achieved a decisive advantage. It was the dead and dying who were left in possession of the field.

It had been during this last desperate push of pikes that the earl of Lindsey was shot through the thigh. The veteran of the Prince of Orange's service, who in vain had advocated the more compact battle line of the Dutch infantry (just as he had used Dutch engineers to drain the Lincolnshire fens to which he gave his name), now cursed in the bitter cold of the night. Why had the King not heeded his counsel, taking instead that of a 23-year-old thruster? 'If it please God I should survive,' he declared to his son, 'I never will fight in the same field with boys again!'

But it did not please God. Just before midnight the earl of Lindsey, like so many of his Lincoln regiment, joined the 'harvest of death'.

Others on the Royalist side were soon thinking the same as their late general-in-chief, if not directly blaming Rupert then recognizing that Edgehill was not the way to make war. And those who had recently seen service abroad also knew that a continental army would have swept them from the field. It was well that Britain was an island, and the navy capable – and that the Parliamentarian army was no better found than they.

Parliamentarian officers were thinking along the same lines, too. Oliver Cromwell, MP for Huntingdon and a captain of horse, had arrived too late on the field at Edgehill to see action, but he had been able to see well enough what had happened. He wrote at once to his cousin John Hampden, one of the Parliamentary leaders:

Your troopers are most of them old decayed servingmen and tapsters; and their [the Royalists'] troopers are gentlemen's sons, younger sons and persons of quality; do you think that the spirits of such base and mean fellows will ever be able to encounter gentlemen who have honour and courage and resolution in them? You must get men of a spirit that is likely to go on as far as gentlemen will go, or else I am sure you will be beaten still.

Cromwell may have been a puritan, but he was a puritan gentleman. However, he also recognized the weakness of the Royalist cavalry. Their lack of discipline had let slip a thorough victory at Edgehill, and he sensed it would not be the last time that Royalist *élan* would turn into unruliness. They could be countered by disciplined troops.

At Edgehill something profound had been, if not born, then certainly conceived:

Thank Heaven! At last the trumpets peal
Before our strength gives way.
For King or for the Commonweal—
No matter which they say,
The first dry rattle of new-drawn steel
Changes the world today!*

The change that Kipling wrote of two and a half centuries later was not merely the overturning of the constitution but the dawning of the realization that war could no longer be

---

* From 'Edgehill Fight' – one of the 'songs' written for C. R. L. Fletcher's *A History of England* by Rudyard Kipling (1911).

made in the old feudal way; that there must be system and discipline, and thus (eventually) a regular, professional army. For although there would be another two years' inconclusive fighting (during which Cromwell would rise to lieutenant-general) before Parliament grasped the nettle and raised an army in which enlisted men received proper training and regular pay, and the officers were selected and promoted on professional merit, Edgehill was the genesis of the 'New Model'.

And when Parliament did at last grasp the nettle it did so resolutely and without too much scruple: puritan ministers might teach the Gospel, but it was the likes of Carlo Fantom, a Croatian (the Croats were famed for their irregular light cavalry), who would teach the sword. 'I care not for your cause,' he boasted: 'I come to fight for your half-crown and your handsome women.'

Fantom was indeed a notable ravisher, and would soon change sides for the promise of more half-crowns. But the Royalists, on that occasion at least, proved to have the greater principles and eventually hanged him – for ravishing.

The New Model Army would not be especially large, however – 22,000 men and 2,300 officers, two-thirds infantry to one-third cavalry (about the number, indeed, of the British infantry today) – but it would be superbly disciplined, equipped and trained. And for the first time a British army would wear a true uniform – red. Cromwell was certain of the type of man he wanted to lead such troops, too: 'I had rather have a plain, russet-coated Captain, that knows what he fights for, and loves what he knows, than what you call a Gentleman and is nothing else.' Out went the officers who had attained their ranks as MPs, and in came those who had proved themselves capable. While theirs was not a vast army, the Parliamentarians believed that professional quality would make up for lack of numbers, although ironically Charles himself was never able to muster many more troops than they.

Curiously, attempting to raise more troops was in fact the only Royalist response to a war that was not going their way. Charles and his generals did little to change their tactics, nor did they develop any sound military strategy. There were some loyal and able supporters, such as the estimable Sir Ralph Hopton who raised a formidable little army in Cornwall and almost captured the earl of Essex. Hopton's *Maxims for the Management of an Army* (1643) would have served the New Model admirably, with the terse injunction to 'pay well, command well, hang well!' But paying well became increasingly difficult for Charles, and commanding well, to his mind, remained synonymous with birthright, while capital punishment was no deterrent to a man who evaded service in the first place.

By the beginning of 1645, as the New Model was being readied for action, the Royalist forces were spread thinly about the country in a patchwork of sieges and counter-sieges, none of which was vital, and none of which promised a decision. The arrival in the field of the New Model could easily tip the scales irrecoverably in Parliament's favour, and Charles's more astute advisers urged him to attack before it was fully formed. But they urged in vain: Charles's want of military strategy, especially the planning of campaigns, was as great as his want of political instinct.* For without operational art war becomes a set of disconnected engagements, relative attrition the only measure of success or failure. Charles conflated sieges with sovereignty: not only were towns and cities the source of the money and arms with which war was made, they were key elements of his realm. The Parliamentarians, on the other hand, were not viscerally

* Campaign planning – what is known today as 'operational art' – involves the setting of military objectives to achieve the strategic aim, and the tactical employment of forces on the battlefield to achieve those objectives.

THE
SOULDIERS
CATECHISME:

Composed for

The Parliaments Army:

Consisting of two Parts : wherein
are chiefly taught :

1 *The Iustification* }
2 *The Qualification* } *of our Souldiers.*

Written for the Incouragement and Instruction of all that have taken up Armes in
this Cause of God and his People; especially the common Souldiers.

---

2 Sam. 10. 12. *Be of good courage, and let us
play the men for our people, and for the Cities of our God, and the Lord do that which
seemeth him good.*

Deut. 23. 9. *When the Host goeth forth against
thine enemies, then keepe thee from every
wicked thing.*

Imprimatur. JA. CRANFORD.

Printed for J. Wright *in the* Old-Baily. 1644

The *vade mecum* issued to every man of the New Model Army.

connected with borough or shire; they were intent only on defeat of the King. Raising the New Model Army was therefore a strategic stroke of huge significance – a move of war-winning potential, for an army that could not be beaten was, self-evidently, able to dictate the course of events in the field, and it was only in the field, now, that the political issue could be settled. Those of Charles's advisers who advocated attacking the New Model before it reached its full effectiveness had grasped this essential strategic fact. Unfortunately for the Royalist cause, Prince Rupert, by now general of the army, had not. He proposed instead to recover the north of England and join forces with the Royalists in Scotland.

As the crow flies, from Edgehill where the war began in 1642 to Naseby where effectively it ended in June 1645 is but 30 miles. And in three years of fighting little appeared to have changed in the design for battle, although at Naseby it was the Parliamentarians who would form up on a ridge, with the infantry in the centre (five large regiments in the front line and three in reserve) under the New Model's admirable serjeant-major-general Sir Philip Skippon. Commissary-General Henry Ireton had command of the left wing of five and a half regiments (predominantly horse), and on the right was his future father-in-law, Cromwell, with six and a half regiments. As at Edgehill the artillery, a mere eleven guns, was 'penny-packeted' in the intervals between the infantry regiments, and just as at Edgehill they would play little part in the battle, the first salvoes going high and the two sides soon too closely engaged for the guns to be safely used. But if the deployment for battle was the same, the soldiers of the New Model were very decidedly not. Properly regimented, well armed, well drilled, well motivated and well led, they would be a match for the *élan* with which the Royalists could undoubtedly still fight. And unlike the King's, the cavalry of the New Model could charge home and then rally quickly to exploit its shock action.

Poor King Charles: he had taken Rupert's advice to rally the Scots and recover the north of England, but in truth all chance of this had been lost – and with it Rupert's reputation as a commander – in July 1644 at Marston Moor outside York. Rupert had only evaded capture, indeed, by hiding in the corn; sadly his dog 'Boy' failed to do likewise, and was captured and summarily shot. At Naseby, Charles must have known that the tide was running ever more strongly against him, and Rupert that the dashing cavalry charge no longer decided matters. And now they faced the New Model for the first time. Nevertheless, morale was by no means low, for they had stormed Leicester a fortnight before (and dealt with the defending garrison brutally), drawing Parliament's troops north from the siege of Oxford, Charles's de facto capital.

If only they had had the high ground. However, it was perhaps as well that the New Model was drawn up *behind* their ridge, for, well trained as they were, many a man had not seen battle, and there was still the touch of majesty in the Cavalier ranks as they formed up below – the drums, the trumpets, the colours streaming, the morning sun glinting on armour. And in the midst of the great, proud panoply rode Charles himself, sword drawn. He would personally command the reserve of infantry (the King's and Prince Rupert's regiments of foot) and his Life Guard of Horse: whatever else might be said of this king, he did not lack courage. Jacob Astley, his splendid serjeant-major-general, who had prayed on his knees before Edgehill, was deftly deploying the battle line, while 2,500 horse under Rupert and his brother Prince Maurice champed and pranced on the right wing facing Ireton's stolid troopers. And over on the left Sir Marmaduke Langdale, high sheriff of Yorkshire, who had come south at the head of 1,500 horsemen known for both hard fighting and in-discipline, was taking unruly post on the flank of the mile-long line opposite Cromwell's grim-faced professionals.

Rupert was intent on attacking, for all that doing so uphill would be no less risky than the Parliamentarians had found it in the opening battle of the war. But with the sharp spur of his ignominy at Marston propelling him, Rupert now led his Cavaliers in a charge straight at Ireton's men, with Astley's infantry beginning their more measured advance behind him. The clash of mounts was violent, and Rupert's first line was checked. Then into the mêlée galloped the second line under the earl of Northampton, and the unexpected happened: Ireton's men broke.

The Royalist cavalry swept on after them, baggage-bound exactly as at Edgehill, not stopping till it reached Northampton 15 miles away, giving Ireton a chance to recover – if not quite as quickly as on the New Model's training grounds of East Anglia. The Royalist advance would now depend on Sir Marmaduke Langdale's men neutralizing the other flank of the New Model's cavalry – the flank commanded by Cromwell.

The Royalist infantry, too, was making progress. As Sir Edward Walker, secretary of the King's war council, later recalled: 'Presently our forces advanced up the hill, the rebels only discharging five [artillery] pieces at them, but [these] overshot them, and so did their musketeers. The foot on either side hardly saw each other until they were within carbine shot, and so only made one volley; ours falling in with sword and butt end of muskets did notable execution.' In this push of pike the Parliamentarian foot began to give way, their commander, Skippon, himself badly wounded by a musket ball in the chest (though he would not quit the field).

And now was the moment when Langdale's cavalry might have tipped the scales. Cromwell knew it, and coolly held his ground rather than shifting to support the wavering infantry, though he could see men in red coats throwing away their arms and making off. Up the slope began the Yorkshire horse. On this flank, however, the ground was broken, and they could hardly get into a gallop, still less keep formation. Cromwell judged the moment perfectly: down the hill he took his 'Ironsides', as soon they would be known from Rupert's lament that they cut through anything, and saw off Langdale's hearties in short order. Unlike Rupert, however, Cromwell had his men in hand: he sent two regiments in pursuit, and with the rest he turned against the Royalist centre while those of Ireton's men who had at last rallied attacked on the other flank.

The Royalist foot fought like tigers. Rupert's regiment of 'Bluecoats' stood their ground to the last, their ensign, who would not yield, killed by the Parliamentarian commander Sir Thomas Fairfax. Charles tried to lead his Life Guard of Horse to the rescue. 'Sire, would you go on your death so easily?' cried the earl of Carnwath, seizing his bridle and forcing him to halt. But the tide of Parliamentarian foot and horse was overwhelming, dragoons now pouring fire into the Royalist flanks. With no help from Rupert's cavalry in sight, those of the infantry that could get away now broke and ran.

The aftermath was bloody and inglorious. Fairfax's troops hunted down the fugitive Royalists and put to the sword even those who surrendered. Coming on the baggage train they hacked to death a hundred camp-followers, believing them to be 'whores and camp sluts that attended that wicked army', or else Irish, or both, though they were in fact simply the innocent distaff side of Charles's Welsh regiments, who paid highly for their inability to protest their virtue in English.

At Naseby, although the New Model had not performed uniformly well, they had been able to rally, and thereby proved the superiority of professional troops. And now there was no time for Charles to make good the deficit. His cause was all but finished, though there would be another three years of bloody, pointless skirmishing before the war in Britain was over.

But when peace came, so did the obvious question: what to do with the New Model? Was a standing army any more lawful, affordable or expedient than it had been before the war? How would Parliament control it?

The place of the army in the state, whether republic or monarchy, remained a fundamental and problematic issue – and by no means, of course, one for these islands alone. In Britain it would not be settled for another half-century; in France, not for two centuries; in Germany, three. The problem was that by 1649, when the execution of King Charles apparently settled the nature of the state, the New Model Army, though its officers were 'professionals' and its rank and file in regular pay, had become thoroughly imbued with puritan zeal, and thus politicized. After 1649 it became the means of imposing a political and religious vision on the civil population – as well as conducting the brutal suppression of Catholic (and therefore Royalist) Ireland, where war dragged on until 1653. The decade that followed the execution of Charles I was to see the nightmare that Englishmen had so far only heard of from across the Channel – martial law, the 'rule of the major-generals'. And the scourge of foreign wars, albeit largely naval ones, returned too.

In fact Cromwell's missionary zeal and the rectitude of his intentions were by 1658 (when he died) so mired in the cruelty and increasing absolutism of his means that many a man formerly sympathetic to Parliament began thinking himself no better off than his fellows on the Continent who had endured the clash of armies in religion's name for thirty years. Country gentry and town merchants alike wanted peace, order, lower taxes – and fewer soldiers. Indeed, the legacy of the Commonwealth was to be a hearty dislike of soldiers and a renewed mistrust of standing armies.

Cromwell's son Richard now succeeded his father as 'Lord Protector', and a power struggle began – among generals, among politicians, among leaders of the religious sects, between Parliament and generals, between generals and the army. Little wonder, therefore, that even those who had fought against the old King were soon looking to a return of the old order – the King in Parliament – which meant the return of a properly elected parliament and, of course, of the King himself.

But to restore the Stuart king would take soldiers, and a man with vision, integrity and grip to lead them. In Scotland commanding the army of occupation was Lieutenant-General George Monck, an old professional (he had seen much action in Dutch service) who had begun the war a Royalist. Perhaps fortuitously he had figured little in the fighting in England, serving in Ireland until 1643 and then in January the following year being taken prisoner in Cheshire. Refusing the offer of his liberty on condition he changed sides, he had spent the next three years in the Tower, until in November 1646, at the end of the first Civil War – the defeat of the Royalists in England – he finally took an oath of allegiance to Parliament, whereupon he was made major-general and commander in Ulster. In 1650 he took part in the invasion of Scotland, and after the shattering defeat of the Scottish Royalists at Dunbar Cromwell promoted him lieutenant-general and commander-in-chief north of the Tweed. Over the next eight years he had earned a name for firm but fair government and loyalty to his troops – not least over their pay, which was always heavily in arrears. He had not benefited personally from confiscated Royalist or Church assets, and so had no financial stake in Cromwell's Protectorate. Above all – at least in the eyes of the exiled son of Charles I, the king-in-waiting – he was not a regicide: he had not signed Charles I's death warrant, nor even been a member of the High Court of Justice which condemned him. He was known to all, indeed, as 'honest George Monck', and to some as no more than a simple soldier, almost a bumpkin, an image given force by his rich Devon accent and enormous bulk. He had nothing to fear from the return of the King.

By the summer of 1659 Charles II was trying to make contact with Monck through the general's brother, a clergyman in the most Royalist of counties, Cornwall, and through Viscount Fauconberg, a grandee of Royalist-inclined Yorkshire.* Charles offered titles and baubles, but the general would not yet commit himself, publicly at least. And so when in the depth of winter honest George Monck gathered his troops at Coldstream, a tiny 'border toon' whose name but for this assembly few would know, he had confided his intentions to no one. Some of his men may have had thoughts of their own – Monck for Lord Protector, indeed – but for the most part they were content to follow him in the hope of getting their promised arrears of pay. Before leaving Edinburgh he had paraded and addressed them directly: 'The army in England has broken up the Parliament, out of a restless ambition to govern themselves ... For my part, I think myself obliged, by the duty of my place, to keep the military power in obedience to the civil.' It was indeed a statement of fundamental doctrine. No general since the Restoration has tried to overawe Parliament, let alone break it up. Few, probably, have even thought of it.

On 1 January 1660 Monck crossed the Tweed, the border between his command in Scotland and that of northern England, and began his march on London, just as Julius

---

* After the Restoration the Reverend Nicholas Monck was made provost of Eton and bishop of Hereford, such was the King's gratitude and General Monck's favour.

Caesar had crossed the Rubicon and begun his march on the capital of the Roman republic – *alea iacta est*. Unlike Caesar, however, Monck did not burn his boats, for the Tweed was not quite the legal barrier that the Rubicon had been, and neither was it a Gallic torrent. There was, indeed, a good bridge, although 6,000 men could not cross it expeditiously, and so many of them waded through the icy stream. Was their general for King, or was he for Parliament? No one but Monck truly knew.

Steadily south marched these 'Coldstreamers', and all opposition, real and imagined, melted away before them. 'The frost was great, and the snow greater; and I do not remember that we ever trod upon plain earth from Edinburgh to London,' recalled John Price, Monck's chaplain. But unlike some later epics of winter marching – in Saxony, or in northern Spain – when the British army's discipline faltered, this long haul saw the morale of Monck's Coldstreamers increasing with every mile, their reception in town and village warm, the church bells ringing joyfully. 'They were certainly the bravest, the best disciplined, and the soberest army that had been known in these latter ages: every soldier was able to do the functions of an officer,' wrote Gilbert Burnet, later bishop of Salisbury. Doubtless he exaggerated a little, but perhaps not too greatly.

Far from being a winter's march by which the army was almost broken, indeed, it was a march by which – almost literally – the British army was made.

His Excellency GEORGE MONCK Generall of all the Forces in England Scoland & Ireland &

If not the father of the British Army, then certainly the midwife.

2

The Return of the King

Blackheath, south-east London, 29 May 1660

THE KING'S THIRTIETH BIRTHDAY WAS THEATRE AS GOOD AS ANY monarch of the *ancien régime* could have wished for. Charles Stuart, as still he was officially known, had stepped on to English soil – or, rather shingle – four days earlier for the first time in nearly a decade, and now he rode ceremoniously on to the 'bleak heath', which since Roman times had served as a marching camp, to take possession of England's army. At Dover he had knelt momentarily and thankfully on the beach, to be greeted as he rose by Monck, and from there had made his steady progress via Canterbury. Having knighted the former Roundhead general in the dilapidated cathedral he rode on through cheering crowds in the lanes of Kent to this vast and grassy parade ground where 30,000 troops, with Monck at their head, waited to salute their new sovereign. The sun shone, although the silence must have seemed at least a trifle forbidding to Charles, for most of these men had fought his father in the war. Some might even have connived at his execution.

'You had none of these at Coldstream,' muttered one of Monck's officers as the glittering royal party came on parade. 'But grasshoppers and butterflies never come abroad in frosty weather!'

Charles did indeed cut a fine figure – tall, 'black and very slender-faced', in a doublet of silver cloth, a cloak decorated with gold lace, and a hat with a plume of red feathers. And his brothers the duke of York and duke of Gloucester were hardly less gorgeously arrayed. Three men in their prime, the very image of Cavaliers, and behind them the Life Guard of eighty troopers, 'gentlemen's sons' as Cromwell had dubbed them ruefully – exiles all, and as glad to see their native country again as was the King himself. Paradise lost, and now regained.

Sir George Monck, made captain-general by Parliament after his march from Coldstream, broke silence with the command to 'Take heed and pay attention to what you hear!'

The ranks of the former New Model Army braced as he read out a declaration of loyalty to His Majesty on their behalf.

At the signal, pikemen and musketeers gave loud cheers, raised their hats and their weapons, and shouted, 'God save King Charles the Second!'

Monck silenced them as swiftly with a hand held high. 'Lay down your arms!'

Thirty thousand men in the pay of the Commonwealth bent the knee and laid down musket and pike.

'Retire!'

They turned about, marched a few token paces, halted, and faced front once more.

'To your arms!'

Drums beat as each pikeman and musketeer ran to his mark.

'In the name of King Charles the Second, take up your arms, shoulder your matchlocks and advance your spears!'

Again they bent the knee; and took up musket and pike as soldiers of the King.

If there was a precise moment when the British army was formed, this was it, marked (though not with this purpose in mind) by this highly symbolic act. Yet it goes unheeded today except by the in-pensioners of the Royal Hospital at Chelsea, who celebrate each year the return of the King, parading with sprigs of oak leaves in their hats to commemorate Charles's escape after the battle of Worcester, when he hid in an oak tree, the 'Royal Oak', in the grounds of Boscobel house in Staffordshire. 'Oakapple Day' is founder's day at Chelsea, and the nearest that any in the Queen's uniform come to celebrating the founding of the army itself.*

But Monck's 30,000 were vastly more than the King needed. Indeed, they were vastly more than Parliament (even a Cavalier parliament) was willing to pay for. Honourable Members had intended the return of the King, but also of the *status quo ante* – the position before Charles's father had dismissed Parliament and tried personal rule. They had no wish now to provide the new King with the means to coerce them out of the power they had earlier wrested from the Crown. The army was therefore to be disbanded. And at once.

It would at least, and at last, be paid. Parliament had voted two-thirds of the arrears (which amounted to the impressively precise sum of £835,819 8s 10d† – testimony to the diligence with which the pay serjeants even then did their work), and Charles would find the balance.

But some soldiers there would still have to be: the

---

* The Royal Hospital was founded by royal warrant in 1681. It was built on royal land, and in part with royal money. The Royal Hospital Kilmainham served the same purpose in Dublin.
† About £90 million in today's retail prices, and £1.25 billion in terms of average earnings.

monarch needed close protection, and there were coastal forts to be garrisoned. Two regiments would therefore be retained as well as the garrisons: the King's Life Guard, and Monck's regiment of 'Coldstreamers', which Charles creatively called his 'guards' to circumvent Parliament's objections to a standing force. Otherwise there was to be a return to former practice: if need arose for an army (as all prayed it would not), the militia would provide – the sort of men who had stood at Edgehill 'in the same garments in which they left their native fields; and with scythes, pitchforks, and even sickles in their hands'. That scratch forces had proved no use to either side in the late war was conveniently forgotten. In Scotland, too, there was retrenchment. Now a separate polity again, released from Cromwell's forced integration with the Commonwealth, it retained only Lord George Douglas's Regiment, which had been raised in 1633 for service with the French – Les Royal Écossais – and had served continuously abroad. Other regiments transferred *en bloc* to the Portuguese service for the war of independence from Spain.

Parliament's hopes for better (and cheaper) times soon proved delusory, however. In the autumn and winter several conspiracies came to light. Charles and his commander-in-chief the duke of Albemarle, as Monck was now styled, were therefore able to convince Parliament that a larger body of guards was needed to guarantee law and order. A permanent establishment of four regiments was duly authorized at a cost of £122,407 15s 10d a year, roughly 10 per cent of the annual supply voted to the Crown. And so the army came into official being (though it had never actually disbanded) by Royal Warrant on 26 January 1661, consisting of: the King's Regiment of Horse Guards (later called simply the Life Guards); the King's Regiment of Horse (later the Royal Horse Guards, or 'Blues'), formed from Cromwell's old Life Guard of Horse; the 1st Regiment of Foot Guards (after Waterloo called the Grenadier Guards), formed from Lord Wentworth's

Regiment, which had been raised at Bruges in the Spanish Netherlands as bodyguard to Charles in exile; and Monck's own regiment of Coldstreamers, the Lord General's Regiment of Foot Guards (after Monck's death in 1670 called simply the Coldstream Guards).* These were all to be stationed in London. Around the country in addition were some twenty-eight garrisons of various sizes, none of them large – a sort of nascent Territorial Army – at a further cost of £67,316 15s 6d a year. The small but longstanding garrisons in the West Indies and North America were to be financed largely by local taxes; and the Irish establishment remained entirely separate. To tide things over while these new arrangements were put in place, Lord George Douglas's Regiment of Royal Scots was brought to England from France.

Charles and Monck had thereby laid the foundations of today's regimental system, and on these there would now be steady and continuous building. In 1661 Charles contracted a marriage with the Infanta Catarina (Catherine of Braganza), second surviving daughter of John IV of Portugal. Her dowry included sugar, a deal of plate and jewels, bills of exchange worth £20 million in today's terms, the rights to free trade with Brazil and the East Indies, and the port-colonies of Bombay and Tangier. Bombay was a prospering trading post, needing a mere 400 men to secure it (Charles sold it to the East India Company a few years later), but Tangier was a different prospect. It held a strategic position at the entry to the Mediterranean, constantly menaced by Moors and Barbary pirates, and needed an altogether bigger garrison – the 'Tangier Regiment of Horse' and a regiment of foot which Charles was soon speaking of as 'Our Dearest Consort's, the Queen's Regiment'.

---

* Wentworth's regiment took precedence over Monck's, although it had been formed later, by virtue of its earlier allegiance to the Crown. The Coldstream Guards make their point nevertheless by their regimental motto, *Nulli secundus* ('Second to none').

Still Parliament was content enough, for only the Guards were seen in the streets. The unregimented garrisons were tied to their forts, the foreign garrisons were largely self-financing, and the few regiments such as Douglas's were in effect mercenaries, paid for by the French, the Dutch or the Portuguese, although those in the French service were suspected of being a closet Royalist reserve for nefarious purposes – perhaps for coercing Parliament, or bringing back Catholicism. Events soon began to bring many of these 'foreign' troops home, however. When the Second Anglo-Dutch War broke out in 1664 the English regiments of the Anglo-Dutch Brigade quit Flanders for London and were at once re-formed as a third regiment of foot, called the Holland Regiment; other soldiers were taken into the Admiralty Regiment,* which later became the Royal Marines. A third Dutch War (1672–4), in which England joined the French in a costly attack on the United Provinces,† came to an abrupt end at the insistence of Parliament, alarmed not least by the expense, and consequently the Anglo-French Brigade was dissolved and its men re-enlisted in the regiments at home (Lord George Douglas's Regiment, the Royal Scots, were brought officially on to the English establishment in 1678).‡ The

* Formed in 1664 as the Duke of York and Albany's Maritime Regiment, it consisted of naval infantrymen under the Lord High Admiral rather than the commander-in-chief.

† The exact title was 'The Republic of the Seven United Netherlands', or 'of the Seven United Provinces' – commonly referred to as the United Provinces, or the Dutch Republic, or Holland. Holland was in fact but the dominant province – and the richest.

‡ Eventually the Royal Scots would be given seniority as the 1st Regiment of Foot in the infantry of the line (of battle), and the Queen's Regiment would be the 2nd. The distinction between Guards and line regiments was simply that of purpose: the Guards were 'Household troops', for the personal protection of the monarch, from which they derived a certain cachet; the line was the workaday muster of regiments that did the job of fighting. When the King took to the field, however, the Guards went with him and fought as a sort of ultimate reserve – or so it was in theory, for Charles did not actually go on campaign.

regimental acquittance rolls must have been a nightmare for the pay serjeants.

And then in 1684, the year before his death, Charles at last abandoned his dear consort's troublesome and expensive dowry, Tangier, and so the home establishment gained a further – fourth – regiment of foot (renamed the Duchess of York and Albany's Regiment, later the King's Own) and a third of cavalry – or rather, mounted infantry, for the Tangier Horse were re-designated the Royal Dragoons. In Scotland, too, there were changes. Charles's forces north of the border were always few – never more than 3,000 – but there was a notable birth, a regiment of dragoons known later as the Scots Greys, and known better still today, musically at least, as the Royal Scots Dragoon Guards.*

In Ireland, unusually, there was peace, except that, being Ireland, it was a peace with constant breaches requiring an active constabulary. At the Restoration the Cromwellian garrison there had immediately declared for the King and quietly acquiesced in a severe reduction that left about 7,500 men in tiny, scattered posts, unregimented till 1672. It was a largely Protestant army, neglected, in arrears of pay, and miserable in their alien barracks. Catholics engaged instead for foreign service, later coming to be known as the 'Wild Geese', though occasionally they found themselves serving under unexpected orders – as, for example, when the Third Dutch War ended and Viscount Clare's regiment of infantry transferred from the Stadtholder's service to the English establishment as the 5th Foot.

By the time of Charles's death, therefore, in the curious, haphazard way that the regimental system continues to

---

* The Scots Greys were formed in 1681 from three separate troops by Lieutenant-General Tam Dalyell, whose namesake and descendant, the Labour MP and author of the intractable 'West Lothian Question', was a National Serviceman in the Greys in the 1950s.

develop even today, there had grown up an English standing force (army is perhaps too grand a description) that had no common system of organization or drill, but plenty of fighting experience. And varied fighting at that, from 'continental warfare' (sieges, counter-sieges and field battles) learned under the great French marshal Turenne and the impressive Dutch general Schomberg to the wholly irregular warfare of Tangier. England was becoming accustomed to the fact of a standing army 'in time of peace', even if in law it did not exist.

Parliament had its qualms, however. The regiments were officially part of the royal household, raised by royal warrant; Charles had always been careful to refer to them as his 'guards and garrisons', as indeed had Parliament. But there were practical problems arising from this de facto existence which could not be ignored for much longer, as the number of soldiers just continued to grow. For the army's internal discipline on active service abroad was regulated by the 'Articles of War', first issued in 1663, but these had no legal force before the English courts. They were, so to speak, the rules of a private club. This did not matter much in the case of relatively minor military offences, where some suitable punishment such as stoppage of pay or fatigue duties could be imposed; but the serious cases, of which desertion was the most prevalent, had to be represented as felonies before a civil court – and the wary courts were reluctant to become involved in military discipline. Parliament wished it could continue to turn a blind eye to the operation of military justice, for to regularize it would be to recognize in law the existence of an army, but with the numbers of troops constantly growing it was increasingly difficult to do so.

Not that Parliament wanted to see the back of them altogether, for soldiers were proving useful to the common good. Cavalrymen patrolled the roads to apprehend highwaymen and footpads; they escorted valuable merchant convoys and searched for contraband. And in the days before the constabulary, foot soldiers guarding the royal palaces and the theatres were a welcome presence of law and order on the streets of London. When from time to time the mob erupted into violence towards property, they assisted the magistrates in putting down the riot, while in other more peaceful ways the soldiery mitigated their unpalatable image – building roads, repairing bridges, fighting fires and the like. Charles personally supervised the efforts to put out the Great Fire of London in 1666, using his guards to pull down buildings to create firebreaks. But Parliament knew it must soon address the existential question: who 'owned' the army? Events rather than debate would force the issue.

Although Charles himself sired many children, poor Queen Catherine produced no heir, so that when the King died in February 1685 his brother James, duke of York, though a Catholic, ascended the throne. He was immediately challenged by one of his Protestant but illegitimate nephews, his 36-year-old namesake James, duke of Monmouth and Buccleuch.

Monmouth had made a name for himself as a military commander: in Flanders he had been brave if erratic, and after Monck's death in 1670 he had been made commander-in-chief for the Third Dutch War. Charles, however, had steadfastly refused to legitimize him. Nevertheless, Monmouth believed he had both the rightful claim and, critically, the support of the Whigs in Parliament, although it was probably the encouragement of the earl of Argyll, who intended declaring for him in Scotland, that tempted him to act. From the Netherlands, where he had lived in exile since the 'Rye House Plot',* in May 1685 Monmouth sailed

---

* Named after the house in Hertfordshire where the plotters (it has never been precisely established who they all were, and how complete were their plans) intended to assassinate Charles and his brother on his way back from the races at Newmarket.

challengingly for England, but with an 'army' of a mere hundred or so, evidently putting his trust in that persistent but false principle of war, *hope* – in this case that the men of the West Country would rise in his support. Indeed, his campaign plan had the merit of simplicity: like a ball of snow rolling down a long hill, his army would grow in size with its own momentum, and sweep all before it in the march on London.

Monmouth's three ships were, for a reason never adequately explained, unmolested by the Royal Navy in their leisurely passage through the English Channel and along the south coast. He landed at Lyme Regis on 11 June, and at Taunton ten days later was proclaimed king – truly a provincial folly if ever there was one. Local men, by all accounts, 'flocked to his support'; it was an apt verb, since they would soon be lambs to the slaughter. First, however, the reckless independence that has frequently characterized the county of Somerset worked to his advantage. There was indeed a 'snowball effect': 8,000 men had soon joined his march – but towards Bristol, not London. This diversion proved costly, for a troop of Life Guards, hastily dispatched from Whitehall, were able to turn them back from the city in pretty short order. Further deflated by the news that the earl of Argyll had been arrested, it was a dispirited and now diminishing host that trudged back into the Somerset Levels, Alfred the Great's famed fastness. But this time the low-lying marshland proved no refuge, and by the first week of July Monmouth lay cornered in the town of Bridgwater, his army having dwindled to perhaps 3,500 pitchforks.

Monmouth had really not chosen his moment well, for with England at peace abroad there were regiments to hand at home, and it was now – if there had ever been any doubting it – that the superiority of disciplined troops told. During the night of 5–6 July, in a desperate throw of the dice, and perhaps to buy time to escape by sea, he decided on a surprise attack against James's forces from the least expected direction – across the soggy wastes of Sedgemoor. Unfortunately for him, this approach was not only un-expected; it proved too difficult for undrilled men to manage, and they lost both direction and order. Inevitably, too, they were detected by the cavalry vedettes (mounted sentries), and James's troops, under command of Lord (John) Churchill, later duke of Marlborough, stood to arms efficiently in the darkness. As dawn revealed the wretched-ness of their situation, the few rebel cavalry fled, leaving the foot soldiers to face the music unsupported. The battle was a short one. In open country and without cavalry support, Monmouth's army was helpless – and was utterly destroyed. With much irony, which the wits would soon be tattling, Monmouth's sheep were butchered by 'Kirke's Lambs', as the former Tangier Regiment was popularly known, after their colonel, Sir Piercy Kirke, and the emblem of Catherine of Braganza – a paschal lamb – emblazoned on their colours.* It is said that soldiers of Kirke's descendant regiment are still banned from the public houses of Bridgwater today.

Although in the end the Monmouth Rebellion was little more than a peasant revolt in the long tradition of hopeless causes, it could hardly have been better staged to demon-strate the need for a standing army, for the militia had proved too slow to muster and too lumpen in manœuvre. In the bloody aftermath of the rebellion, James was able to raise nine new regiments of foot, five of horse and two of

---

* The infantry carried colours (flags), and the cavalry standards or guidons, as a rallying point for troops and to mark the location of the commander. Later they were emblazoned with battle honours and other distinctions to make them recognizable to their followers. Once con-secrated they achieved almost mystical importance as the soul of a regiment, and extraordinary efforts would be made to save them from capture. The term 'serving with the colours' came to mean service with the regiment rather than being a reservist.

dragoons, paying for them with money formerly voted by Parliament for the militia. Thus by 1688 the army stood at 24,000, with a further 10,000 in Ireland: the largest number since Cromwell's New Model.

Not surprisingly, Parliament grew ever more alarmed. Not only was money they voted being misappropriated, but James was now appointing Catholics to key positions – in Ireland replacing virtually all Protestant officers by Catholics. With the birth of a son to James's ultra-Catholic second wife Mary of Modena after fifteen years' barrenness, the 1688 equivalent of the men in grey suits (known with retrospective admiration as the Immortal Seven) moved to secure a Protestant succession, inviting James's nephew and son-in-law Prince William of Orange to take his place. 'Nineteen parts of twenty of the people . . . are desirous of change,' they assured him.

And despite the failure of Monmouth's earlier endeavour, 'Dutch William' managed to assure himself that the venture was politically feasible. Being a seasoned soldier, he also knew how to make sure it was militarily feasible. Not for him three ships and the hope of scythes and pitchforks: on 1 November he left Antwerp with a fleet of 200 troop transports escorted by forty-nine warships. The Royal Navy once again failed to intercept, pleading contrary (no doubt Protestant) winds, and William landed barely four days later at Torbay with 11,000 foot and 4,000 horse, including 4,000 English and Scots from the Anglo-Dutch Brigade.

At once he began a methodical but direct march on London, and the West Country, seemingly willing to overlook the blood-letting after the Monmouth rebellion, rallied to his side. Encouraging news of support in the midlands and the north also came, as did deserters from James's army; further backing came from William's sister-in-law Anne, James's own daughter, and her husband Prince George of Denmark.

James moved determinedly to intercept the invader, ordering the Huguenot but ultra-loyal commander-in-chief, the earl of Feversham, to concentrate his troops on Salisbury Plain, while decanting the Court from London to the cathedral close in the city itself. At this stage, militarily at least, the outcome was not a foregone conclusion. James's army outnumbered William's, and despite rumours and fears of disaffection the regimental officers were loyal enough – though this was perhaps less to do with personal commitment to James (the bulk of them thought his Catholicizing odious) than to the investment they had made in their commissions.

And here chance played its part; or, to put it another way, James's record caught up with him. The 4th Regiment of Foot was commanded by Colonel Charles Trelawny, whose brother, the bishop of Bristol, along with six other bishops, had been put in the Tower the previous year after protesting against James's Declaration of Indulgence (allowing freedom to practise religion according to conscience – or, depending on point of view, freedom publicly to espouse and promote Catholicism). The bishops' incarceration had provoked the biggest wave of popular support the Church of England has ever seen.* Trelawny's regiment, veterans of fighting the Moors at Tangier and Monmouth's men at Sedgemoor, were in outposts at Warminster on the western edge of the Plain, and therefore the first infantry likely to make contact with William's army. Their disaffected commander's lieutenant-colonel – the officer in executive command – was Charles Churchill, younger brother of (Lord) John Churchill, the hammer of Monmouth's rebellion and now Feversham's second-in-command. Treason, as Voltaire famously said, is a

---

* Indeed, two centuries later the feeling in Cornwall, whence the Trelawny family came, was still strong, inspiring the famous Parson Hawker of Morwenstow to write what became the 'Cornish national anthem' – 'The Song of the Western Men' or, simply, 'Trelawny'.

matter of dates; exactly when John Churchill decided to throw in with William has never been established, but one person who would certainly have known was his brother Charles.

Trelawny and his lieutenant-colonel, Charles Churchill, now declared for William, and almost to a man the regiment followed. The news of this and other defections ('desertions', had the outcome been different) unnerved James, who abruptly ordered a withdrawal to London despite the urging of John Churchill and Feversham to stand and fight. But if the order was precipitate, in its fear it was self-fulfilling: by the time James's army reached Hounslow Heath, where it camped, there were fewer than 8,000 men still in its ranks. Days of trying to 'fix' and to horse-trade followed, but on Christmas Eve, with the Dutch at the gate, James fled the country. The 'Glorious Revolution' was accomplished with barely a shot fired – in England, at least.

Ironically, the Restoration army whose very *raison d'être* had been to secure the Stuart monarchy was in the end the death of its male line (though 'Jacobitism' would flicker and occasionally flare for half a century and more) and, with it, of all religious ambivalence in the English state. There was now an unequivocally Protestant monarch – two, indeed, for William was to reign coequally with Mary. And with such a settlement England's fledgling standing army might have been reduced – had it not been for two things. First was Ireland: James's Catholic army did not intend to surrender the country without a fight. Second was the Netherlands, or rather King William's Dutchness: he was first and foremost the Stadtholder, and there was unfinished Dutch business on the Continent – war with Spain, France and Austria. To 'Dutch William', England and Scotland were a source of manpower for these wars, and he had no intention of reducing the army. On the contrary, he had every intention of making his Britain a continental power.

3

Glorious Revolution

1689–1702

'WE HAVE NOTHING TO EQUAL THIS!'

King George III's astonishment, dismay even, on first seeing Blenheim Palace half a century after it was finished was perhaps not surprising. By comparison, his own royal residences must have seemed provincial, whereas Blenheim ranked in scale and magnificence – as still it does – with the great baroque palaces of Europe.

From the laying of the foundation stone in the winter of 1704, the house was called, at Queen Anne's wish, Blenheim – the gift of a grateful nation, of a grateful sovereign indeed, to John Churchill, first duke of Marlborough, after his victory over the French at the village of Blindheim in Bavaria in August that year. Marlborough had engaged Sir John Vanbrugh to build a suitable memorial in Woodstock Park in Oxfordshire opposite the old royal palace, and so Blenheim became, uniquely, a 'palace' rather than a 'house'.

'Palace' suited both the significance of the victory at Blenheim (as Blindheim was soon known in English) and

Marlborough's own ambition, as well as his capability. He was, quite simply, the greatest figure of British military history, an accolade he shares only with the duke of Wellington. And Blenheim's architecture, in the exuberant, even florid, Italian baroque style – encapsulated this greatness, and rubbed in, so to speak, the victory over Louis XIV, the French 'Sun King', who for decades had been master of Europe. A vast lake was later created before the palace,* spanned by a bridge nearly 400 feet long, beyond which an enormous victory column raises the duke, swathed in a Roman toga, far above even the most royal of visitors as they approach. Inside the house itself the grandness of the state rooms, with their magnificent tapestries, marble and wood-carving, together with the sheer vastness of the whole, almost defy description. Ironically, the baroque style had originated as a statement of the wealth and power of the Catholic Church, inspired by the Counter-Reformation of which Louis was so ardent a disciple. At Blenheim that style made the same statement, but on behalf of the Protestant commander-in-chief of what had become the leading Protestant nation of Europe. And in a further historical twist, as if to reaffirm Blenheim's place in the continuing struggle for Europe, it was here, 170 years after the great battle that had given the palace its name, that the most illustrious of Marlborough's descendants was born – Winston Churchill. The victory at Blenheim did indeed bear fruits for more than one century.

Baroque did not live long in England, however. Its exaggerated style was altogether un-English, a thing of the Continent; and the Hanoverians, when they succeeded to the throne on Queen Anne's death in 1714, were duller even than their native subjects when it came to public art. That

---

* 'Capability' Brown was brought in by the second duke to make an English pastoral landscape, and most of the baroque gardens – a mini-Versailles – were destroyed.

Blenheim had been built in such a style was a statement that Marlborough had come of age – and, through its captain-general, that the British army had come of age too. George Monck and John Churchill were both sons of West Country gentry; both were raised from penury to ducal dignity by loyal and capable military service to the Crown (only one other English soldier, Wellington, would reach such heights after them). But George Monck, duke of Albemarle, with his pension of £500 a year, had lived at the 'Cockpit lodgings' in the old palace of Whitehall. The British army had come a long way, from an affair of the country gentry to one of the great institutions of state – and in the space of just fifty years.

But before the palace was built, before even the battle was fought, the business of the army's legal status and its control had had to be settled. The 'Glorious Revolution', as James's dethroning became known, gave Parliament its chance to resolve the matter more or less for good. Hitherto the army had functioned as a department of the royal household, funded by indirect and opaque means. After Monck's death in 1670 the army's day-to-day affairs had been run by a lower tier of secretaries, each managing a sub-department such as pay, medical services and judge-advocacy, of whom the secretary at war was the most important. This quaint title (as opposed to the secretary of state *for* war, the later cabinet post) derived from the original function of the appointment – that of secretary to the commander-in-chief when on campaign. And since after Monck's death neither Charles nor James had appointed a commander-in-chief except when the army took to the field, these civilian secretaries had acquired both experience and increasing power. Through the various legislative instruments of the Glorious Revolution, notably the Bill of Rights of 1689 (and ultimately the Act of Settlement of 1701), and by an annual Mutiny Act for the disciplining of the army, Parliament at last established both de facto and de jure control. Thereafter, to keep more men under arms than Parliament actually voted 'supply' (funding) for was unequivocally illegal.

Having sorted out the 'ownership' of the army, Parliament was now more or less content to concede its 'government and command' to the prerogative of the Crown. Thus although the army's strength – indeed its very existence – depended on the consent of Parliament, exercised through the annual estimates and the Mutiny Act, all promotions, commands, honours and awards, organization and training, and the maintenance of discipline, were the business of the King, who in the usual course of things was his own commander-in-chief. Even today, an officer is promoted to major-general and above with the express approval of the Queen as commander-in-chief.

The decade and a half that followed the Glorious Revolution was a time of deeper modernization, however. Every act of the King, including but by no means only military acts, was taken on the advice of a minister who was personally responsible both to the Crown and to Parliament. Marlborough made the army what it was in the field, but without William Blathwayt, secretary at war from 1683 to 1717, described by the diarist John Evelyn as 'very dexterous in business ... [having] raised himself by his industry from very moderate circumstance', he could scarcely have achieved what he did. Just as Pepys had earlier served the navy as Secretary to the Admiralty, so Blathwayt became the army's general administrator and principal staff officer, in effect founding the War Office. He became in addition 'Secretary to the Forces', responsible to the King for the detail of the army's 'command and government', and also to Parliament for the exercise of financial control and for guarding against military encroachment on civil liberties. The happy coincidence of great general and great minister has more than once been the making of success: Marlborough and Blathwayt were the prototype combination.

One result of this early separation of the civil and military functions in the army's government was the distinctively apolitical character of the men in uniform, at least in comparison with their counterparts on the Continent. Before the Glorious Revolution the army had been a political affair because it had been the decisive instrument of internal politics. 'The standing army were a body of men who had cut off his [Charles II] father's head ... had set up and pulled down ten several sorts of Governments,' wrote Clarendon in his history of the Civil War. He might have added, had he lived, 'and chased their King, his own brother, from his realm'. Now the standing army was very firmly under control of the King in Parliament, at the hands of a civilian secretary at war. A century later, the great Whig statesman Charles James Fox would write that 'the theory of the constitution consists in checks, in oppositions, one part bearing up and controlling another'; and the army had become a model of constitutional theory and practice. When the duke of Wellington was commander-in-chief in the 1820s he remarked (without complaint, for there was no greater advocate of constitutional control than he) that he 'could not move a corporal's guard from London to Windsor without obtaining the authority of the civil power'.

William of Orange and Mary Stuart, Anne's elder sister, had ascended the throne jointly – or, more correctly, ascended the *thrones*: the King in England as William III, in Scotland as William II and in Ireland as William I (in the north of which, then as now, he was known affectionately by the Protestants as 'King Billy'). Above all things, William was a warrior. His life had been defined by conflict with Spain over the Spanish Netherlands, and with Louis XIV who coveted Flanders and who, despite buying millions of tulip bulbs each year for Versailles, despised the Dutch as merchants as well as disapproving of them as republicans and utterly detesting them as Protestants. Seeing his new kingdoms as a source of money and troops with which to prosecute war with Spain and France in his struggle to preserve the United Provinces as an independent, Protestant polity, William made doubly sure that what passed for an army in Britain was shaped for war on the Continent. This meant, initially at least, bringing in Dutch experts, of whom the most intriguing was Frederick (later duke of) Schomberg. William appointed him Master General of the Ordnance and commander of the army in Ireland, charged with putting down the rebellion (or loyal resistance, as James's mainly Catholic adherents there saw it).

Born German (von Schönberg), Schomberg had become a marshal of France and then, when Louis XIV revoked the Edict of Nantes, provoking the exodus of Huguenots, had thrown his lot in with the House of Orange (Charles II had even tried to poach him to take over from Monck). Bringing in an outsider did not endear William to home-grown senior officers. Not least of the noses put out of joint was that of John Churchill, who might have expected the honour himself after committing himself to the Revolution at the critical moment. An earldom went some way towards pacifying James's former deputy commander-in-chief, however, and after a stint in Ireland (not long enough to tarnish his reputation, mercifully, for the campaign, if not as bitter as Cromwell's, was nevertheless bloody), the new earl of Marlborough was sent to the Continent.

At first there was not much opportunity to display his talents: Marlborough commanded a mere 8,000 men under the German-born General Waldeck in laborious siege operations along the principal waterways, the economic and strategic arteries of the Low Countries. Battles were rare, for non-conscript armies represented a significant investment, too valuable to risk in unpredictable actions; and commanders sought victory instead through the advantage of fortress lines or by campaigns of manœuvre. Marlborough was, however, seeing how continental warfare was done and making calculations for the future – not least when it was right to avoid battle, and when it was right to fight. He already had some experience, if on a small scale, and had seen enough fighting to understand the elements of battle; and the colonel of the sister regiment in which he had cut his teeth, Monck, must have been a powerful example of generalship and statecraft to him in those early days of the Restoration. Indeed, Monck's own reflections on the soldier's art, *Observations on Political and Military Affairs*, published in 1671, the year after his death, contained much that spoke to Marlborough in the pursuit of personal success: 'A General is not so much blamed for making trial of an ill-digested project, as he will be for the obstinate continuing in the same. Therefore the speediest leaving of any such enterprise doth excuse the rashness which might be imputed to the beginning.'

William's war with France had by the time of the Glorious Revolution become a much wider affair. There were other princes who had had enough of Louis XIV's territorial aggrandizement, for it went beyond merely acquiring defensible frontiers. Louis wanted his own man on the throne in Spain when the ailing Charles II breathed his last, and wanted also to make sure the 'right' man was elected Holy Roman Emperor. In 1686, therefore, an unlikely alliance had united Protestants and Catholics alike against France. Holland, the Rhenish states, Prussia, Denmark and the hitherto traditionally pro-French Sweden joined with Catholic Austria, Spain, Portugal, Bavaria and Savoy in the 'League of Augsburg' against the Sun King. France's invasion of the Rhineland in 1688 had given William his opportunity to land in England (for Louis would otherwise have come to James's aid), and so now Britain joined the League. This 'Grand Alliance', as it then became known, would stand against Louis, the embodiment of absolute monarchy, on and off for the next twenty years.

Marlborough knew Flanders well; he had first served there *against* the Dutch. But his light was now pushed even further under the bushel by the arrival of King William in person after defeating the Irish rebels and James's Franco-Irish army at the battle of the Boyne. Indeed, the light was almost extinguished, for the following year, 1692, he was dismissed from his lieutenant-generalship and briefly imprisoned in the Tower of London on charges of Jacobitism and mis-appropriation of funds (not *wholly* trumped up), having made the mistake of being too vocal in his criticism of Dutch favourites.

In Marlborough's temporary absence the war against France did not go well, though the number of troops rose steadily. By 1692 there were 40,000 British soldiers in the Low Countries, and by the end of the war five years later there were 56,000 (the same number as in the British Army of the Rhine in 1990 when the Berlin Wall began to crumble) – and this from a population of at most six million in the British Isles as a whole, less than a tenth of that today. And although the Scottish and Irish contingents remained officially separate from the English, in Flanders they increas-ingly operated as a single order of battle. A distinctly 'British' character to the army was beginning to emerge.

The Guards were forging their own character, too (the Scots Guards, originating before the Commonwealth as the King's bodyguard in Scotland, had been placed on the English establishment in 1686). Proximity to the sovereign, 'public' (guard) duties at the royal palaces, and a con-sequential emphasis on smartness of uniform and precision in foot drill lent them distinction. And alongside the general increase in infantry and cavalry, 'specialists' were appearing. Grenadiers – troops trained to prime and hurl grenades – had been introduced in 1678, and by 1690 every infantry regiment had a company of them. They were hand-picked, the taller and stronger of the rank and file, and acquired a certain cachet through their 'bishop's mitre' hats (which allowed an over-arm throwing action that the tricorn would have hampered) and their honoured place on the right of the line or at the head of the regiment when it marched in column. Fusiliers made their entrance too, men armed with the new *fusil*, a prototype flintlock musket (in which a piece of flint produced a spark to the initiating charge when the trigger released a cocked hammer), rather than the matchlocks carried by the rest of the infantry. The first regiment to carry the fusil, numbered seven in the line of battle,* was known as the 'Ordnance Regiment', its job being to escort the artillery (which still belonged to the Board of Ordnance), for there was less chance of accidental ignition of the gunpowder in the ammunition waggons than there would have been had the escorts carried the matchlock. Later they would be known as the 7th Regiment of Foot, or the Royal Fuziliers.†

There were changes in the rest of the line too. The proportion of pikemen to shot had been steadily reducing as the musket became more efficient, but it was the introduction of the bayonet in 1670 that signalled the end for the pike, for the musketeer could now truly be his own pikeman. The first bayonets had been plug-ended, slotting into the muzzle of the musket, with obvious limitations. Indeed a dramatic

---

* The number in the 'line' fixed their seniority (usually by date of raising): the lower the number, the higher the seniority.
† How fusilier (with or without a z) regiments subsequently acquired their cachet is baffling: escorting anything except the sovereign has always been *infra dig*. But fashionable they were (until the 2007 infantry reorganization reduced them dramatically), if for no better reason than the coloured feathers they wore in the head-dress (or pure white in the case of the late, incomparable, Royal Welch Fusiliers, as well as the Royal Fusiliers (City of London Regiment) until the RRF was formed in 1968, and the too-short-lived Royal Highland Fusiliers).

setback in the field, at the 'battle' of Killiekrankie in the Highlands during the Glorious Revolution, had demonstrated just how disastrous those limitations could be. Two English regiments were routed having fixed bayonets too soon, thereby rendering themselves unable to fire. A new pattern was therefore designed – the 'socket' bayonet, with a 'cranked' hilt fitting over the muzzle to carry the blade to one side, allowing the musket to be reloaded with the blade fixed. It was so effective that at the end of the Nine Years War the pike was withdrawn from service altogether, only the half-pike ('spontoon') or halberd remaining in the hands of NCOs and some officers, more as an instrument of correction than a weapon.*

The change from pike to bayonet changed the character of the infantry, promoting aggression rather than mere proficiency with unwieldy arms. Indeed, the bayonet would become more than just a weapon: it would acquire iconic status as the symbol of infantry spirit. The 'push of pike' had always been a lumbering affair, a clumsy locking of horns; but the bayonet charge achieved rapid results. The infantry could now generate 'shock action', which hitherto had been the role of the cavalry. At Minden in 1759 six British regiments of infantry would put the French cavalry to flight at the point of the bayonet. 'Fix bayonets!' now, as then, is as much a statement of resolve as it is an order, and though it is by no means the historical preserve of the British army, the British infantry have long made an art of it.

And it was at Steenkirk (Steenkerque or Steenkerke), one

---

* The 'Royal Corps of Halberdiers', the regiment of Evelyn Waugh's *Sword of Honour* trilogy, are entirely fictitious. There never was a formed regiment of halberdiers in Britain's armies. The Swiss Guard famously used the weapon to cover the flight of the pope during the 1527 sack of Rome, dying to a man. NCOs were still carrying the spontoon at Waterloo, but it had been abandoned by the time of the Crimean War (one of the few examples of progress during those forty years of military atrophy).

of the least-known but hardest-fought battles in the army's history, that the bayonet was properly blooded. The battle was fought some 30 miles south-west of Brussels on 3 August 1692, beginning before dawn and continuing until late afternoon – an uncommonly long fighting day in the experience of most Englishmen – with 15,000 allied troops under William's personal command pitted against a much larger force under the capable and experienced duc de Luxembourg. It began well, for the French were not expecting the allies to attack, not least because of William's dissembling tactics. Having achieved his immediate object in the days before – the capture of the fortress of Namur – he had camped nearby and made a pretence of resting. But since the early hours of 3 August his troops had in fact been moving stealthily towards the duc de Luxembourg's camp, and as dawn approached his advance guard was within striking distance of the sleeping French. But then the alarm was raised, and a furious, piecemeal fight began.

The main body of William's force quickened to the sound of the fight but got stuck in thick woodland. By nine o'clock they had managed to get some guns forward, but the French made good use of the respite to form a strong battle line. Not until midday did William's own line start to form properly, by which time the advance guard of British and Danes had been in action for the best part of nine hours. William nevertheless ordered an attack, which broke the first French line with the bayonet.

The duc de Luxembourg was too experienced a soldier to be unnerved by a poorly supported attack, however, and stood his ground as the British and Danes ran out of steam. Seeing the loss of momentum, the Dutch lieutenant-general Count Solms ordered up his cavalry, which was still with the main body, but they too found it difficult to move over the bad roads and heavy ground, and ended up blocking the way for the infantry. Cursing Solms and all the other Dutch generals in the usual way an infantryman curses everyone outside his own regiment, some of the British now pushed to the front, bayonets fixed, whereupon Solms ordered them to clear the way for his cavalry. William's counter-order for the cavalry to halt only appears to have made things worse, and by early afternoon the allied advance was at a standstill. The French now counter-attacked, forcing the allies to abandon the field. Five British regiments were completely destroyed, and their commander, Major-General Hugh Mackay, was killed (as was Solms himself). The survivors blamed the Dutch for incompetence, though the retreat was ably covered by General Ouwerkerk, William's second-in-command. Allies are never so 'useless' as in a defeat.

Although Marlborough was not at Steenkirk – indeed, he was not long out of the Tower – the battle told him just as soberly as it told the army in the field what the face of battle with the French would be like: hard pounding, heavy casualties, and certain defeat if things were mismanaged. The events at Steenkirk were soon public knowledge and remained vivid in the public consciousness even after the victories that followed. In his 1759 novel *Tristram Shandy*, Laurence Sterne recounts the soldiers' memories of the battle when Captain Toby Shandy, invalided from the army after the subsequent French siege of Namur, regales his old servant, the former Corporal Trim, with memories of the fighting, Parson Yorick adding his own views:

'Had count Solmes, Trim, done the same at the battle of Steenkirk', said Yorick, drolling a little upon the corporal, who had been run over by a dragoon in the retreat, –he had saved thee; –Saved! cried Trim, interrupting Yorick, and finishing the sentence for him after his own fashion, –he had saved five battalions, an' please your reverence, every soul of them:–there was Cutt's,–continued the corporal, clapping the forefinger of his right hand upon the thumb of his left, and counting round his hand,–there was Cutt's,–Mackay's,–Angus's,–Graham's, –and Leven's, all cut to pieces;–and so had the English life-guards too, had it not been for some regiments upon the right, who marched up boldly to their relief, and received the enemy's fire in their faces, before any one of their own platoons dis-charged a musket,–they'll go to heaven for it,–added Trim.*

Discharging a musket had at least been easier at Steenkirk than at Edgehill, for by then most infantrymen, not just the Fusiliers, were carrying the flintlock. But it was still a novel weapon, its potential not yet realized. Under Marlborough, in the hands of infantrymen who were acquiring a reputation for pugnacity and discipline not enjoyed since Elizabeth I's day, the flintlock musket would become the instrument of the British habit of victory – the deadly volley and then the bayonet charge. Three centuries later, in the Falklands War, the Scots Guards would run at the Argentinian defenders of Tumbledown Mountain with the same instinct for cold steel as their forebears had at Steenkirk. And other regiments since then, in Iraq and Afghanistan, have pressed home just as vigorously with the bayonet. The battle of Steenkirk pointed to the enduring future for the infantry: their job then was to close with the enemy and kill him, and it remains so today.

But Marlborough's greatest victories were ten years and more away. In 1692 he remained *persona non grata* with William, and perhaps even more so with the Queen. His rehabilitation was painfully slow, but when Mary died in December 1694 the thaw began, for the childless William was increasingly conscious of the succession; and Mary's

---

* *The Life and Opinions of Tristram Shandy, Gentleman* was published in nine volumes, the first two appearing in 1759, and seven others following over the next ten years. It reached the big screen in 2005 as *A Cock and Bull Story*, with Steve Coogan as the eponymous gentleman.

sister Anne was a staunch supporter of Marlborough, not least through her close friendship with Sarah Churchill, his wife. Nevertheless, it was not until 1698, a year after the formal ending of the Nine Years War, that the earl of Marlborough was restored to his lieutenant-generalcy and to the Privy Council.

And here Marlborough's story might have ended, his name that of a general even less well known today than Monck – but for 'events'. The Treaty of Ryswick had brought the war to an inconclusive end: the underlying conflict between Bourbons and Habsburgs remained unresolved, and the peace was therefore an unstable one. Yet Parliament expected retrenchment, and so the army was reduced once more – to a paltry 7,000 on English soil, with 17,000 in Ireland and the overseas garrisons. Thus was set the pattern of premature disbandment that has continued ever since – what in recent years has become known sardonically as taking the 'peace dividend'. And excessively premature it was too, for almost as soon as the regiments were disbanded Charles II of Spain, who was so interbred even for a Habsburg that he was called *El Hechizado* (The Bewitched), died childless. It was Nature's blessing, said some – though not a blessing on Europe, for at a stroke the future of the Spanish Netherlands was back on an agenda that could be settled only by war.

William's health was deteriorating too. And if he is remembered with little affection outside Northern Ireland, perhaps he should be given credit for deciding that Marlborough, with his influence over the soon-to-be Queen and his undoubted soldier's laurels, should take centre stage in the unfolding drama. William sent him to The Hague as ambassador-extraordinary and commander of English forces in the Netherlands to arrange a new coalition against France and Spain. The embassy bore fruit in September 1701 with the Treaty of the Second Grand Alliance, signed by

England, the Holy Roman Empire and the Dutch Republic. Six months later William died after a fall while riding in Richmond Park, his horse stumbling on a molehill. At once the Jacobites were toasting the mole, 'the little gentleman in black velvet', and dreaming of an early restoration of the rightful Stuart king. Indeed, as Marlborough's greatest biographer and descendant, Sir Winston Churchill, put it in his *History of the English-Speaking Peoples*, William's passing 'opened the trapdoor to a host of lurking foes'.

Most pressing of those foes was the alliance of France and Spain, a formidable military combination, its gold, men and ships prodigious, and its leader, Louis XIV, wielding such absolute power and lust for *La Gloire* as to blind him to the cost of war. And England was now without a monarch of experience. Marlborough, however, was both willing and able to act, so that Count Wratislaw, the Imperial ambassador in London, was soon writing to the Emperor that 'The greatest consolation in this confusion is that Marlborough is fully informed of the whole position and by reason of his credit with the Queen can do everything.'

This credit now paid a handsome personal as well as public dividend: in recognition of his diplomatic and military accomplishments, Anne appointed Marlborough Master General of the Ordnance, knight of the garter and 'Captain-General of Her Majesty's Armies at Home and Abroad'. Now at last, with the power both to organize the nation's land forces and to direct their employment, Marlborough was able to show what an English general might achieve, and of what feats of arms British troops were capable.

4

Corporal John

The north bank of the Upper Danube, Bavaria, 12 August 1704

IT WAS A SIGHT TO BEHOLD: TWO GENERALS IN FULL-BOTTOMED wigs sweating up the steps of the church tower in the little village of Tapfheim. They had met only two months before – the duke of Marlborough (elevated two ranks to the pinnacle of non-royal nobility by Queen Anne in reward for his diplomatic and military achievements and to give him the necessary standing with foreign heads of state) and Prince Eugene of Savoy, President of the Imperial War Council and de facto commander-in-chief of the Holy Roman Empire. Together they commanded an allied army of 52,000 men: 160 squadrons of cavalry, 66 battalions* of

---

* 'Battalion' is largely interchangeable with 'regiment' at this time, except in the cavalry, which never used the term. 'Regiment' tended to be used of the complete 'administrative identity', whereas 'battalion' described the 'tactical identity', a unit composed of a variable number of companies, fighting as an entity, and today under a lieutenant-colonel. Regiments could raise additional battalions in time of war, and nowadays the infantry comprises principally what are called 'large' (multi-battalion) regiments rather than the post-1950s single-battalion 'county' regiments.

infantry and 60 guns. If only 16,000 of these were truly 'British' (English, Scots and Irish), their red coats were as conspicuous as their capability, for these were now seasoned soldiers, the core of Marlborough's military machine.

The sight that greeted the two military leaders of the Grand Alliance when they reached the top of the church tower was ripe indeed: a mere 2 miles to the south-west, before the village of Blindheim (spelled 'Blenheim' in the dispatches), were the tents, horse lines, artillery and baggage of the 56,000-strong Franco-Bavarian army: 147 squadrons of cavalry, 84 battalions of infantry and 120 guns.

Marlborough was fifty-four years old. He had heard the sound of the guns almost yearly, and yet he was still to command in a major battle. His opponent Marshal Tallard, the former ambassador at the Court of St James, was fifty-two (both Wellington and Bonaparte were forty-six at Waterloo) and one of the most experienced battlefield commanders in Europe. Prince Eugene of Savoy was forty-one; he had spent half his life on active service, with the brilliant success of the battle of Zenta seven years earlier to his name. For Tallard and Eugene the Upper Danube was, if not familiar, then certainly not *terra incognita*: France lay not 100 miles to the west, Austria not 100 miles east. But Antwerp, the English base-port where Marlborough's regiments would have to re-embark if the campaign went badly, lay 350 miles north-west. Britons had indeed marched, in Winston Churchill's ringing words, 'where Britons never marched before'. Monck's men had marched the same distance from Edinburgh to London, but no English soldier had marched so deep into the territory of a continental enemy, and with a flank exposed to the most powerful army in the world. It was as prodigious a feat of imagination as it was of organization. But in the decade since the first faltering steps into the world of continental campaigning (at Dutch William's command and under his Dutch generals), the British army had changed in marked degree. Marlborough had been one of the engines of that change, as he would now be its principal beneficiary.

To begin with, the army had grown to an unprecedented size. In addition to the 'guards and garrisons' at home and in Ireland, from 1702 Britain's treaty contribution to the Grand Alliance was 40,000 men, of whom half were to be 'subject troops' – men raised within the kingdom(s) – and the rest hired from continental princes. Later there was 'the Augmentation' – 20,000 men, a third of them subject troops, half the bill footed by the Dutch. Eventually Marlborough would have more than 30,000 red-coated subject troops at his disposal in the Low Countries. Nor were British eyes directed only to the south-east; from 1707 Britain would take the fight to the Bourbon Philip V, the French placeman on the Spanish throne, and there would be almost as many redcoats in Portugal and Spain.

This fourfold increase in the post-Ryswick figure could not have been achieved without the proprietor-colonel system in which some notable – a man of standing in the county, but not necessarily with military experience – would be contracted to field a regiment of specified strength. In 1704 an infantry battalion comprised some 800 men organized in twelve 'battalion companies' of sixty, and one grenadier company of seventy. They were equipped by the Board of Ordnance but clothed, fed and paid by the colonel. His return was his own pay, which was not great, plus any legitimate surplus from the public money he was given for the maintenance of his regiment. Day to day, the regiment was commanded and trained by a lieutenant-colonel. And since the proprietor-colonel frequently had other interests such as offices of state or, on campaign, a superior appointment (command of a brigade or a position on the staff), the lieutenant-colonel usually commanded on operations too. Charles Churchill, Marlborough's brother, by now a lieutenant-general and a key member of the duke's high

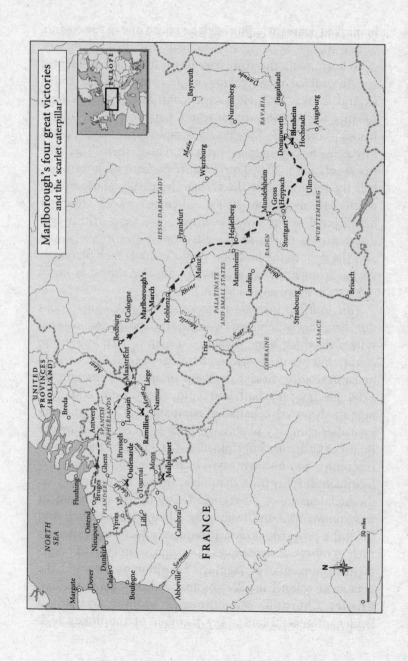

command, was still colonel of the 3rd Foot, a regiment with which, though it was to fight at Blenheim, he had only the most perfunctory contact. Most of his proprietorial interest would have been handled by a London agent, a figure who would grow in importance to the whole system of army administration. The lieutenant-colonel, the major (the lieutenant-colonel's second-in-command) and the company officers (a captain assisted by two or three ensigns) bought their commissions from the Crown but were paid direct by the agents on the Crown's behalf. Officers were able to buy and sell their commissions more or less at will, and this became big business for the agents. Purchase was in fact the usual route to promotion to the next rank up, although commissioning and promotion without payment was sometimes conferred for men with the right connections, or for distinguished service.

The origins of 'purchase' (the whole system of buying and selling of commissions) are uncertain, but throughout Europe in the Middle Ages mercenary troops were raised on the expectation not of pay but of profit; not just of booty but of legitimate trading in the wake of conquest. Those who wished to hazard for profit in this way were expected, in effect, to buy a share in the undertaking, and since the profit was shared out among the various ranks pro rata, it followed that the price of each rank – each 'share' – would be different. The term 'company' for such a body of mercenaries possibly derives from this military–mercantile arrangement. Bearing in mind the Crown's penury after the Restoration, purchase was a useful additional source of income to fund the unconstitutional army. In effect, therefore, an officer purchased an annuity, though he could always recover his original investment by 'selling out'. If he wished to purchase a greater annuity – if he wished for higher rank – all he had to do was pay the difference between the price of the higher rank and his own, as long as there was a vacancy in the higher rank. If his own regiment had no vacancy the aspiring officer would have to buy into a regiment that did, and regimental mobility was a feature of the British army until late in the nineteenth century, when purchase was abolished.

Those who could not afford to purchase promotion had to put their faith in the fortunes of war, slow and unpredictable as they were. If an officer died on active service, his commission was forfeit to the Crown. The next senior officer was promoted in his place without payment, which in turn created another vacancy, and so on down the ranks to the lowest, where seniority on the list told. This gave rise to the black-humoured toast 'To a bloody war, and a sickly season!' If purchase seems incomprehensible as a concept today, it is as well to remember that almost every public office in Stuart times was obtained in this way. Even William Blathwayt had paid for his post as secretary at war.

One exception was general officers, those beyond the 'field' ranks (as the regimental officers, captain to colonel, were known). Whereas field officers purchased their commissions at a remove through agents, generals were appointed personally by the Crown. And while it was not always undiluted merit that determined promotion to general rank, money did not change hands. Generals were concerned with the direction of campaigns and battles and with their logistic arrangements. In the Middle Ages and 'early modern period' men of proven worth were appointed to senior command on campaign in an ad hoc fashion: there was no permanent establishment of generals, the rank being frequently honorific. Regiments were 'brigaded' – grouped in brigades, the term that superseded 'tertia' after the Civil War – and commanded by the senior regimental commander. The 'brigadier' had no formally allotted staff, only the officers he chose from his own regiment to relay his orders in battle, for he was not administratively responsible for the regiments in his brigade, only for their actual control during the fighting. But by 1704 these arrangements were becoming more formalized, though brigades were still not permanent commands. The rank of brigadier-general had come into use, but it was a temporary one given to the senior officer within the equally temporary grouping of regiments. Indeed, a brigade might be commanded by a major-general if that was the rank of the senior 'proprietor-colonel',* though he was still referred to (confusingly to modern ears) as 'brigadier'. Thus Colonel Archibald Rowe of the Scots Fuzileers, a proprietor-colonel who actually took his regiment into the field, was appointed to command one of the three brigades at Blenheim as Brigadier-General Rowe, while a second brigade was commanded by Major-General James Hamilton (earl of Orkney), colonel of the 1st Foot (Royal Scots), and a third by Major-General James Ferguson, colonel of the 26th (the Cameronians). All three led the brigades in which their own regiments were mustered, these regiments being commanded in their absence by the lieutenant-colonel. There were no 'divisions' – groupings of brigades – at this time: brigades were mustered into 'columns', usually for administrative and security purposes on the march, although just occasionally – as at Blenheim – divisions were employed as entities for battle itself.†

Altogether then, Marlborough's army marching along the Danube was a far more professional affair than any fielded

---

* An officer would remain colonel of his regiment – an appointment – long after promotion to general officer. In time, the colonelcy became honorific, and regiments would hope to have a senior general as colonel for his influence and prestige – and in some cases for his deep pockets.
† Later in the eighteenth century the term 'division' came into use. The divisions were typically each under command of a major-general, and in Wellington's Peninsular army were themselves grouped into corps under a lieutenant-general.

before by Britain. Crucially the infantry, with the shorter and lighter flintlock musket (not needing a 'rest' on which to lay the barrel to take aim) and socket bayonet, were a great deal handier than their predecessors with whom William had made war in Flanders only a decade earlier. The flintlock's calibre was also smaller than the matchlock's and the ball therefore lighter, while no less lethal, allowing the infantryman to carry more rounds. The new weapon simply permitted a greater rate of fire. At Edgehill it had been one round in two minutes; now it was three or even four a minute in the best-drilled regiments.

And Marlborough had made sure this innovation was exploited to the utmost. Fifteen years earlier, when first appointed to command the army in Flanders, he had written to the secretary at war, Blathwayt, to 'desire that you will know the King's pleasure whether he will have the Regiments of Foot learn the Duch [*sic*] exercise, or else to continue the English, for if he will I must have it translated into English'. What he meant was the standardization of volley fire, for the practice had generally been for each rank (perhaps two companies in line – up to 200 men) to fire as a single entity, with the rank behind firing the next volley while the first rank reloaded, and so on. Since this involved 'dressing' (the realignment of the rank about to fire after it had stepped forward of the previous front rank) there was inevitably a hiatus between volleys. In Marlborough's 'Duch system', however, the companies were subdivided into 'platoons', each firing independently, so that a rolling fire could be kept up.*

Musketry was now therefore a decisive force on the battlefield, where before it had been more often than not a hazard,

---

* The private soldier did not fire his musket except at the express command of an officer – a key ingredient of what was (and is) called fire discipline.

John Churchill, first duke of Marlborough, 'Captain-General of Her Majesty's Armies at Home and Abroad'.

an irritant, and secondary to the *arme blanche*, as the knightly sword and lance or the 'puissant pike' were known. The British infantry became hugely adept at this system of fire control: time after time their disciplined volleys won the day in Marlborough's battles. And they would continue the ascendancy in the later continental wars of the eighteenth century – and then most spectacularly of all at the hands of that master of the tactical battle, the duke of Wellington. Tight, platoon-based fire control is still today the hallmark of the British infantry. Its deep-seated importance in the collective subconscious of the army is demonstrated in the annual Queen's birthday parade on Horse Guards ('Trooping the Colour'), for the drill evolutions through which the Foot Guards are put in that magnificent hour – sharp, precise, emphatic – are the relict of the battlefield drill that got the serried ranks of infantry-men to deliver volleys in whichever direction was needed and in the shortest possible time. No other troops in the world save those of the old Commonwealth look like the British on parade, for their drill comes from a different period and purpose.*

By 1704 the War of the Spanish Succession was in its fourth year. Marlborough had achieved much, but in strategic terms it had been inconclusive. Indeed, the year before had been one of consistent success for France and her allies, par-ticularly on the Danube. Vienna itself was now within French grasp, and the Austrian and Imperial capital's fall would no doubt be followed by the collapse of the Grand Alliance. To hasten it, Marshal Tallard's army of 50,000 were

---

* The 'goose-stepping' of some armies makes a political point, while the more languid pace of those in the Napoleonic heritage – the French and Italian notably (and even to an extent the US) – are more the relict of the old drill which moved large numbers of conscripts about the battlefield in column.

to strike out from Strasbourg (then in French territory) and east along the Danube, while 46,000 Frenchmen under Marshal Villeroi kept the Anglo-Dutch army pinned down at Maastricht. Which left only Prince Louis of Baden's force of 36,000 in the Lines of Stollhofen 30 miles north-east of Strasbourg, and a much smaller force at Ulm, standing in the way of Tallard's march on Vienna.

On learning of this, Marlborough at once saw the imperative. His most recent biographer, Professor Richard Holmes, quotes Clausewitz writing a century and more later that in a coalition war the very cohesion of the coalition is of fundamental importance. Holmes goes on to say that 'Marlborough was, first to last, a coalition general, supremely skilled at holding the Grand Alliance together'. And holding the alliance together now meant that Marlborough had to prevent the fall of Vienna. But the problem was the Dutch: they would simply not let the allied army quit the Spanish Netherlands to march deep into Bavaria. So he devised a ruse, disguising his purpose from The Hague and Versailles alike. His intentions, he wrote to London in late April, were

> to march with the English to Coblenz and declare that I intend to campaign on the Moselle. But when I come there, to write to the Dutch States that I think it absolutely necessary for the saving of the Empire to march with the troops under my command and to join with those that are in Germany ... in order to make measures with Prince Lewis of Baden for the speedy reduction of the Elector of Bavaria.*

And so, in Churchill's ever colourful words, the 'scarlet caterpillar, upon which all eyes were at once fixed, began to crawl steadfastly day by day across the map of Europe,

---

* Bavaria had changed sides after the Nine Years War.

dragging the whole war with it'. The scarlet caterpillar was, like the first polar expeditions, a supreme demonstration of the art of the possible, a masterpiece of organization and planning, and an example for the future. Ever since then, British soldiers, and for that matter politicians, have accepted the idea of hazarding far from the island fastness, on exterior lines* and against an enemy of the first rank. It had become 'no big deal'.

How did the scarlet caterpillar do it? Captain Parker of the Royal Regiment of Ireland described the routine:

> We frequently marched three, sometimes four, days, successively, and halted a day. We generally began our march about three in the morning, proceeded four leagues† or four and a half by day and reached our ground about nine. As we marched through the country of our Allies, commissars were appointed to furnish us with all manner of necessaries for man and horse, and the soldiers had nothing to do but pitch their tents, boil kettles and lie down to rest. Surely never was such a march carried on with more order and regularity, and with less fatigue to both man and horse.

Little wonder the troops called Marlborough, affectionately, 'Corporal John'. His diligence in appointing subordinates and agents to provide such order and regularity – including novelties such as light, two-wheeled carts to carry the tents and camp kettles rather than relying on the troops themselves as beasts of burden – was as inspired as it was uncommon. The simple, but not easy, expedient of breaking camp, marching two hours before first light and halting at nine kept the exertions to the coolest part of the summer day. Eight

---

* Lines of communication through territory held by the enemy or a neutral state – or even uncharted – as opposed to 'interior lines' through territory firmly held.
† One league = two and a half miles.

to ten miles in the day, 40 miles in five days, may seem slow progress, hardly 'the angel guiding the whirlwind' as one essayist described Marlborough's generalship (Churchill's 'scarlet caterpillar' was altogether better chosen). But a march of 350 miles (the route from the assembly area at Bedburg, 20 miles north-west of Cologne, to the Danube at Blindheim) could not have been conducted at any greater speed without considerable attrition in horses, guns *and* men. The non-marching days were filled with work, or 'interior economy', too – oiling, cleaning, making and mending, baking bread. And it worked: fewer than 1,000 men – less than 5 per cent of the force that left Bedburg – fell out along the way. On 12 August, as Marlborough and Eugene spied out the French lines from the top of the steeple in Tapfheim, they could be confident the army was fit to fight.

But if the 1704 campaign was a turning point in the army's strategic confidence, it did not follow that its campaigning ability was forever assured. Supply of the army would remain in the hands of a civilian commissariat for a century and a half, and it took a strong-minded commander to get the commissary officers to answer to him first rather than to the Treasury. It was not until the Napoleonic Wars that a military Corps of Waggoners was formed – an unglamorous organization in uniform but one which earned the increasing admiration of the troops whose biscuit and powder the Waggoners carried. In the long peace that followed Waterloo the corps would be disbanded, for 'Soldiers in peace are like chimneys in summer,' as Queen Elizabeth's adviser Lord Burghley had once remarked – and perhaps none more so than those 'behind the line'. It would take the very public failures of the Crimean War to get supply placed on a regular footing once more. Thereafter, tellingly, the newly formed Army Service Corps would be the only branch of the army to avoid criticism in the Boer War. If not the certainty, then the expectation of success had taken root on that march to the Danube: if the duke of Marlborough had once made his logistics work, so in the future could others.

From the top of their steeple in Tapfheim, Marlborough and Eugene were able to make a fair estimate of Marshal Tallard's dispositions. The French infantry were bivouacked behind a marshy tributary of the Danube, the Nebel, their flanks secured by the Danube itself and the fortified villages of Blindheim on the right and Lutzingen, itself backed by the rising hills, on the left – in all, a frontage of 5 miles (vastly more than any of the battles of the English Civil War). The cavalry, Marlborough and Eugene supposed, would be posted on either flank in the usual way, but they could see good galloping ground just west of Blindheim, too. Tallard's dispositions did not look as formidable as they might have been, however. In truth he did not believe the allies would seek battle against a more powerful army, especially when they were running short of supplies, as Tallard believed Marlborough was. And so, expecting the allies to retire north as soon as they discovered the presence of superior forces, he had merely taken the customary defensive precautions of the marching camp. It was a fatal assumption: in the heat of that high summer's afternoon, Marlborough and Eugene decided to attack the following morning – Sunday.

Not long after midnight, therefore, the allied army was roused silently from its sleep under the stars (and brilliant they were that night) in the wooded hills between Tapfheim and Munster to the east. At two o'clock they broke camp and crossed the Reichen stream beyond the village, which Marlborough's pioneers had been clearing of bosky obstacles all evening, and on across the Kessel stream. In eight columns of double brigades they advanced undetected by Tallard's patrols, a considerable feat of field discipline almost impossible a decade earlier with glowing slow-matches and clanking tin cartridges. Across the Kessel, on the favourable open ground which Marlborough and Eugene had observed from the church tower, they formed up in line of battle, but with the cavalry (the only arm in which the allies were superior) in the centre, and the infantry on the flanks. Marlborough intended delivering such a violent attack on Blindheim that Tallard would have to reinforce the village or risk his flank being turned. In reinforcing, he would have to weaken the centre, and it was there that the allied cavalry would strike the decisive blow. Meanwhile Marlborough's guns were being hauled up as stealthily as possible along the Munster–Hochstadt road.

In essentials Marlborough's artillery was little different from Cromwell's. It was principally a siege train still, with a few lighter field pieces to thicken up the infantry's firing line. Marlborough had begun trying to make the heavier guns more manœuvrable, but since the cannon, horse-teams and drivers belonged to the Board of Ordnance he had had only limited success.* Had the artillery possessed the handiness even of Wellington's a century later, he might have had the guns forward sooner at Blenheim en masse to wreak havoc with Tallard's ponderous dispositions, but the dominance of field artillery in major battles would have to wait for the emergence of Bonaparte, the supreme gunner, and his belief that 'It is with artillery that war is made' (although the British army would not fully subscribe to that view until the First World War).

Marlborough could expect some spirited fire, nevertheless, for his gunners had gained much skill with the three types of ammunition now carried in their limbers. 'Canister'

---

* Queen Anne had, however, recently made him Master General of the Ordnance as well as captain-general (commander-in-chief), and so at Blenheim the artillery (and engineers) were at least obliged to follow his orders, though it would be some years before he got the sort of guns and teams he wanted.

(commonly but inaccurately called grapeshot, which was the naval equivalent) consisted of little more than a gun barrel full of scrap metal that was fired point-blank in the face of the oncoming infantry or cavalry – a desperate, last-minute device requiring as much nerve as skill. The standard projectile, however, was the solid, round iron shot which battered down the walls behind which the infantry took cover, or scythed through their ranks in the battle line – the more ranks, the more casualties. The bigger guns could send shot half a mile, though Marlborough's main piece, the sixteen-pounder (the weight of the shot it threw) was at its best at 500 yards. But the most sophisticated type of shot, the one that most tested the gunner's skill, was explosive shell. It looked like solid shot but was hollow inside and packed with gunpowder. A fuse inserted in the touch-hole was ignited by the firing of the gun's main charge, and the round would explode among the enemy, the shell case fragmenting into lethal shards. The gunner's skill lay in trimming the fuse to just the right length: too short and the shell would explode in flight; too long and it would hiss harmlessly past its target. The best gunners could trim the fuse to make the shell explode above the infantry – 'air burst' as it would later be known – though shell was at its most effective if the enemy's lines were stationary. As soon as they began to advance, solid shot was the most serviceable – much faster to load, and the gun itself requiring the least 're-laying' (an artillery piece is 'laid' not aimed).

But at Blenheim Marlborough expected the heavy work to be done by his cavalry and infantry; the shock would come not from artillery but from surprise – the impudence of attack against superior forces, and the audacity of attack at dawn.

And shock it was too. The comte de Mérode-Westerloo, one of Tallard's cavalry commanders, was still asleep when the curtain of his camp bed was pulled aside by his groom –

not with the customary cup of chocolate but with the news that the allies were pouring on to the field. 'I rubbed my eyes in disbelief,' he recalled afterwards, pulling on clothes and rushing to the horse lines to harry his equally sleepy troopers into the saddle, as signal guns called in the foraging parties.

By eight o'clock the French artillery were managing to send solid shot towards the allied ranks as they formed up, but neither the infantry nor the cavalry, despite the encouragement of Mérode-Westerloo, were in any shape to counter-attack – not even when Marlborough's engineers came forward to lay bridges across the Nebel. Towards ten o'clock the English brigades on the left, under Major-General John 'Salamander' Cutts (so-called for his remarkable ability to survive fire), were able to get across the stream to begin the assault on Blindheim; but they then had to wait, and under increasingly troublesome artillery, until Eugene, who had the harder approach march, could get into position. There being nothing else to do but bear the cannonade, and it being Sunday, Cutts ordered the drums to be piled, altar-like, the colours laid on them, and the regimental chaplains to conduct divine service – the prayer probably a deal more earnest than usual.

At midday, with Eugene now forward and Blindheim ablaze from well-directed shell fire, the assault got properly under way. Heavy fighting quickly spread the length of the 5-mile line, which even without the thick pall of smoke soon shrouding the battlefield would have been a challenge to Marlborough's capacity for control. Yet in his first test as a major battlefield commander he showed the same sure judgement as Wellington at Waterloo in choosing where best to place himself both to observe and to influence – how to 'read the battlefield'. He did so with considerable luck, too, for there were close shaves: at one point a round shot struck the ground between his horse's legs and showered him with earth – just as at Waterloo a cannon ball took off the leg of an officer sitting astride his horse next to Wellington. But Marlborough also relied on picked men to bring him intelligence of the battle. It was standard practice to have a suite of aides-de-camp to relay orders, but ADCs acted as messengers and interpreters of instructions, and they were mounted on the fleetest horses. Marlborough's two dozen 'running footmen' were dismounted, making them less likely to become casualties – hand-picked young officers and NCOs who not only carried orders but observed for themselves how the battle was going.* In the Second World War, Field Marshal Montgomery copied Marlborough's system, with young men in jeeps 'battle reporting'.

It took most of the afternoon to accomplish, but on the allied right Eugene was able to keep the Elector of Bavaria's troops fixed while on the left Cutts managed to tie down virtually all of Tallard's infantry in and around Blindheim. The centre ground therefore became a free-manœuvre space for the cavalry, as Marlborough had planned. He was still taking no chances, however, and took the unusual step of placing a line of infantry between his two lines of cavalry.

Tallard's cavalry fell into the trap, charging and breaking through the first line of allied cavalry, only to be met by withering volleys from the infantry. The matter was now decided in Cromwellian fashion by the second line of cavalry charging home with the sword, Marlborough himself at their head. The shaken French, instead of counter-charging, pistol-volleyed then fled – all of them, including the vaunted Maison du Roi, the French Household Cavalry. The French infantry beyond were quickly surrounded and cut down, and Tallard's flanks, suddenly isolated, collapsed.

The unthinkable now happened: Tallard's army began

---

* They carried gold-, silver- or bronze-tipped staffs to indicate their status.

streaming from the field. A century later Robert Southey, the poet laureate, would sum it up in five memorable words: 'It was a famous victory.'

Famous, complete and *bloody*. The French lost 30,000 killed, wounded and missing, a huge 'butcher's bill' by the standards of the time. The allies lost 10,000. The spoils were colossal too: guns, powder, shot, horses, waggons, coaches, coin, plate, wine, beer, 30,000 tents, 10,000 sides of beef, and no doubt a *vivandière* or two. There was even a French marshal's baton, with the marshal holding it.

'Marshal Tallard is in my coach,' wrote Marlborough on the back of a tavern bill to his duchess that very evening: 'I have not the time to say more but to beg you will give my duty to the Queen and let her know that her Army has had a glorious victory.'

It is hardly possible to overstate the significance of Blenheim, which planted ambitious notions of what could be done in the minds of officers and soldiers alike. The strategic–logistical brilliance of the march to the Danube, the tactical daring of the night approach to contact, the unorthodoxy of the infantry–cavalry deployments, the all-arms coordination, and the sheer aggressive use of the infantry in particular – all these set the standard for British troops to measure themselves against in the future. If any one battle could be said to have made the British army, it was Blenheim.

And what effect it had throughout Europe! Vienna, and thereby the Grand Alliance, was saved. Indeed, it was re-invigorated. France had had *La Gloire* knocked out of her: the French monarchy had never suffered such a defeat since its rise to pre-eminence in the century before. No such crushing victory had been won by any general since Gustavus Adolphus, 'the Lion of the North', had stamped all over the Baltic and the Rhine eighty years before.

'Five years were to pass before the French armies redeemed their reputation,' wrote the late David Chandler, the most devoted of Marlburian scholars: 'and, for the first time for centuries, England had assumed the military leadership of Europe.'

5

Chimneys in Summer

The United Kingdom of Great Britain, and Ireland, 1713–43

IN HIS DECLINING YEARS THE DUKE OF MARLBOROUGH WOULD walk the corridors of Blenheim Palace, gazing at the paintings and the huge Flemish-woven tapestries which commemorated his greatest victories – Blenheim, Ramillies (1706), Oudenarde (1708), Malplaquet (1709). And once when he came to Sir Godfrey Kneller's grand equestrian portrait of his youthful self in knightly armour, charger *en levade* eager for the off, he was observed to pause and with a heavy sigh declare, 'This was once a man.'

He might also have sighed and added, 'This was once an army,' for with the Peace of Utrecht in 1713 came swingeing cuts. At its peak, the year of Malplaquet (a battle whose butcher's bill came as a savage reminder that victory could not always be bought cheaply), the army stood at 70,000 excluding the old overseas garrisons, with a further 12,000 on the Irish establishment. By 1714 it had been slashed to 26,000, nearly half of whom were in the new garrisons of Gibraltar, Minorca and the West Indies. Nor were the cuts carried out with much justice (a feeling not unknown in more recent cuts): the regiments to be disbanded were chosen not by seniority, determined by their date of raising, but by their perceived loyalty to the Hanoverian succession. For on the death of Anne in August that year, George, Elector of Hanover, had become King. His mother the Electress Sophia had been named Anne's successor by the 1701 Act of Settlement, which stipulated a Protestant monarch, but she had died in June. Sophia was Prince Rupert's sister, Charles I's niece, and there was Stuart blood in her veins and in her son George's – but not as much as in those of James II's son, Prince James Edward. 'Jacobite' claims to the throne would therefore plague the country for the next half-century. The army had within living memory deposed one king, if not two: it was hardly surprising, perhaps, that the Whig government therefore wanted to make sure the proprietor-colonels and their regiments did not depose a third.

Of eighty-five regiments of foot on the British establishment at the time of Utrecht, only forty-two survived.* When Marlborough was fighting at Blenheim there had been thirty-eight regiments of horse and dragoons (including those on the Scots and Irish establishments); fewer than half were left after Utrecht. Lord Burghley's lament seemed most apt: George I's government could see little cause for chimneys now that summer was come, and all the continental fires had been put out.

The retrenchment would cost dearly, however, if not immediately. The essayist, spy and novelist Daniel Defoe, reviewing the snakes and ladders progress of William's wars in the 1690s, had argued against the reductions after the Peace of Ryswick (1697) on the grounds that it had taken, and would take again, three years and 30,000 lives to make a

---

* The 6th Foot was one of the first to go, though they had been raised as early as 1674.

new army, describing English recruits as 'the worst raw men in the world and the best once got over it' (rather as the duke of Wellington famously described his recruits as 'the scum of the earth'). Sixty years after Utrecht, on the eve of the American war, which would see for the first and only time the complete defeat of British arms, the proto-economist Adam Smith wrote of the folly of those reductions in his *Inquiry into the Nature and Causes of the Wealth of Nations*:

The soldiers of a standing army, though they may never have seen an enemy, yet have frequently appeared to possess all the courage of veteran troops and the very moment that they took the field to have been fit to face the hardiest and most experienced veterans. In 1756, when the Russian army marched into Poland, the valour of the Russian soldiers did not appear inferior to that of the Prussians, at that time supposed to be the hardiest and most experienced veterans in Europe. The Russian empire, however, had enjoyed a profound peace for near twenty years before, and could at that time have very few soldiers who had ever seen an enemy ... In a long peace the generals, perhaps, may sometimes forget their skill; but, where a well-regulated standing army has been kept up, the soldiers seem never to forget their valour.

Indeed, Smith also cites the attack on Cartagena in Central America (in 1741, during the 'War of Jenkins' Ear') as an example of the same bravery; but since this was a disastrous episode in a thoroughly inglorious campaign, described by one historian with charitable understatement as a war of probing and learning, it serves more to illustrate his point that generals forget their skill and that the army forgets how to organize itself to fight. For forgetting it was: the knowledge was all there when the Peace of Utrecht was signed. Indeed, it is remarkable how much of the British army of today is recognizable at the end of Marlborough's wars.

And Marlborough had shown London – the world, indeed – what a supreme strategist and tactician could achieve. All future generals had therefore to measure themselves against this standard, and not just in the estimation of the capitals of Europe: the sobriquet 'Corporal John' was an expression of the army's, the common soldier's, appreciation of Marlborough's attention to administrative detail. The soldier could now have expectations, too. If there were setbacks, or murdering battles like Malplaquet, a commander's reputation for being careful of men's lives and attentive to their welfare buoyed up morale. They saw and appreciated that there was science in Marlborough's fighting: he brought them to battle in as good a condition as possible, with his logistics on a sound footing and with the ingenious help of his increasingly capable engineers. Once battle was joined he did not rely solely on brave men's breasts, throwing them needlessly against the fire of the enemy. Instead he co-ordinated the attacks of the infantry with those of the cavalry, and supported them with increasingly mobile artillery. In other words – those of today – he fought an 'all-arms battle'. Robert Parker, one of his infantry captains, wrote of him: 'It cannot be said that he ever slipped an opportunity of fighting when there was any probability of coming at his enemy; and upon all occasions he concerted matters with so much judgement and forecast, that he never fought a battle which he did not gain, nor laid siege to a town that he did not take.'

For their part, Marlborough's soldiers, his infantry especially, were ready to endure gruelling marches and the bloodiest fighting. Why? In part because of their commander's reputation for success. Good administration played a large part too, but there was something else. It was not fear of the lash, for that instrument was used sparingly compared with Wellington's day a century later (when the Russian, Swedish, Austrian and Turkish armies had regimental executioners). There was nothing of the revolutionary, patriotic fervour that made Napoleon's armies march the length and breadth of Europe, for a dynastic war was not a nationalistic war. Neither pay nor loot was great: the one was always in arrears, and could dwindle to next-to-nothing after all the so-called 'off-reckonings', the stoppages of pay for clothing and subsistence; and the opportunities for the other were few and far between – and severely punishable. In his *The Seven Ages of the British Army* (1984), the late Field Marshal Lord Carver, never a sentimental man (having been a Royal Tank Regiment officer), suggests that the morale of Marlborough's men derived from

> the feeling of being a member of a family, in which the opinion of his fellow soldiers, of a community based on sharing common dangers and hardships, exercising mutual responsibility for the lives of comrades-in-arms, mattered more than any consideration of national, moral or personal factors. For the cavalry and the infantry, this centred on a soldier's regiment – of horse, dragoons or foot.

This feeling of being a member of a sort of family was and remains today what the regimental system was about, though it was not devised as such by careful military calculation. It grew out of the peculiar financial arrangements of the day, and to help recruiting to what was, after all, scarcely an attractive proposition. As the irrepressible Dr Johnson averred, 'No man will enlist for a soldier or go to sea who can first contrive to get himself in prison.' The regiment was, in fact, an accidental act of genius.

But as well as the sense of belonging, there was another factor emerging (or re-emerging) – the British soldier's sense of his own, individual superiority. 'Re-emerging' because it had been there at Agincourt, when the English bowmen had famously gestured in derision with their drawing fingers;

and it had certainly been true of Elizabethan fighting men, soldiers and sea-dogs alike. After the victory at Blenheim, Marlborough had honoured Marshal Tallard in very eighteenth-century fashion by inviting him to review the allied army, and the civilities are revealing:

> *Marlborough*: I am sorry that such cruel misfortune should have fallen on a soldier for whom I have the highest regard.
> *Tallard*: And I congratulate you on defeating the best soldiers in the world.
> *Marlborough*: Your Lordship, I presume, excepts those who had the honour to beat them.

It would be easy to dismiss this as mere banter between elevated gallants. But it was what the men in red coats had come to believe: that as British soldiers they could both take it better and dish it out harder, that their volleys were more withering, their bayonet charges more fearsome. Forty years after Blenheim, the British statesman and man of letters Philip Stanhope, fourth earl of Chesterfield, wrote: 'That silly, sanguine notion, which is firmly entertained here, that one Englishman can beat three Frenchmen, encourages, and has sometimes enabled, one Englishman, in reality, to beat two.' And on the ridge at Waterloo, when the red-coated descendants of Marlborough's regiments faced for the first time the Grande Armée with Bonaparte himself in command, they did so with all the assurance of men who could not be defeated: 'However proletarian and semiliterate he may have been,' suggests the Italian Professor Alessandro Barbero in *The Battle*, 'the English soldier, well nourished with meat and beer, stimulated with gin, and convinced of his own racial superiority to the foreign rabble he had to face, was a magnificent combatant, as anyone who has ever seen hooligans in action at a soccer match can readily imagine.'

And so by the end of Marlborough's wars the British army had acquired the habit of success. It was a habit born of superior generalship, which knew how to get advantage out of the innate fighting quality of the troops, and also of the psychological advantage of the regimental system. Such was the dividend that was in large measure squandered in the wholesale disbandments of 1714.

As the century advanced, both Whigs and Tories sought to disengage permanently from the continental strategy, harking back instead to the Elizabethan vision of the 'sceptred isle' which Sir Walter Raleigh had once advanced with the compelling logic that 'Whosoever commands the sea, commands the trade of the world, commands the riches of the world, commands the world itself.' There emerged, therefore, what appeared to be an appropriate formula of ends, ways and means: the *ends* were command of the world, which would leave Britain free of foreign interference; the *ways* were, principally, command of the riches of the world through trade; the *means* were command of the sea and possession of colonies. In other words, British military strategy was to be maritime – 'blue water' – not continental.

Where did this leave the army? In many respects back in square one. MPs who still had a horror of the notion of a standing army could rail in Parliament against the annual estimates, which were still couched in tentative, even apologetic, tones as 'a number of troops not exceeding . . .' By the end of the eighteenth century's second decade the number was in fact a paltry 8,000 in England and Scotland, the same as in Charles II's day, with once again an entirely separate Irish establishment. And with no war to fight on the Continent, soldiers – 'the brutal and licentious' as Kipling would much later and with heavy irony versify them – were scattered about the country once again in small detachments, billeted on inn-keepers and the like with mutual dissatisfaction, and winning even fewer friends when enforcing public order, which too often looked like party political work. It were better, argued men in Parliament and shire hall alike, to place the nation's defence in the hands of the militia once more, which was in turn in the hands of the county authorities. No one appears to have identified the possibility of treating the two as complementary, with the militia as a partially trained reserve for the regular army, offering some sort of administrative framework for recruiting and the basis for rapid expansion. All this would have to wait for another 200 years, until the decade before 1914.

The cuts might have gone even further after 1714 had there not been a sharp reminder that Utrecht did not guarantee universal peace. Indeed the provisions of the treaty at once precipitated a crisis, for Prince James Edward Stuart (the 'Old Pretender') had been sheltering the while in France, which along with Spain and the papacy had recognized him as King, but the treaty required Louis XIV to expel him, which in 1715 he decided to do – unfortunately by warship to Scotland.

In anticipation of James's landing the Catholic Highlands and the more Episcopalian parts of the Lowlands rose in 'feudal' loyalty, and a scratch Jacobite army under the earl of Mar had soon captured the ancient capital, Perth, without opposition. The great fortress of Stirling Castle remained in government hands, however, and after a month's indecision in which the Jacobites first took and then lost Edinburgh (Leith citadel), and with James still at sea, Mar made the fateful decision to divide his force and invade England. Committed only to restoring James to the throne of Scotland, many Highlanders deserted, but a Northumbrian Jacobite squire, Thomas Forster, had assured Mar that there would be armed support for him in Lancashire, a strongly recusant county. And so over the border they poured, 2,500 undrilled men in homespun tartan, like the cattle-reiving 'blue bonnets' of centuries past, led by the inveterate but wholly unmilitary Lowland Jacobite, Viscount Kenmure.

At Carlisle they met with the usual anti-reiving opposition, not so much from the under-garrisoned castle as from the Cumberland militia armed with pitchforks. These they easily brushed aside, but few recruits had come their way on the march through the Border country, and the long, onward trudge through the rainy autumn fells of Westmorland disheartened many a Highlander, though he usually thought of rain and rough pasture as home. Lancaster received them sullenly, but on 9 November at Preston 1,500 recruits rallied to the cause.

The new Whig government's response to the threat had been swift. Regiments, principally of dragoons, were hastily re-raised. But the response was somewhat paranoid, too. In the contrary West Country suspected Jacobites were arrested, though many of them were really no more than residual 'non-jurors' – clergy and officials whose conscience had prevented their repudiating the original oath of allegiance to James II in order to take a second to William – and probably no more than moderately Jacobite in sympathy. A good number of troops were also moved to Oxford in the fear that the university's traditional Royalism would somehow transmute to Jacobitism. Despite these diversions, however, government reinforcements were soon arriving at Preston, from Scotland as well as the south, including the Cameronians, a viscerally anti-Catholic regiment and veterans of Marlborough's four great victories. Opposition to the Act of Union of 1707 which bound England and Scotland together as the Kingdom of Great Britain with one parliament, and thereby made the Scots and English armies officially into one, had undoubtedly fuelled support for James among the Scots; but the Act also made for a more efficient government response.

And so in Preston the skirmishing began, with buildings

set alight and a little cannonading to keep the Jacobite outposts busy. After a few days there was street fighting – noisy,
destructive, time-consuming, but to no purpose or effect.
With reports of government reinforcements arriving by the
hour, the Highlanders rapidly began losing heart, and on 14
November Viscount Kenmure's 'army' surrendered un-
conditionally. The 'bag' numbered barely 1,500, of whom a
third were English (though the majority of the Lancashire
men had fled). Although fewer than twenty Jacobites had
been killed, 200 government troops had been killed or
wounded – more than enough to frighten London; and
ministers were not inclined to be magnanimous. Kenmure
was tried for treason, and beheaded.

Meanwhile at Sheriffmuir south of Stirling the earl of
Mar, his forces swelled to 12,000, clashed inconclusively with
a government force half the size led by the doughty duke of
Argyll, who had served with real distinction under
Marlborough. Neither side was left in possession of the field,
but Mar, dithering as before, retreated towards Perth, leaving
Argyll to lick his wounds and bring reinforcements from
Stirling.

At last in December James landed, north of Aberdeen. But
he was at once severely cast down: besides the bleak granite
prospect of that city, which cannot have cheered him much
after Paris, the news of Preston and Sheriffmuir was hardly
encouraging. Beset by a raging fever, he did little to inspire
his confused and demoralized troops. He briefly set up court
at the old royal palace of Scone outside Perth, and ordered
the burning of villages to hinder the duke of Argyll's
advance, which even the deep snow was failing to do.

But James's counsellors had also lost heart. The earl of
Mar, not for nothing nicknamed 'Bobbing John' for his rapid
switching of allegiances, ordered a retreat eastwards to the
coast on the pretext of finding a stronger position. James saw
that the game was up, however, and like his father before him

abandoned his throne without a fight, taking ship for France on 4 February and in customary Stuart fashion leaving a message for his fatally loyal Highlanders to 'shift for themselves'. A century later, when tartan sentimentality was gripping even the Prince Regent, Sir Walter Scott romanticized the whole sorry affair in *Rob Roy*.

The Fifteen, as the campaign became known (to distinguish it from the second Jacobite rebellion of 1745, which would press the government much more sorely), and a minor but potentially more successful rebellion (or rising, according to taste) in 1719 had given London a fright, but it also taught the army a good deal about 'counter-insurgency' and the occupation of hostile territory. The Irish general George Wade was appointed 'Commander in Chief of His Majesty's forces, castles, forts and barracks in North Britain', and told to pacify the Highlands.

Wade had seen what it took to pacify his own country. He had also served in Flanders with Marlborough, in Spain and in the West Indies, and he had studied the Roman army of Agricola's day which had brought the *Pax Romana*, of a sort, to Caledonia. He would now apply what he had learned in a grand scheme worthy of the legions themselves, building firm bases in the Great Glen at Fort William and Kiliwhimin (later Fort Augustus) and at Inverness (Fort George), with fortified barracks at strategic points such as Ruthven midway between Perth and the Highland 'capital'. They can all be seen today, in various states of repair – the finest of them, Fort George, rebuilt east of Inverness after 1745, with its vast ramparts and massive bastions one of the most formidable artillery fortresses in Europe, and still garrisoned by regular troops (it is also home to the regimental museum of the Queen's Own Highlanders). Wade linked these strong points by a beacon chain and constructed 250 miles of metalled road requiring forty new bridges (including one over the wide Tay at Aberfeldy) to move troops and artillery rapidly to anywhere that trouble threatened. And in case these measures failed to contain rebellion within the Highlands, he also reinforced the great strongholds of Stirling and Edinburgh, and engaged Vanbrugh's assistant at Blenheim, Nicholas Hawksmoor, to build a huge barracks–fortress at Berwick to guard the eastern route into England. It remains today in all its military elegance under the patronage of English Heritage, and houses the regimental museum of the King's Own Scottish Borderers, the regiment now incorporated, along with the Queen's Own Highlanders, into the Royal Regiment of Scotland (though Berwick is still an English city).

And while south of the border the militia was increasingly neglected, in Scotland Wade organized the 'Highland Watches' under the local gentry to prevent fighting between the clans, deter raiding and enforce the new disarmament laws. In 1739 these watch companies would be mustered into a regiment of the line, the 'Black Watch', so-called for their dark-coloured 'government' tartan, and for many years they were the only troops allowed to wear tartan of any kind. Wade also mounted a 'hearts and minds' campaign, opening schools in the Highlands funded from forfeited Jacobite estates.

But it would not be quite enough. A Jacobite romanticism, not wholly the retrospective invention of Sir Walter Scott, and a genuine attachment to the 'Old Religion' (Catholicism) would mean a good many more Highland men slain – in the Forty-five, when James's son, 'the Young Pretender' Charles Edward, sailed from France and raised the Stuart standard in the heather.

For the time being, however, retrenchment remained the policy. As soon as the threat of the Fifteen was past, the army's entire establishment was cut again – to 18,000, the lowest figure since James II's reign. Britain was still unquestionably a major European power, but it was a power based on residual memories of victories past, not the capability for victories future. An army of 18,000 was to prove wholly inadequate for a nation of the first rank, even when employed dextrously in collaboration with the ever-strengthening Royal Navy. It was not enough for Britain's new wars of trade, nor its unfinished business with Spain; and certainly not for yet another continental dynastic contest – this time of the Austrian succession.

6

The Mousetrap

Dettingen, Bavaria, 27 June 1743

KING GEORGE II, WHO SUCCEEDED HIS FATHER TO THE BRITISH throne and to the electorate of Hanover in 1727, had three passions: the Queen, music, and the army. Caroline of Brandenburg-Ansbach was probably the most intelligent woman a British king has ever had the fortune to marry. By relentlessly championing Robert Walpole as 'prime minister' she played no small part in consolidating both the Hanoverian succession – which in spite of the ghastly Stuart alternatives was by no means universally popular – and constitutional monarchy itself. Nor was she merely a power behind the throne: when the King was away on state business in Germany, as he frequently was, Caroline had vice-regal authority. A contemporary verse ran:

> You may strut, dapper George, but 'twill all be in vain,
> We all know 'tis Queen Caroline, not you, that reign.

They had eight children, and despite – perhaps even because

of – his several mistresses, George was devoted to her. When she was dying, in 1737, she urged him to take another wife. 'Non,' he replied resolutely: 'j'aurai des maîtresses!' And he had a pair of matching coffins made with removable sides, so that when he followed her to the grave (twenty-three years later) they could lie together again.

George inherited his passion for music from his father, whose protégé Handel he continued to champion: he is famously credited with the custom of standing during the 'Hallelujah chorus', and Handel composed the anthems for Caroline's funeral, as he had for the coronation. But like his cousin Frederick William I of Prussia (*der Soldaten-König*), George believed the army to be the first and noblest occupation of a king. He certainly took little interest in government, which was ably if corruptly (in modern eyes; the eighteenth century was on the whole more tolerant) conducted by Walpole. It all worked rather well.

George was also physically brave. He had fought at Oudenarde, the third of Marlborough's great 'quadrilateral' of battles (alongside Blenheim, Ramillies and Malplaquet), and would sometimes parade in his old battle coat. 'And the people laughed, but kindly, at the odd old garment, for bravery never goes out of fashion,' wrote Thackeray a century later.

Whenever George dealt with army business he took off his habitual Court brown to put on more military red. He put on red as much as he possibly could, indeed, loving to interfere in the army's business, although he scarcely considered it interference, for despite the measures enacted by Parliament after the 'Glorious Revolution' the limit of the royal prerogative was still unclear. Like his father, George laboured manfully to standardize drill – what Wolfe, the hero of Quebec (1759), complained of as 'the variety of steps in our infantry and the feebleness and disorderly floating of our lines' – though it would be many years before there was a truly common system. He championed the Royal Military Academy, which opened at Woolwich in 1741 to teach gunnery and engineering (a permanent corps of artillery had been formed at Marlborough's urging in 1716, becoming the Royal Regiment of Artillery in 1727). He regulated the price of commissions, abolished the trade in the regimental proprietor-colonelcies and sought to advance able officers, keeping a book in which he made notes on their capabilities and appointments. If he had had his way, he might also – like his cousin Frederick William – have introduced compulsory military service. It was not surprising that when in 1743 the army found itself once more in Marlborough's old stamping ground, Bavaria, George insisted on taking to the field at its head.

Britain had in fact been at war with Spain since October 1739. By the Treaty of Seville ten years earlier, Britain had agreed not to trade with Spanish colonies in the Caribbean and South America, and to verify the working of the treaty the Spanish were permitted to search British vessels. While boarding the *Rebecca* in 1731, the Spanish coast guard severed the ear of her captain, Robert Jenkins, or so it was claimed. British merchants, determined to penetrate the Atlantic trade, used the incident as a *casus belli* against Spain in the Caribbean (though tardily to say the least, hostilities not beginning for a full seven years). Jenkins exhibited his pickled ear to the House of Commons, and the entirely predictable outrage forced a reluctant Walpole to declare war. Thus began an episode of Caribbean skirmishing – the 'War of Jenkins' Ear' – that yielded very mixed results.

The gravest unintended consequence of the skirmishing was the slide into the much greater affair of the War of the Austrian Succession. In 1740 Emperor Charles VI died, leaving the crown to his daughter Maria Theresa. Frederick II – later 'Frederick the Great' – had succeeded to the throne of Prussia earlier the same year, and had lost no time in exploiting the questionable legitimacy of female succession by invading Silesia, defeating the Austrians at the battle of Mollwitz. He was joined by Charles Albert of Bavaria, rival claimant to the Habsburg lands, and almost as a matter of course by France. Britain – or rather George, for Walpole was against a continental entanglement – backed the old ally, Austria, fearful that Prussia would not stop at the borders of Hanover. Britain and France came to blows not by declaring war, therefore, but as auxiliaries of their respective German allies. With Spain taking France's side, the war quickly began to look like a continuance of the War of the Spanish Succession – an affair as old as King George's Oudenarde coat.

Militarily, however, two things had changed. Prussia astounded everyone by the quality of its army – not so much the cavalry, which bolted at Mollwitz (Frederick's reforms had yet to touch them), but the infantry, which could fire at the rate of five rounds to the Austrians' three. The lesson was at once driven home to every prince in Europe: a standing army, albeit one made up of conscripts, would beat an improvised army even twice its size. By contrast, the French, who had remained a considerable power even after the run of defeats at Marlborough's hands, were uncertain, even ponderous, in the field.

In 1742 the Prussians, having got what they wanted – Silesia – withdrew from the war. But two French armies had managed to reach Prague and Vienna, and a third was keeping watch on Hanover from east of the Rhine. The situation looked bad on the map, until all three French armies were obliged to retreat in the face of a revitalized Austro-Hungarian counter-offensive. To hasten their return to France, a British army assembled in Flanders comprising four troops of Household Cavalry, eight regiments of horse and dragoons, three battalions of Foot Guards and twelve of the line – some 16,000 men under the septuagenarian Field Marshal John Dalrymple, earl of Stair.

Stair was a true Marlburian, his age a measure more of experience than of disability. At nineteen he had fought at Steenkirk, and he had been in each of Marlborough's four great battles. His campaign plan for the autumn of 1742 could indeed have been designed by Marlborough himself: he proposed to combine with the Austrians in a bold thrust towards Paris along the valley of the Moselle. George now demurred, however, reverting to the pretence that Britain was not at war with France as such. Nothing happened for months as the army watched the seasons change about them and felt the winter's bite in their Flanders billets. But, unusually for troops confined for so long, they fared well. One of the effects of Walpole's perpetual retrenchment had been the emergence of a corps of experienced junior officers, for since there were few regiments to command and consequently little promotion, there were a great many captains with long service and substantial know-how. These proved invaluable in the hastily expanded army, which emerged from winter quarters in uncommonly good health and spirits – or, as Stair put it, 'with great modesty and good discipline'. Marlborough would certainly have approved. Indeed, he had set the standard.

Opposing forces abhor a vacuum. Notwithstanding the delusory state of non-war, as the three French armies resumed their retrograde march towards the Rhine the combined English–Hanoverian–Austrian army in Flanders, now 44,000 men, was drawn east across the Lower Rhine towards Frankfurt. In mid-June King George arrived – with a vast baggage train, including 600 horses (which severely clogged the roads), and his younger son, the 22-year-old Major-General the duke of Cumberland – intending to take personal command.

Though George had the advantage of ten years on the earl of Stair, and had fought in the same battle in his Oudenarde coat thirty-five years earlier, he did not, alas, have the old field marshal's instinct for campaigning. Against Stair's advice, he now posted his army on the north bank of the Main at Aschaffenburg, 30 miles upstream from Frankfurt, hemmed in by the Spessart Hills to the north. The French, even without *La Gloire*, were not ones to miss an opportunity and quickly cut his lines of communication, isolating the allied army from its magazines and depots at Hanau just east of Frankfurt. After a week the army was showing signs of starving, and George decided to withdraw north-west back to Hanau.

The French marshal, the duc de Noailles, was exactly midway between George and Stair in age (the three may well, indeed, hold the record for combined age in command). Withdrawing south around Frankfurt, Noailles was quick to see his chance, and despite enjoying only a 50 per cent numerical superiority he at once split his force, sending some 28,000 men under his nephew, the relatively youthful (54-year-old) marshal the duc de Gramont, to block the allied withdrawal in the bottleneck between the village of Dettingen and the Spessart Hills. Meanwhile, five brigades would hook south to cross the Main at Aschaffenburg and attack the allied rear, enabling the bulk of the French artillery to enfilade the allied main body from south of the river. On 26 June, with some justification, Noailles boasted that he would have the allies 'dans une souricière' – in a mousetrap.*

George got the army in motion by daybreak the following morning, leading with his own cavalry, followed by that of the Austrians, and then the British and the Austrian infantry, with his best troops – the Guards and Hanoverians – as

---

* A body of troops or a defensive position is 'in enfilade' if fire can be directed along its longest axis, thereby inflicting the greatest damage – from which the verb 'to enfilade' is derived. The allied ranks would therefore have roundshot bowling the length of their lines from the left as they advanced.

rearguard, followed by the artillery and baggage. By seven o'clock the advance guard had reached Klein Ostheim, 4 miles west of Aschaffenburg and halfway to Dettingen. Beyond the village the cavalry halted to let the infantry catch up, but the French batteries south of the Main opened a raking fire from which there was little shelter. Sam Davies, a major's servant in the 3rd Dragoons, recounts in a letter to a tapster friend at the White Hart in Colchester how he was sent to the rear with the other servants and led horses:

We stayed there till the balls came flying all round us. We see first a horse with baggage fall close to us. Then seven horses fell apace, then I began to stare about me, the balls came whistling about my ears. Then I saw the Oysterenns [Austrians] dip and look about them for they dodge the balls as a cock does a stick, they are so used to them. Then we servants began to get off into a wood for safety, which was about four hundred yards from where we stood. When we got into the wood we placed our-selves against the largest trees, just as I had placed myself, a 12-pounder came, puts a large bough of the tree upon my head, the ball came within two yards of me, indeed it was the size of one of your light puddings, but a great deal heavier.

By now the *souricière* was discovered, and the earl of Stair, stung by George's assumption of command and dismayed by his tactical ineptitude, decided that, in his words, 'it was time to meddle'. He began deploying the army in three lines: the front line with British and Austrian troops, the support line British and Hanoverian, and the Guards in the reserve line on higher ground to the rear. But it took all of three hours – as long as it had taken the Royalist infantry to form up at Edgehill. Marlborough's regiments would probably have managed it in a quarter of the time.

At midday Marshal Gramont, thinking the allied main body must have eluded him and that he was facing instead the rearguard, advanced across the Beck stream and likewise drew up in two lines and a reserve. George, brave as ever, if lacking an eye for the tactical situation, began urging his men forward, waving his sword and shouting encouragement in his thick German accent, doubtless to mystifying effect all round. With the enfilading fire of the French artillery south of the river, and no proper order, the advance was uneven. And then when the infantry opened fire on the Maison du Roi (the French Household brigade) it was dangerously premature, ragged and wholly ineffectual – except, it seems, for the effect on some of the allies' horses: George's in particular, which suddenly took hold of its bit and bolted rearwards, its rider only managing to pull up in a grove of oak trees where a company of the 22nd Foot (later the Cheshire Regiment) was sheltering. Evidently the unexpected royal visit went well, for regimental legend has it that George rewarded them for their warm reception with a sprig of oak leaves, which in time became their cap badge.

The infantry of the Maison du Roi now advanced, Marshal Gramont believing he had the advantage. By this time, however, the allied regimental officers had got their battalions in hand, and the front line was soon volleying by platoons in the old Marlburian drill. The Garde Française staggered to a halt, and then hastily withdrew behind the cavalry of the Maison, who in turn charged the allied left. However, they had the misfortune of falling on the 23rd Foot (later the Royal Welch Fusiliers), one of the regiments kept in being after Utrecht, and better drilled than most. The cavalry of the Maison du Roi were seen off rudely by a volley and a hedge of bayonets.

'Our men were eager to come into action,' one of the 23rd's officers wrote afterwards:

We attacked the Regiment of Navarre, one of their prime regiments. Our people imitated their predecessors in the last war gloriously [an early example of consciousness of regimental heritage], marching in close order, as firm as a wall, and did not fire until we came within sixty paces, and still kept advancing; for, when the smoak blew off a little, instead of being amongst their living we found the dead in heaps by us.

During the following charge of the 3rd (King's Own) Dragoons against a great mass of French horse, the duke of Cumberland was severely wounded in the leg. Some said that his mount, like his father's, bolted, though towards the French not away, but this seems unfair: few riders in a cavalry charge, then as later, would have been wholly in control. A typical charge would start calmly enough at the walk, the riders knee-to-knee. The trumpeter would sound 'Trot' once the line had cleared its own side's guns and pickets, and then 'Gallop', when the line would buckle and bow as riders struggled to keep the 'dressing'. Finally the commanding officer would point his sword and cry 'Charge!', from which point all semblance of control would be lost for the final 50 yards, the noise of pounding hooves so great as to drown all shouted commands, trumpet calls and even the sound of firing.

While the cavalry were battling on the flanks, a hard infantry fighting match had developed along the whole length of the line. Here and there the sudden shout 'Cavalry!' would throw up a tight square of bayonets until the danger was past and the volleying could resume. Riderless horses on both sides barged through the ranks to add to the picture of chaos. And all the while the French guns south of the Main kept up their raking fire, answered hardly at all by the allied artillery, who found it extraordinarily difficult to come into action in the growing confusion of 'the mousetrap', and even harder to get up close to the infantry.

It had been thirty years and more since the British had fought in formed lines against regular troops, and if the general officers were rusty the infantry, as at the desperate fight at Steenkirk, were relearning what the bayonet and resolution could do. But although it was the cavalry that kept the French horse busy, and the bayonet that almost literally steeled the infantry's resolve, the day was won by dogged volleying, which grew steadier with the practice. According to Lieutenant-Colonel Russell, commanding the 1st Foot Guards,

> excepting three or four of our generals, the rest of 'em were of little service ... our men and their regimental officers gained the day; not in the manner of Hide Park discipline, but our Foot almost kneeled down by whole ranks, and fired upon 'em a constant running fire, making almost every ball take place; but for ten or twelve minutes 'twas doubtful which would succeed, as they overpowered [outnumbered] us so much, and the bravery of their maison du roy coming upon us eight or nine ranks deep.

Towards the middle of the afternoon, seeing they could make no progress, the French began to quit the field, leaving behind 5,000 dead and wounded. Pressed by the allied cavalry, their retreat soon turned into flight, and many *mousquetaires* drowned in the press to get back across the bridge of boats west of Dettingen, especially among the Garde Française, who had attacked first and taken the most casualties, but who by all accounts tried to cross the bridge with indecent haste. It is ever the fate of Guards regiments to incur the scorn of those more workaday regiments of the line if they appear not to live up to their advance billing: so many of the Garde fell into the river that the line dubbed them 'Les Canards du Main'.

But the allies, who had been under arms since the early hours and were exhausted by the best part of a day's fighting, failed to follow up and turn defeat into rout. Besides, though

Edgehill was a century behind them, the fear of loosing the cavalry and regretting it was still strong. And the French in their Gallic obstinacy might even now turn on them with renewed vigour, for their artillery was still in place and protected by the waters of the Main. Only the most seasoned battlefield commander could have judged it aright – a Marlborough, or later a Wellington. Indeed, at the culmination of Waterloo the 'Iron Duke' would throw all caution to the wind and urge the line forward: 'Go on, go on! They won't stand!' But King George, for all his bravery, was no such judge. He flatly refused to pursue at all, even in the days that followed. And so, while the allied army restocked its canteens and cartridge cases at Hanau, Noailles limped back to France unmolested.

Dettingen, though a worthy feat of arms, was ultimately therefore of no strategic significance. It blooded a good many green men and subalterns, however, and reminded the field officers – if they had ever forgotten it – that in a bruising fight they could prevail by superior musketry. It showed George and his general officers that their military system was lacking; and it would be the last time a British monarch commanded in the field. But Dettingen, for all its insignificance in the strategy of the War of the Austrian Succession, was seen increasingly as a model of British fighting spirit, above all in the infantry. When at the end of the battle the King playfully chided the commanding officer of the Royal Scots Fusiliers, Sir Andrew Agnew of Lochnaw, for letting French cavalry break into his regiment's square, Agnew replied drily: 'An it please Your Majesty, but they didna' gang oot again!'

'Dettingen' is a name habitually given to recruit platoons in the Army still; and for as long as anyone can remember there has been a Dettingen Company at Sandhurst, so prized is the occasion as an example to officers. And the battle was something of a watershed in the making of the army, for it had been a close-run thing – perhaps only a matter of ten or twelve minutes, as Colonel Russell of the Foot Guards had reckoned: it would not do in future to pit too many scratch troops against veteran Frenchmen, even Frenchmen without the *élan* of Marlborough's day. In London the battle was celebrated as a famous victory, Handel promptly writing a Te Deum to mark it. But the red-coated regiments had been lucky: the French had not been on form. How long would it be before they regained it?

7

A Family Affair

Drummossie Moor, Inverness, 16 April 1746

WITH SO SMALL AN ARMY, AND THAT OCCUPIED ON THE continent, the threat of invasion was a business for the Royal Navy. Yet twice – during the Monmouth rebellion and the 'Glorious Revolution' – the navy had been unable or unwilling to intercept a hostile fleet. Although the House of Hanover had reigned in Britain for thirty years, there was still a 'King Across the Water', and now that full-blown war with France was inescapable the spectre of French-backed Jacobitism came back to haunt the Court. Hasty militia reform measures were enacted to provide the belt in case the navy's braces failed.

The French were quick off the mark, assembling a substantial fleet in February 1745, though February was not a month known for the best of weather in the Channel. In fact, a storm soon scattered their warships and sank some of the transports even as the invasion force of 10,000 was embarking. With that puzzling yet characteristic lack of maritime self-confidence, the French promptly abandoned the venture.

But storms and His Majesty's ships could do little to prevent 'invasion' from within. With the predictability of the cuckoo's return in spring, war with France heralded another attempt by the Stuart pretender to seize the crown. Indeed, the French invasion plan had been predicated on English Jacobite support. Almost unbelievably, Louis XV's master of horse had supposedly toured southern England after Dettingen listening to disaffected Tories' proposals. As February's Protestant wind was blowing away the invasion plans, Prince Charles Edward Stuart, the Old Pretender's Italian-born son – 'Bonnie Prince Charlie' – had been in Paris, waiting to cross the Channel to take the crown as regent. All had seemed lost; but he had received word from the Highland Jacobites that if he would land with just 3,000 French troops they would raise the clans to his cause.

Louis XV must have seen the venture as a diversion, now, with only the faintest possibility of success, for he gave Prince Charles Edward every encouragement but no troops. Eventually, by private enterprise – indeed, by the promise of privateering* – Charles Edward found a ship of the line, the *Elisabeth*, and a frigate, and with 700 volunteers from the Irish brigade in French service he sailed in July for Scotland. Intercepted by the wonderfully aptly named HMS *Lion*, the *Elisabeth* turned back for France; nevertheless Charles Edward, blithe as ever, was able to land in the Hebrides, though with only seven men.

The Hebridean clan chiefs were not impressed. 'Go home!' said Macdonald of Boisdale when Charles presented himself.

'I am *come* home,' replied Charles in his Italian-accented English.

Macdonald of Sleat and Macleod of Macleod refused

---

* Under authority of 'letters of marque' private captains were allowed to sink or capture enemy merchant ships and seize their cargo without threat of being treated as pirates – except, of course, by the enemy.

point blank to raise their clans, thinking the venture a romantic folly.

Undaunted, Charles rowed to the mainland and in the middle of August raised his standard at Glenfinnan, at the head of remote Loch Shiel west of Fort William, to which other Macdonalds, and Camerons and Macdonnels – about 1,200 men in all – rallied.

The Forty-five was under way.

But if it is occasionally portrayed as a willing, spontaneous rising to defend the Highland way of life, the picture is false. The Highlanders followed their clan chiefs out of feudal duty or through the habit of decades of inter-clan brigandage. Some chiefs, like George Mackenzie, third earl of Cromartie, resorted to the simple threat of killing the menfolk of any village that refused to assist him and burning their houses. It was a fearsome enough Highland host that marched towards Perth – as legend has it, in colourful plaid and carrying the famed broadsword (though at this stage most were carrying sickle and pitchfork) – but it was no embryonic Scots Nationalist army. To deal with them were about 3,000 untried government troops scattered about the various 'Wade garrisons', the bulk of them south of Perth. Lieutenant-General Sir John Cope, Wade's successor, who had commanded the cavalry with some distinction at Dettingen, gathered up as many as he could, raised fresh recruits, and marched north to intercept the Highland army. Hearing reports of greater numbers than expected, however, he made for haven in Fort George at Inverness, and with no formed bodies of government troops before him, Charles and his army were able to slip out of the Great Glen and have a free run to Perth and beyond – ironically via Wade's own roads. On 15 September they brushed aside two regiments of dragoons in the outskirts of Edinburgh and the following day entered the city, where for the second time his father *in absentia* was proclaimed James III and VIII.

Cope, who had by now learned of the evasion, marched the 100 miles from Inverness to Aberdeen to embark his troops for Dunbar, intending to retake Edinburgh from the south or to fall back on the border fortress of Berwick as the situation dictated. Marching on the capital, his force was surprised and given a terrible drubbing at Prestonpans. Having not put up much of a fight his troops ran away, the gallant Colonel Gardiner of the 13th Dragoons (a regiment raised in the Fifteen and blooded at Preston) being pulled from his horse and hacked to death while trying to rally the infantry – an episode with which Scott makes full play in the first of his historical novels, *Waverley*. In that story, the eponymous Captain Edward Waverley, recently commissioned into Gardiner's regiment, but having joined the Jacobite cause in confused circumstances,

> cast his eyes towards this scene of smoke and slaughter, [and] observed Colonel Gardiner, deserted by his own soldiers in spite of all his attempts to rally them, yet spurring his horse through the field to take the command of a small body of infantry, who, with their backs arranged against the wall of his own park (for his house was close by the field of battle), continued a desperate and unavailing resistance. Waverley could perceive that he had already received many wounds, his clothes and saddle being marked with blood. To save this good and brave man became the instant object of his most anxious exertions. But he could only witness his fall. Ere Edward could make his way among the Highlanders, who, furious and eager for spoil, now thronged upon each other, he saw his former commander brought from his horse by the blow of a scythe, and beheld him receive, while on the ground, more wounds than would have let out twenty lives.

Cope was court-martialled but eventually exonerated. And since history and Scottish literature have generally given

Cope a bad press, it is worth noting what American professor Martin Margulies has written of the court-martial proceedings:

> Anyone who scrutinizes it closely can only conclude that the Board was correct. What emerges from the pages is not, perhaps, the portrait of a military genius but one of an able, energetic and conscientious officer, who weighed his options carefully and who anticipated – with almost obsessive attention to detail – every eventuality except the one which he could not have provided for in any case: that his men would panic and flee.

The Highland army now stood at 6,000, tempting a march on London, for Charles was ever hopeful of a French landing and English Jacobite support. But at first he delayed, enjoying the pleasure of his Holyrood court but also having difficulty persuading his council of war that the English as well as the Scots crown was theirs for the taking. For the Scots had not rallied to his standard in the numbers expected. Many who in principle held Jacobite sympathies refused to throw in their lot with so unreliable a man as James Edward or his untried son – and certainly not at the risk of forfeiting their estates. Many more, largely Presbyterian, simply loathed the Stuart religion. Charles only carried the vote to march over the border by falsely declaring that he had Tory assurances of a rising in England.

Having resolved to march south, the Highland army waited until November before striking out. It was hardly the best month to begin a campaign, and the delay allowed London to withdraw more troops from Flanders (including Hanoverians, and Dutch and Hessian mercenaries), sending them to Newcastle under the veteran 67-year-old Lieutenant-General Henry Hawley, rumoured to be a bastard son of George I. By November Hawley had assembled 18,000 men.

Learning of this, Charles Edward took the western route into England instead, as had the Jacobite army in the Fifteen. And despite the old Roman road which ran from Newcastle to Carlisle, Hawley was unable to intercept them. The Jacobite march rates were consistently impressive throughout the campaign, and a new road would be built after the rebellion to allow troops to get from one side of the country to the other faster.

In Lancashire, however, the promised Jacobite support failed to materialize, the Lancastrian recusants proving even less ready to rally to the cause than they had been thirty years before. Just 200 deluded souls from Manchester joined the colours – and these were Anglicans not Catholics. About this time, too, the long-awaited French landing took place, but not in the strength that Charles had boasted of – just 800 men of Louis XV's Écossais Royaux* and the Irish Brigade. And they landed not on the English coast but in Scotland.

Still marching fast, by early December Charles was at Derby. When the news reached London there was a run on the Bank of England, whose collapse was prevented only by the chief cashier's presence of mind in ordering payments to be made in sixpences to gain time. But at Derby came the habitual Stuart rift with advisers. Charles was all for marching on, but his generals, the capable Lord George Murray in particular, argued that it was futile: 3,000 men had deserted, there were more government troops covering the capital than first supposed, three armies were converging on them (there were in fact only two, the other the product of clever deception), and – perhaps more dismaying still – Murray had learned that Charles had deceived him about both English and French support. Back they turned for Scotland.

They paused to fight a rearguard action near Penrith, the

***

* Catholic (on the whole) Scots who preferred foreign service to Hanoverian – like their Hibernian co-religionists, the 'Wild Geese'.

last field battle on English soil, then left the Manchester Anglicans to hold the castle at Carlisle, a hopeless cause if ever there was one: government troops took the place in short order (visitors can still see the grooves in the walls of the dungeons where the Jacobites licked desperately for moisture). But still Hawley's men could not catch the main body. By Christmas Charles's army was in Glasgow, and in early January they had managed to restore themselves enough to lay siege to Stirling castle. Stirling proved no easier a nut to crack than in the Fifteen, however, and as government reinforcements began arriving from the north, Murray decided it was time to turn and meet their pursuers.

Accounts of the battle at Falkirk on 17 January are varied, and historians disagree over the detail – including the direction of the government attack – but what is not in doubt is that with around 6,000 men Lord George Murray inflicted a sharp reverse on Hawley's superior numbers. It was raining hard, and without practice of late the government musketry had lost its edge. On top of this, some of the rawer troops showed scant inclination to meet steel with steel. Hawley, indeed, was furious: 'Such scandalous cowardice I never saw before . . . The whole second line of foot ran away without firing a shot.'

The duke of Cumberland now hastened north and took personal command. This was 'family' business, and King George could have no greater reassurance in the final solution to the Stuart problem than having his son at the head of the army. Cumberland had taken command after Dettingen, and although he had been worsted at Fontenoy two years later, he had profited a good deal from his time in the field. He knew that the answer, as in Germany, lay in discipline and well-delivered volleys. The army would need to retrain and apply the lessons learned – a response that was to become a characteristic of the British army after a tactical setback.

The duke of Cumberland – a considerably more able general than
his youth or appearance suggest.

But first Cumberland had to secure the Lowlands against a resurgence of the Highland host, and then disengage. Leaving a force to guard the passes south into the central Lowlands, and now augmented by 4,000 Hessians recently landed at Leith, he allowed the Jacobites to fall back on the Great Glen, and in early February marched to Aberdeen, resupplied along the way by sea. And for six weeks, in the relative comfort of that granite city, he drilled his regiments.

At the beginning of April he was satisfied that they were ready for battle, and on the eighth they began marching for Inverness, the Jacobite depot, by way of Banff and Cullen. It was a less direct route but more secure, the right flank guarded by the sea from which, again, the army could be supplied by the Royal Navy. And very deftly did the 24-year-old Cumberland handle his force of 9,000 infantry and cavalry: using his artillery to concentrated effect, he forced a crossing of the Spey, a tricky river and the only place the Jacobites had any real chance of holding up the advance, and reached Nairn on the fourteenth. It was a march of 100 miles, with a contested river crossing, in six days – and with the army in as hale a condition at Nairn as they had been on leaving Aberdeen, it was an impressive rate of advance.

On learning there were redcoats in Nairn, Charles left Inverness (and his supplies) and marched east with 5,000 men to block their further advance. They skirmished with Cumberland's outposts, but little more, and that afternoon the Highland army bivouacked in the grounds of Culloden House a dozen miles west of Nairn while Charles and his lieutenants rode over nearby Drummossie Moor (or Muir) to see if it would serve as a defensive position. Besides spectacular views (in good weather) of the Moray and Beauly Firths and the mountains beyond, the place had nothing to recommend it. Indeed its wet, springy, even boggy moss made the going for man and horse difficult, and since a quarter of Charles's men were armed with broadswords only, the advantage would be to Cumberland's regulars, whose powder and shot did not depend on good going. Lord George Murray objected to 'so plain a field' and urged a different course: leave Inverness uncovered, he begged the prince, and play to the Highlanders' strengths by drawing Cumberland's men into the hills south of Nairn. Charles, with predictable Stuart self-assurance, ignored him.

The following day, the fifteenth, was the duke of Cumberland's birthday. To celebrate, and with the imminence of battle in mind, he provided extra beer and spirits for the army to drink his health – and beef. Indeed, his men had eaten beef every day thanks to the victualling ships coasting the Moray Firth.

The Jacobite army was not so fortunate, their supply meagre in the extreme. A local antiquarian writing in the early nineteenth century describes their condition – the very antithesis of the Marlburian ideal to which the duke of Cumberland had sensibly adhered:

> The scarcity of provisions had now become so great, that the men were on this important day [of battle] reduced to the miserable allowance of only one small loaf, and that of the worst kind. Strange as the averment may appear, I have beheld and tasted a piece of the bread served out in this occasion, being the remains of a loaf or bannock, which had been carefully preserved for eighty-one years by the successive members of a Jacobite family. It is impossible to imagine a composition of greater coarseness, or less likely to please or satisfy the appetite: and perhaps no recital, however eloquent, of the miseries to which Charles's army was reduced, could have impressed the reader with so strong an idea of the real extent of that misery as the sight of this singular relic. Its ingredients appeared to be merely the husks of oats and a coarse unclean species of dust, similar to what is found upon the floors of a mill.

Morale could not have been high in the ranks of Charles Stuart's army, therefore. The hardy Highlanders may indeed have found the short and unwholesome rations familiar enough, but they were scarcely sustaining. A third at least of the Jacobite army were Lowlanders, too, or men from the good farming and fishing country of Aberdeenshire and Angus – or Irish and Scots regulars in French pay: their table did not usually consist of sweepings. But Charles had one more idea. His army had caught the government troops napping at Prestonpans the year before: they might do the same at Nairn, where 'drunk as beggars' the redcoats would be sleeping even more soundly. 'Heigh! Johnnie Cowp, are ye wauken yet?' ran the Jacobite taunt after Prestonpans; they could surely pull off the surprise one more time?

They set off after sunset, about eight o'clock, though not all the foragers had returned, and for six hours they stumbled east in the moonless night. At two they halted, still with 4 miles to go for Nairn. But Murray knew there was not enough time left to get into position before Cumberland's army would stand to for the dawn, and so he ordered his men to turn back. Before the night march the Jacobite army was half-starved; when it mustered on Drummossie Moor next morning it would be half-starved and exhausted.

And now it began to rain – driving, sleeting rain from the east.

Cumberland's army was roused from its well-nourished sleep at four o'clock into a night even darker because of the cloud; but the rain was no more welcome to men with fuller bellies if they could not now get the cooking fires going: 'A very cold, rainy morning,' wrote one private soldier of the Royal Scots, 'and nothing to buy to comfort us, but we had the Ammunition loaf.' The 'ammunition loaf' ('biscuit', that is, bread double-baked) was a good deal better than the Highlanders' nothing, however –

and the weather would be at their backs, too, making it a good deal easier to keep their powder dry.

Cumberland was keen to press his advantage, and before first light he had his men on the march – three brigade columns, each of five battalions, with the cavalry on the left and the Highland auxiliaries scouting ahead. Twice they halted to form line, and twice they resumed the march when the scouts reported false alarm, until towards the middle of the morning, seeing the Jacobite army hastily forming on the moor, Cumberland ordered his regiments to deploy for battle.

Because the broadsword was not a weapon that rested easily in waiting hands, Cumberland expected the Jacobites to attack quickly. He therefore began deploying in three lines, the 'continental' way, but seeing Charles's dispositions – two lines, guns in the centre and on the flanks, the Scots and Irish regiments of the French army, with the horse, in the second – he reduced his reserve to just one battalion and strengthened his first and support lines (to eight battalions and six), and thereby managed to extend on the left to anchor his flank on a low stone wall running towards the Jacobite line.

Training told. 'On our approach near the enemy, the army was formed in an instant,' ran the account in the *Gentleman's Magazine*, which though *parti pris* seems to have given a fair picture of events. The gunners were quick about things too, with pairs of three-pounders run forward between the battalion gaps in the front line, and the mortars just in rear. Cumberland now spoke to every battalion, 'yea, almost to every platoon', continued the magazine enthusiastically: 'Depend, my lads, on your bayonets; let them mingle with you; let them know the men they have to deal with!'

But before the enemy were allowed to mingle, they would take a deal of shot. Cumberland pushed the 8th Foot (Wolfe's regiment) forward on the left flank to use the cover of the low wall to enfilade the Jacobite right, while sending dragoons and the Campbell militia round behind the wall and through the grounds of Culloden House to be ready to take them in the flank.

The Jacobite line should have numbered 6,000, of which the front rank of 4,000 Highlanders were to have made the irresistible charge, but a third of them were still missing after the fiasco of the night ramble. Colonel John O'Sullivan, the Jacobite chief of staff and one of the few professionally trained officers, therefore had to weaken the support line to strengthen the front. It did little for the morale of men who knew already that they were outnumbered.

With their dispositions thus completed in the early afternoon, the two commanders – both 25 years old, both with the future of their royal houses in their hands – could survey the battlefield in silence for a few minutes more. But whereas Prince William Augustus, the duke of Cumberland, could do so with confidence all round, it would not have served for Prince Charles Edward to compare his military credentials with those of his adversary – any more than it would for him to compare his army to the government's in anything save courage. The sleeting rain continued to lash the faces and bare limbs of the Jacobite front rank – weary, wasted men, the bravest of whom could not but have been in some part overawed by the regular ranks of red which stood 600 yards distant with the rain on their backs and who 'kept dry our flintlocks with our coatlaps'. Drummossie Moor was indeed 'so plain a field'.

At one o'clock Charles Edward's artillery (twelve guns) opened fire – a pathetic, ineffective ripple of roundshot, 'extremely ill-served and ill-pointed', wrote Cumberland in his dispatch. His own gunners answered, whereupon the Jacobite gunners ran.

Accounts of Cumberland's cannonade differ (it lasted between fifteen and thirty minutes) but it played havoc with the exposed Jacobite lines. It was not so much the physical damage – there were not that many guns, the rate of fire was slow and the soft ground absorbed the impact of solid shot – but its galling effect on half-disciplined troops unused to standing in line under fire was marked.

It appeared, also, to have paralysed Charles Edward. He hesitated for what seemed an age before giving the order to charge, then urged his front rank not to throw away their muskets in their ardour to close with the tormenting guns. At first the Macdonalds would not budge, however, angry because they had been placed on the left of the line instead of the honoured right. Clan Chattan (Lady Mackintosh's Regiment) were first away, but boggy ground forced them to veer right, into the line of the following regiments, so that the weight of attack was towards the wall on the government left.

Despite the rain, Cumberland's redcoats were able to get off several good volleys, the product of their six weeks' training at Aberdeen. The artillery now switched to canister, with terrible effect, 'the men dropping down by wholesale', as Cumberland would record.

Even dropping down wholesale, however, numbers spoke as a great mass of Highlanders crashed into the two regiments on the left of Cumberland's line, principally on the grenadiers of the 4th Foot ('Barrell's Regiment') commanded by the marquess of Lothian's younger son, Lord Robert Kerr. The 4th, who had been the first regiment to go over to William in 1688 under the lieutenant-colonelcy of Marlborough's younger brother, was perhaps the most battle-hardened regiment in the army. At Aberdeen they and the rest of the infantry had been trained to thrust obliquely with the bayonet, not at the man coming head on but into the right flank of the man on his left, for the Highlander carried the *targe* (round shield) as well as the broadsword, and with these he had broken the government ranks at

Falkirk. But such a drill, if it were possible at all, required the greatest discipline. The 4th held their ground, but at a price: 117 were killed or wounded where they stood, including Kerr himself who received the first charging Cameron on the point of his spontoon seconds before another cleft his skull in two. 'There was scarce a soldier or officer of Barrell's . . . who did not kill at least one or two men each with their bayonets and spontoons,' wrote Cumberland. Nevertheless the sheer weight of Highlander numbers drove the 4th back on to the support line.

Unlike at Falkirk, however, the support line stood steady, and those Highlanders who broke through Barrell's were promptly shot down 'like rabbits', principally by Semphill's Regiment (the 25th Foot) of Edinburgh and Border Scots. Those Highlanders who survived seemed stunned. They 'never seed the English fight in such a manner, for they thought we were all Mad Men that fought so', wrote one of Barrell's afterwards.

Meanwhile Wolfe's 8th Foot had 'wheeled in upon them; the whole then gave them five or six fires with vast execution', as one of his captains later recalled. Having thrown away their muskets despite Charles's entreaties, the Highlanders could make no reply but hurl dirks, stones and even clumps of earth in the fury of impotence – or sense of betrayal?

On the Jacobite left the Highlanders never made contact with the government line. Nervous of the cavalry to their flank, knee-deep in water and with little support from the second line, the Macdonalds simply failed to press home the attack. Cumberland, coolly surveying the wretched scene, now moved his cavalry forward on the right, 'upon which, rather like Devils than Men they broke through the Enemy's Flank, and a Total Rout followed,' he wrote in his dispatch with understandable satisfaction.

It was soon finished on the left and in the centre too,

which suddenly gave way 'in the greatest hurry and confusion imaginable; and scarce was their flight begun before they were out of our sight', continued the dispatch. None was so fast in that flight as Charles Edward himself – covered only by the steadiness of the Scots and Irish regulars of the French legion.

By two o'clock the Forty-five was over. It remained only for the Highlands to be 'pacified', as after the 1715 rising. But this time there would be 'no more Mr Nice Guy': the fugitives were hunted down, and many of them were dispatched on the spot (there had been little quarter given on the battlefield itself). Since they would have been hanged for treason anyway, it probably seemed only summary justice. And the heart of the rebellion was torn out, too: Jacobite estates were even more thoroughly destroyed or confiscated than after the Fifteen, and penal laws were introduced to destroy, in effect, the clan system – of which the banning of tartan and the Highland (bag)pipes were but the least brutal measures. The duke of Cumberland would receive an honorary degree from Glasgow University, Handel would compose yet another work to celebrate a Hanoverian victory (with an anthem in his especial honour – 'See the Conqu'ring Hero Comes!'), but north of the Highland line Cumberland would gain the name 'Butcher'. There would be no more Jacobite risings, however.

As for the military lessons learned, the Forty-five was yet another demonstration of the superiority of well-trained regulars, the devastating effect of volley fire, the winning factor of logistics, and the 'force multiplier' (as it would be called today) of the Royal Navy's close support. These lessons the duke of Cumberland would take with him to the Horse Guards (via two more seasons' campaigning in Flanders), where in 1748 he was installed as commander-in-chief.

8

The Perfect Volley

Three continents, three years: 1756–59

AFTER DETTINGEN AND CULLODEN THE WAR WITH FRANCE stumbled on. It spread to India, where both countries had growing trading interests and increasing numbers of troops to protect them. It spread to North America, where along the Ohio, Missouri and St Lawrence rivers colonial volunteers and British regulars clashed with French troops and their native Indian auxiliaries. Finally in 1748 the Peace of Aix-la-Chapelle brought an inconclusive end to operations in Europe, though skirmishing would continue intermittently in both India and North America.

But the fighting in three continents had underscored the lessons of Dettingen and the Forty-five that disciplined volleys of musketry and handy artillery carried the day in the field battle; that sound logistics were crucial; and that co-operation with the navy could be the key to success. The prime importance of firepower and logistics the army had understood in Marlborough's time; but, Cumberland having shown the way, cooperation with the Royal Navy was only now emerging as a vital principle in the strategy of Britain's expanding trade and the colonies.

There was a further lesson, but it was perhaps too ambiguous yet for the army to embrace. 'Light troops' – troops that fought more as individuals than as parts of the volleying machine – would be indispensable in 'frontier' warfare (what would later be known as 'small wars') in which the enemy did not himself fight in formed ranks. And as the continental armies grew ever more rigid in their formations on the battlefield, so there would be opportunity to dance around them with troops who could move in ways other than the strictly regulated adjustment of column and line. The Austrians were already using light troops, mainly horsed irregulars from the wilder margins of the empire – Pandours, Croats and, of course, Hungarian hussars – though principally in reconnaissance and raiding beyond the battlefield. Frederick the Great was experimenting with his *jaegers* – men in green, hunters from the German forests, and deadly with the rifle, a weapon with which to take careful aim rather than having to rely on sheer weight of shot in the volley.* The French had their *chasseurs*, too, who had helped defeat the duke of Cumberland at Fontenoy. But the opening British volley at Fontenoy had felled 700 officers and men in the French front rank, and that devastating spectacle was etched deep in the British military mind. It would be a full decade, indeed, before the British would employ light troops – in America – and three decades

---

* The 'rifle' – originally 'rifled musket' – was a weapon with a spiral of grooves in the barrel ('rifling') which imparted spin to the round as it was fired, thereby giving greater accuracy especially over longer range. The round – 'bullet' – was specially shaped (usually cylindro-conoidal) to engage with the grooves. The disadvantage of the rifle over the unrifled musket was its slower rate of fire: since the bullet had to be a tight fit in the barrel it had to be forced in rather than (as with the musket ball) merely dropped in.

before a British army fielded light infantry on the Continent.

'King George's War', as the War of the Austrian Succession was known in England (for many believed that George had subordinated the country's interests to those of Hanover), did, however, put paid to the pretence of 'guards, garrisons, augmentations', and any other euphemisms, for the duration of hostilities. 'Our liberties are in no danger from our standing army,' Henry Pelham, the new prime minister, assured Parliament in the early 1740s, 'because it is commanded by men of the best families and fortunes.' In other words, the country need no longer fear being coerced – militarized, indeed – as in Cromwell's day, for the interests of its military leaders were exactly those of the country as a whole. Britain would not be, as Voltaire said of Prussia, an army with a country attached to it.

With the Peace of Aix-la-Chapelle, therefore, not nearly so many battalions were disbanded as after Utrecht; nor did the coming of summer see the chimneys blocked up as usual, or – worse – pulled down. Indeed, in 1751 the duke of Cumberland formally reorganized the infantry into 'regiments of the line of battle', each known by its number rather than by the colonel's name; and for the first time an army list was published giving details of commissions, regiments and the seniority of general officers. But an army in a country whose strategy was maritime not continental was never going to be very big: if more troops were suddenly needed they would have to be raised from scratch, and paid off afterwards. The line of battle – of the infantry in particular, but of the cavalry too – would still be rather elastic.

The duke's modest reforms, including an attempt to standardize drill, proved timely, for in 1755 the continued skirmishing in North America produced a backwash of war in Europe. At first it scarcely came to shots – except for poor Admiral Byng, who failed to press home the battle with a French fleet sent to capture Minorca and was consequently put in front of a firing squad. Voltaire's Candide famously refused to go ashore at Portsmouth when told that the British liked to shoot an admiral from time to time – 'pour encourager les autres'. But Byng's sentence sent a stark message to admirals and generals alike. War was no longer, if ever it had been, a casual affair: the nation expected victory. And encourage it certainly did. One of the admiral's descendants, General Julian Byng, who planned the first successful tank attack, at Cambrai in 1917, and whose Third Army held on more tenaciously than others during the massive German offensive of 1918, was always acutely mindful of his ancestry.

Paradoxically, the resurgence of war on the Continent came from the very policy of avoiding war on the Continent. For in Britain's grand maritime strategy trade and the colonies were the paymaster, and their protection the first priority after home defence. But France's military force and economic potential continued to be a threat to the North American colonies and to the Royal Navy's hard-won control of the North Atlantic. So in order to stop France from pouring reinforcements into Canada and building more warships, Britain adopted an indirect approach: she would distract France by drawing her troops into war nearer home. It might not have passed for 'just war' with St Augustine, but it had its strategic logic.

In May 1756, therefore, having made subsidy treaties with Hesse-Kassel and several smaller states, Britain concluded an alliance with Prussia. A few months later France obligingly tightened the strategic noose around her own neck by an alliance with Austria: Empress Maria Theresa's obsession with revenge on Frederick the Great over the loss of Silesia would bind Prussia into the fight more surely than any treaty. Ironically, too, Britain could now relax a little about the Scheldt estuary (the perennial fear being that it would fall into French hands, offering a prime site from which invasion barges could bear down on the Thames); for, the Austrian Netherlands now being the domain of a French ally, Louis XV could hardly invade them. And so in September, with the harvest gathered in and the depots restocked, the Prussians marched into Saxony, an Austrian ally, and the War of the Austrian Succession began again, albeit with the line-up of belligerents changed. This time it would be called the Seven Years War.

The elastic British army was expanded once again – to 34,000 at home and 13,000 in the colonies – and the militia were put on alert. Ten new infantry regiments were raised, including two of Highlanders who were allowed to appro-priate all the romantic trappings of pipes and tartan denied their clansmen at home. Existing regiments were augmented by second battalions, which became regiments in their own right two years later. Four regular regiments were raised in America, including the impressive Royal Americans made up largely of Swiss and German immigrants, natural *jaegers*, who would be the progenitors of the army's light infantry and rifle regiments. Two thousand extra men were recruited for the artillery and engineers. And this was just a beginning. By 1762, when the war seemed to offer great opportunities in both Europe and the colonies, the home and colonial establishments stood at 112,000 officers and men, with an additional 24,000 in Ireland – considerably more than in the army of today.

The Seven Years War thus saw the consolidation of all that had gone before in the army's development – consolidation *and* innovation, for, besides the free-shooting Royal Americans, the army's eleven regiments of dragoons each raised a light troop: lighter men on lighter horses, more agile, more versatile, capable of independent action – like the Pandours and Croats, but with the discipline of red coats. They were at once successful on the Continent, and soon whole regiments of 'light dragoons' were being raised.

The embryonic light infantry, on the other hand, were as yet used only in the wilderness of upper New York colony and Pennsylvania, where strictly drilled lines of muskets made little impression, especially with the Huron auxiliaries on whom the French were increasingly relying. It was savage frontier warfare, and the lessons were hard learned; never more so than in July 1755 on the Monongahela River near what is now Pittsburgh. Major-General Edward Braddock, a Scot who had fought with the Coldstream Guards in Flanders, advanced into the forest with a force of 2,200 regulars and colonial militia, including a young Colonel George Washington, intending to capture Fort Duquesne and cut the line of French outposts between the St Lawrence and Ohio rivers. In a scene out of the 1992 film *The Last of the Mohicans* (James Fenimore Cooper's novel is set during this war) he was ambushed by just 900 Frenchmen and Indians. When his men broke ranks to take cover, the old Coldstreamer cursed them back into line, whence they poured volley after volley into the neutral trees. Braddock, as brave as he was stubborn, had five horses shot under him before a bullet lodged in his lungs. His dying words were at last perceptive, however: 'We shall know better another time.'

And so they would, for the following year 'ranger' companies were raised – like *jaegers*, frontiersmen who habitually carried the rifle, and for whom red was the colour of blood rather than a coat. Thus began the notion that camouflage and concealed movement had their place – if only, yet, in the backwoods of America.

Meanwhile in Europe the war to keep the French busy and away from Britain's colonies needed attention. William Pitt (the Elder) became prime minister at the end of 1756, and he was at first reluctant to acknowledge the full implications of the maritime strategy. The *ends* were plain enough: the destruction of French power in America and the West Indies. The *ways* were a maritime–land offensive in North America, and war on the Continent to draw French resources away from the Atlantic and the New World. The *means* were increased naval strength in the North Atlantic, strengthened colonial and regular garrisons in America, and subsidies to the continental allies. The snag was that subsidies alone could not guarantee that the allies would fight. Only British troops fighting alongside them could do that.

Pitt's reluctance to seize the nettle and field a continental army arose from two obvious and perennial fears: first, the fear of invasion; second, the fear of the bottomless pit into which the country would have to pour money once it sent troops to Europe. His was a dilemma that would remain with the army right up to the present day: how to find troops for the Empire (and today for the consequences of empire) – 'light troops', on the whole – while maintaining the 'heavier' capability to fight a first-class army.

Pitt first tried a compromise. Rather than send an army to fight alongside his German allies he would mount seaborne raids on the French coast – 'tip and run' raids. The compromise failed, however, proving to be no more than an irritant to the French, and certainly no diversion, since the landing forces did not stay long enough to draw troops away from Germany or inflict enough damage to justify a permanent redeployment for defence of the coast. In fact all the raids succeeded in diverting were British troops from the two points of decision – the Rhine and North America.

As so often, it was events that forced the strategic hand. In 1757 the French invaded Hanover and defeated the duke of Cumberland (who had by then been made commander-in-chief of the King's German forces). By September they controlled most of the country, and had it not been for the Prussians' victories at Rossbach and Leuthen, George II might have been forced to give up Hanover and make a separate peace. There was no holding back troops from the Rhine now.

But news from across the Atlantic was not encouraging either. The earl of Loudon, commander-in-chief in North America, had led a seaborne operation to capture Louisbourg, which commanded the sea approaches to the St Lawrence. To do so, because the promised reinforcements from Britain had not materialized, he had withdrawn troops from the New York and Pennsylvania frontier. But to no avail, for on arriving off Louisbourg he saw that the French had more warships than he, and so called off the attack. Meanwhile, under their new and capable commander-in-chief, the marquis de Montcalm, the French took advantage of Loudon's absence to capture the strategically important Fort William Henry at the head of Lake George (the historical backdrop to *The Last of the Mohicans*).

And as if to underline Pitt's failure to grasp the nettle, while Loudon was failing to take Louisbourg and losing Fort Henry for want of warships and troops, back across the Atlantic sixteen ships of the line landed 10,000 seasoned infantry at Rochefort, the French Portsmouth, in another raid. It was even less productive than usual, for the port was heavily defended, and so the raiders promptly sailed home again. Had these 10,000 men been in Hanover, the duke of Cumberland might not have been defeated. Had they been in America, Fort William Henry might have been held. And the warships, had they been off Louisbourg, might have overpowered the French squadron. Opponents at the time described Pitt's tip and run tactics as 'breaking windows with guineas'. Two centuries later, with scarce resources spread all over the Mediterranean, and Churchill suggesting yet more Pitt-like schemes, Field Marshal Alanbrooke would write of his frustration with what he called the prime minister's 'stratagems of evasion'.

There was good news from the East, however, which although it would have next to no effect on the strategic aims of the current conflict, would bring to the fore one of the strongest influences on the army's culture and way of war. In India, Colonel Robert Clive ('Clive of India') won a stunning victory over the Nawab of Bengal's army at Plassey, 70 miles north of Calcutta (the Nawab having captured the city the year before and filled its notorious 'black hole' with prisoners). The Nawab had 50,000 men, Clive had only 3,000, fewer than 1,000 of them British, including the only King's regiment in India, the 39th Foot. The others were 'Company troops', raised and administered by the East India Company and commanded entirely by officers commissioned by the Company. The 39th Foot (later the Dorsetshire Regiment) were hired from the Crown, which otherwise had no interest in the subcontinent, for India was a trading opportunity not a colony.

At a distance the Nawab's army – elephants, camels, bullock carts, billowing standards, regiments of horse and foot – had looked like a well-disciplined force, their spears, scimitars, shields and antique matchlocks not discernibly different from the weapons that Clive's own men carried. The Nawab himself rode into the field in a gilded howdah atop a richly caparisoned elephant, with a regalia of swords, fans and umbrellas, as if at the head of a stately procession (not so very different from Marlborough himself, perhaps, who had more than once driven on to the battlefield in a coach). Clive was not going to test even this antique strength, however. Instead he intrigued his way to victory, for in the weeks before he had bribed the Nawab's discontented officers, and their ambivalence on the day unnerved as much as it actually blunted the army's fighting edge.

Fighting there was, however. The battle opened in the humid heat of a June morning, when the Nawab's French-manned artillery began a massive cannonade of the British camp. It actually did little damage, the shot flying high, but gave the impression of much destruction. The Nawab meanwhile paraded his army up and down in a 'great show of noise and futile movements', as one contemporary account has it, for about five hours, until at eleven o'clock the Nawab's battle commander, Mir Madan, launched an attack on the fortified grove in which Clive's force had taken post. But he did so with only about a tenth of the army's strength. Clive's single Royal Artillery battery answered furiously, a fortuitous ball decapitated Mir Madan, and the attack petered out.

Soon after, it began raining heavily. Clive's troops quickly covered their powder, but the Nawab's men did not. When the rain stopped towards the middle of the afternoon, the Royal Artillery reopened fire, while the Nawab's lay useless. Clive then launched a counter-attack, and by dusk, with the Nawab's guns and muskets useless and his cavalry bribed into inaction, the Bengal army was in full retreat.

The Honourable East India Company was now a force to be reckoned with in Bengal, and by degrees would spread its power through the rest of the subcontinent, largely by the skill of its own troops but also with King's regiments hired from the profits of increasing trade and from the taxes raised in the territory annexed in the wake of their victories. India, indeed, would grow so important to the Empire that by the end of the nineteenth century half the army would be either stationed there, or getting ready to go, or just returned; and the other half would be studying its methods.*

The 'oblique' approach to war that Plassey represented would become characteristic of the following two centuries of continual, if usually low-intensity, operations in India and on its frontiers. It was war at a different pace, requiring sophisticated intelligence of local affairs, nerve and immense patience as well as bluff, gold, bullock carts and artillery, together with initiative on the part of junior officers who

---

* It was the 39th Foot (later the Dorsetshire Regiment), however, which claimed the aptly punning motto *Primus in Indis*.

needed as much political as military acumen. From time to time, solid ranks of red discharging regular volleys would be needed, but it is no coincidence that it was in India that red was first replaced by khaki – where cunning, not mere discipline, was needed to defeat enemies who were masters of the first and who disdained the second.

But winning battles on the Ganges Plain, however thrilling to London, would not draw French troops away from either America or Europe. In 1758 Pitt stopped throwing guineas at windows and instead sent troops to Hanover – 12,000 of them, under command of Lord George Sackville and the marquess of Granby – and more ships to America. The Royal Navy's guns were soon covering an exemplary and successful assault landing on Louisbourg, opening the door at last to an offensive in French Canada.

Pitt's strategic courage continued to reap rewards, and the following year, 1759, was the *annus mirabilis* of British arms. At Minden in Westphalia, not far from where the duke of Cumberland had almost lost his name two years earlier, the British infantry had one of their finest hours. In America, the 42-year-old Lieutenant-General Jeffrey Amherst took Fort Ticonderoga, key to movement between New York and Montreal, while the 32-year-old Major-General James Wolfe stormed the Heights of Abraham to capture Quebec. And in November the Royal Navy gained a famous victory at Quiberon Bay off St Nazaire, ending the threat of invasion – and indeed knocking the French navy out of the war.

Minden stands out as a singular and formative achievement. 'I never thought to see a single line of infantry break through three lines of cavalry ranked in order of battle, and tumble them to ruin,' wrote the French marquis de Contades afterwards. Minden has been celebrated on 1 August ever since by the six regiments which distinguished themselves in the charge against cavalry, and has long been recognized by the rest of the army as a prime example of audacious offensive spirit.

The ancient city of Minden, which had been a Prussian possession since the middle of the previous century, sits on the River Weser where it cuts through a forested ridge, the Wesergebirge. After their setbacks of the previous year, a counter-offensive had taken French troops to within 30 miles of Hanover city itself, and from July the fleur-de-lys was flying over Minden. Duke Ferdinand of Brunswick hastily moved his 40,000-strong Hanoverian–Prussian– British army towards the city to halt any further advance along the Weser, and by the end of the month had managed effectively to lay siege to Minden from the north-west.

The French commander, Contades, decided to break the 'siege' by taking the village of Todtenhausen 4 miles north of the city – on which the allied left rested and where Brunswick had garrisoned some 10,000 men, a quarter of his force – believing it too isolated for the rest of the allied line to support. Just to make sure of this, however, he also decided to mount a simultaneous 'fixing' attack on the allied centre so that Brunswick could not switch troops to his left.

Just after dark on 31 July, therefore, Contades led his troops out of Minden so as to be ready to attack at first light. Brunswick learned of this move, however, and at daybreak the allied army was advancing in good order to meet the French – through open country and covered by a mist on the water meadows. Whether reminded of England by the wild roses in the hedgerows, or perhaps just seized with perverse humour, the British infantrymen plucked the flowers as they marched and put them in their hats, like so many regiments of morris men.

Not long after dawn the French frontal attack on Todtenhausen was seen off with predictable dispatch and appalling casualties. The duc de Broglie, commanding the French right wing, at once realized the village was too strongly defended and sent for reinforcements, but the main body was still on the move, and Contades could not oblige him. For an hour or so the battle lapsed into desultory artillery fire, and cavalry skirmishing on the opposite bank of the Weser. But as the sun began burning off the mist Contades saw the allied line deploying steadily in column, and at once turned his guns on them, opening up an especially galling fire on Lord George Sackville's three brigades on the allied right. And soon Contades' cavalry were moving to charge them.

Sackville's regiments did not form square, as they might have done to receive cavalry, but instead deployed into line so as to bring the greatest number of muskets to bear – a highly risky tactic, but potentially devastating to tight-packed horse. And with their hallmark deadly volleying and the brisk support of two agile twelve-pounder batteries, devastate is what they did, repeatedly throwing the French back.

Meanwhile the British cavalry sat motionless. For despite the protests of their commander, the marquess of Granby, Lord George Sackville flatly refused to unleash them for the counter-charge. Why has never been truly established. But the infantry, though badly mauled and doubtless cursing the inactivity of their own cavalry, now decided not to wait for support, and began advancing in line against the French horse.

With the whole of the right flank – three British brigades and one Hanoverian – now in motion, the marquess of Granby could sit still no longer, and ordered his dragoons and dragoon guards (as the former regiments of horse had been renamed) to advance. But even now Sackville halted him, though the duke of Brunswick, himself heavily engaged in the centre, sent galloper after galloper begging him to strike. The infantry therefore continued to advance unsupported. And astonishingly, contrary to all experience and the 'normal usages', the Minden infantry put the French cavalry to flight. At once Contades' left was thrown into confusion, which soon spread to the centre, so that by ten o'clock the entire French line was in full retreat – leaving 10,000 casualties on the field.

Sackville, as if suddenly awakened from a deep sleep, now rode up to Brunswick and asked that he explain his orders to advance. Perhaps he had been genuinely puzzled, but whatever it was that had moved Sackville, Brunswick was dismissive: 'My lord, the opportunity is now passed.'

The British infantry now thought themselves, with some justification, unbeatable. The Royal Artillery, too, had proved themselves agile in action, and their gunnery skilled. The cavalry, on the other hand, had shown themselves not worth their rations; their presence was not, it seemed, necessary for success. Fortunately for the honour of the marquess of Granby and his frustrated troopers, they would soon have the opportunity to prove otherwise.

Six weeks after Minden, following a protracted and debilitating siege, Major-General James Wolfe landed undetected at night with 5,000 infantry and two guns on the north shore of the St Lawrence River below Quebec's 'Heights of Abraham'. Having scaled the precipitous cliffs and overpowered the outposts, his men formed up for battle – inside the French defensive perimeter, if outside the actual walls of the city.

The Royal Navy had taken Wolfe there brilliantly, slipping past the sentries in the dark and finding a narrow strand of shingle on which to disembark the small-boats. 'The heights near by were cleft by a great ravine choked with forest trees,' wrote the American historian Francis Parkman a century later,

and in its depths ran a little brook called Ruisseau St-Denis, which, swollen by the late rains, fell plashing in the stillness over

Major-General James Wolfe: one of the few enduring names of
the wars with France in the middle of the eighteenth century.
Had he lived to command in America, the outcome of the
Revolutionary War might have been different. There is a
magnificent statue of him outside the Royal Observatory at
Greenwich (presented by the people of Canada), unveiled in
June 1930 by the Marquis de Montcalm.

a rock. Other than this no sound could reach the strained ear of Wolfe but the gurgle of the tide and the cautious climbing of his advance-parties as they mounted the steeps at some little distance from where he sat listening. At length from the top came a sound of musket shots, followed by loud huzzas, and he knew that his men were masters of the position.

Masters for the time being, at least. In the first dim light of dawn they had seen the French picket tents and rushed them. The officer had made off, until a musket ball in the heel brought him to a halt, and the picket had soon surrendered. But a handful had managed to get away and raise the alarm.

The marquis de Montcalm moved swiftly – probably too swiftly, for the troops at hand were not his best. Even so, the rest of Wolfe's men, their general with them, had come up by the time the French were approaching, and there on the 'Plains of Abraham' Wolfe's men waited in line, silent, their muskets double-shotted,* as the French came on in column.

And Wolfe let them come on – until, at 30 yards, his centre battalions, the 43rd and the 47th Foot, 'gave them, with great calmness, as remarkable a close and heavy discharge as I ever saw', as one of the 43rd's officers wrote.

The French, stunned, seemed to hesitate, and returned fire only raggedly. Wolfe's line advanced ten paces and fired a second volley.

It was all over. 'They run! See how they run!' shouted one soldier.

But Wolfe had been hit twice, and lay mortally wounded. 'Who run?' he demanded, opening his eyes suddenly like a man aroused from sleep.

---

* One ball loaded and then another on top of it. The range was reduced, the accuracy too, but at close quarters the device had the obvious effect of doubling the strength of the volley.

'The enemy, sire,' one of his staff assured him. 'They give way everywhere.'

'Then', gasped the dying general, 'tell Colonel Burton to cut off their retreat from the bridge.' And he turned on his side and said simply, 'Now, God be praised, I will die in peace.'

Quebec had fallen (to be followed in short order by the rest of French Canada) in even less time than it had taken to put the French to flight at Minden, and both battles had been won by the resolution and disciplined musketry of the infantry. A century and a half later Sir John Fortescue, the first historian of the army, wrote of Quebec: 'With one deafening crash, the most perfect volley ever fired on a battlefield burst forth as from a single monstrous weapon.'

And from then on, until the invention of rapid-fire rifles, it would be the perfection of the infantry's volleying that would carry the day on the battlefield – so long, that is, as the enemy were obliging enough to fight in regimented ranks.

9

Bald-headed for the Enemy

Warburg, at the headwaters of the Weser, 31 July 1760

QUEBEC HAD VINDICATED GEORGE II'S HIGH OPINION OF WOLFE, just as Minden had proved his low opinion of others. 'Mad, is he? Then I hope he will bite some of my other generals!' had been the King's famous reply to a courtier who ventured an unfavourable estimate of the young major-general. Whether or not he would have touched Marlborough's sphere had he lived longer is debatable, but Wolfe had been capable of boldness and tactical 'grip' at a time when British generalship was otherwise not at its best. For generals were having to learn their trade on the job: without a large army in peacetime there was little opportunity to practise except in the service of a foreign prince.

As for Lord George Sackville, his court martial after Minden was unanimous in its verdict: he was 'unfit to serve His Majesty in any military capacity whatever'. He was replaced by his thrusting second-in-command, the marquess of Granby, whose cavalry he had expressly held back at Minden.

Granby's German command would have a stronger British contingent, too, for such was the optimism after the *annus mirabilis* of 1759 that London sent more regiments to the Continent in what became known as 'the glorious re-inforcement'. And Granby was in no doubt that after Minden the cavalry had a debt to settle.

Despite the wonders of 1759, however, in the early months of 1760 the war did not go well. The Prussians had suffered sharp defeats at the hands of the Austrians and the Russians, and the French had pushed the British–Hanoverian–Hessian army back north once more. By July they were threatening Kassel on the River Fulda, only 100 miles south of Minden. Duke Ferdinand of Brunswick, still in command of the allied army, strengthened the garrison at Kassel, his main base, but withdrew the bulk of his force north of the city to allow himself more freedom of manœuvre. A deal of skirmishing for the crossings of the various smaller rivers followed towards the end of the month, while Brunswick's co-commander Karl Wilhelm, the *Erbprinz* (heir apparent) of Hesse-Kassel, occupied nearby Köbecke, intending to attack the main French positions on the ridge north-west of Warburg.

The French here numbered about 20,000 – thirty-one squadrons of cavalry, twenty-eight infantry battalions and twenty-four guns. The *Erbprinz*'s force was slightly inferior in cavalry and infantry – twenty-two squadrons, including two British regiments of dragoons (the 1st, or 'Royals', and the 7th), and twenty-three battalions of infantry, including two British battalions of the 1st Foot Guards, and the 87th and 88th Highlanders – but equal in artillery. In terms of the usual ratio for successful attack, however – three to one – the *Erbprinz* was severely under strength. At last light on 30 July, therefore, having sent a column to take the Desenberg, the hill north-east of Warburg, to distract the French, Brunswick marched west to reinforce him.

By dawn the two generals had met and Brunswick's troops

were not far behind, but there was a distance still to march to the Warburg ridge. The mist was in their favour, however, so they decided to attack as planned, with the *Erbprinz* making a concealed, right-flanking approach. The French commander, the chevalier du Muy, unaware of what was unfolding on both his front and left, mustered his troops without particular urgency along the ridge behind which they had bivouacked.

Fortune continued to favour the allies, who were able to close to the ridge in the late morning without being discovered. At about midday the *Erbprinz*'s outflanking column (including the Guards and Highlanders) burst from the mist and in short order took the Heinberg, the round-topped hill anchoring the left of the French line. Du Muy counterattacked strongly, but the *Erbprinz*'s second column attacked the French in rear and over-ran the guns. There was a brisk infantry fight, and then a charge by the Royal Dragoons decided it: the French began swarming from the ridge.

But du Muy's cavalry (all thirty-one squadrons) on the right of the line had yet to join the battle: a determined charge might yet have thrown back the *Erbprinz*'s men. The marquess of Granby's twenty-two squadrons of horse had now come on to the field, however, and, having summed up the situation with the much-prized cavalryman's *coup d'œil*, Granby attacked at once.

And at speed. Granby himself galloped so fast that his hat and wig flew off, and he went 'bald-headed for the enemy' – providing the inspiration for many a public house name and its bald-pate sign. These were heavy cavalry (dragoons and dragoon guards) – big men on big horses, with straight swords to impale rather than slash: the hooves thundered, the ground shook, and the shock of collision knocked all fight from those French brave enough to stand their ground. For most had just turned their reins and run from the field with the rest of du

Muy's men. The Minden debt had been spectacularly repaid in full.

It had in fact already been repaid in part a fortnight earlier, 60 miles to the south-west at Emsdorff as the allied army was withdrawing on Kassel. The newly raised 15th Light Dragoons – altogether smaller men on smaller horses, carrying lighter, curved swords for cutting in the duel – had repeatedly charged unbroken infantry, cavalry and artillery, against all expectations of what a light regiment could do, and taken 2,000 prisoners and two dozen guns. As reward they were given the right to bear 'Emsdorff' on their helmets and guidons – the beginning of the system of battle honours.

The cavalry was unquestionably the shock arm of the battlefield once more, as Prince Rupert had been certain it should be, and as Cromwell had made it.

A hundred years after Monck had paraded the remnants of the New Model Army at Blackheath, the British infantry stood at last as a formidably large corps as well as a capable one. By 1763 it had grown to four battalions of Foot Guards and 147 of infantry of the line of battle, including twenty-three of Highlanders. It fought in line three ranks deep, and sometimes only two, or in square when attacked by cavalry; and unlike the French it advanced in line too, its effectiveness lying not in numbers and the juggernaut column, but in musketry – firing on orders, as one body. A few commanders, especially those who had seen 'light infantry' in North America, thought there was a place for troops moving and firing on their own initiative under looser control, but for the time being what gave the winning edge in battle was the volley, and it needed iron discipline – unflinching obedience to orders through constant drill.

The Royal Artillery, although still handicapped by having no permanent teams of drivers (who, like carters at harvest-time, were hired only 'for the duration') was also now, in quality at least, a match for that of the continental armies, its guns much handier and able to get about the battlefield in a way that Marlborough would have envied. The duke of Brunswick wrote to one gunner officer after Minden to commend his skill – 'It is to you and your brigade* that I am indebted for having silenced the fire of a battery of the enemy, which extremely galled the troops' – and gave him and his fellow captains generous bounties. British field artillery was coming of age.

With infantrymen who could stand their ground by fire and take the enemy's ground with the bayonet, and with cavalry that could charge home but remain under control, supported by artillery that was handy enough in and out of action to be able to shape the course of the battle, George II's army now had the potential to be as good as any in Europe. All that was required for a successful campaign was their proper handling before and during battle – in other words, generalship.

But if Marlborough had shown the way, disasters such as Sackville's at Minden, Loudon's in America and Sir John Mordaunt's at Rochefort† showed that good generalship was still elusive, and certainly not a precise science. It did repay study, however; many a continental general had been primed in his profession at one or more of the European military academies. But Britain had only Woolwich, a technical college for the artillery and engineers. The odd Englishman had studied abroad, but it was unusual. Some had served with continental armies, but the majority of senior officers

---

* 'Batteries', usually six to eight guns, were grouped into 'brigades' until 1938, when permanent groupings of batteries were introduced and renamed 'regiments'. For this reason batteries of the Royal Artillery still today have a greater sense of independent history than do companies or squadrons in the infantry and cavalry.

† Ironically, Sackville had been a member of the 'board' of Mordaunt's court martial.

had not. Nor was it easy to discern in peace the character-istics of a good general in war. George II had correctly identified Wolfe's ability, but had at first been scathing of Granby's, calling him 'a sot, a bully, that does nothing but drink and quarrel'. Granby was no Marlborough, though he would become commander-in-chief, but he did share Corporal John's common touch, and gained a reputation for generosity towards his own troops which undoubtedly spurred other officers into humanity in a way that would have been strange to the armies of the *ancien régime.*

One of the problems of British generalship lay in the criteria for promotion. The first requirement of an officer in the first half of the eighteenth century was absolute loyalty to the house of Hanover. Better an untried man with a stake in the Hanoverian succession than a proven soldier with un-certain loyalties (and perhaps even closet Jacobite sympathies). John Campbell, fourth earl of Loudon, a key player in suppression of the Forty-five, had afterwards been promoted to command in America. Benjamin Franklin wrote of him: 'On the whole I then wonder'd much, how such a Man came to be entrusted with so important a Business as the Conduct of a great Army; but having since seen more of the great World, and the means of obtaining & Motives for giving Places, & Employments, my Wonder is diminished.'

Likewise Mordaunt, so timorous in the Rochefort raid, was a staunch Whig MP who had commanded the reserve at Culloden and pursued the Highlanders after the battle. And Sackville was the son of the duke of Dorset, Lord Lieutenant of Ireland. These were men who could be relied on politically. Nor were they without courage: Mordaunt had handled his brigade resolutely after the near-rout at Falkirk, and even Sackville had led the infantry from the front at Fontenoy. Yet they had no aptitude for managing a campaign, or perhaps even handling large numbers of troops in a field battle. When the duke of Wellington was asked who should take over from him in the Peninsula were he to fall, to the surprise of many he replied: 'Beresford. He may not know how to lead an army, but he knows how to feed one.'* Unlike in the century before, there were fewer men rising from the ranks of the minor gentry, men who had spent most of their time soldiering, and more who were appointed to command from the aristocracy, whose 'social obligations' could frequently detain them in London or on their family estates. The army was increasingly a gentlemanly pastime, not a means of advancement.

It was also extraordinary just how hard Parliament some-times made it for a general to do his job. The veteran (Huguenot) field marshal Lord Ligonier was both commander-in-chief and Master General of the Ordnance. Just as Marlborough had done when he held both appoint-ments, Ligonier tried his level best to unite the efforts of the cavalry and foot, which answered to his first title, with those of the artillery and engineers, which answered to the second. But Parliament kept the lines of responsibility and the budg-ets of the two departments strictly independent of each other, and this was inevitably mirrored in operations on the ground. It was the same with supply and transport: vict-ualling and clothing were the responsibility of the Treasury, with the same civilian commissary arrangements as in the cen-tury before, while billeting and movement of troops at home remained the business of the secretary at war. The army owned not a draught horse or a waggon of its own, relying instead on civilian hire. It was a perfect system for making sure the army could not threaten the peace of the realm; it was equally im-perfect for making war on the King's enemies. Such was the

---

* William Carr Beresford reorganized and led the Portuguese army, and well. Yet in his battles he almost invariably lost control and had to be bailed out by others – sometimes by Wellington himself.

inheritance of Cromwellian militarism and the Jacobite scares.

Marlborough had overcome the problems through force of personality and willingness to put money into the right hands. Half a century later, few generals had Marlborough's personality and ability to scheme, and fewer still his experience. Duke Ferdinand of Brunswick complained of being unable to take to the field after winter quarters in 1760 because 'I have a monster of a commissariat independent in some respects of me, and composed of several heads independent of each other, each with its own chief or protector in England, but together as ignorant and as incapable as they are avid to line their own pockets.'

Thus the know-how of generalship was largely the preserve of elderly officers schooled somewhat cynically in working the system. And although age did not necessarily make them incapable in battle, as Lord Stair had demonstrated at Dettingen, it did not make for campaign flair either. Even Granby, for all his dash, knew that the art of campaigning did not come easy to him, for 'sudden marches, alarms &tc drive the Commissariat business sometimes right out of our heads'. There were, of course, officers like Wolfe who rose with astonishing speed, but the system as a whole was haphazard. Perhaps, in the end, failure on the Continent was not so calamitous: war there was a diversion after all, and there were always Prussian field marshals and German troops to pull the fat from the fire. And when a raid on a sugar island or a coastal fort went wrong the Royal Navy could usually sort out the mess, landing more supplies or evacuating the army.

But in North America it was different. There, as the early defeats in both wars of the mid-century had shown, mistakes of generalship were not so easily mitigated. The next trial would hammer the point home.

The Seven Years War ended on terms far more favourable than Marlborough's wars, and certainly favourable to the maritime strategy. There was a deal of territory trading around the Caribbean, but the biggest gains in America were in the north and to the west. Before the war, British colonial America had been like a thin layer of sandwich filling between the Atlantic, whose control was uncertain, and the French who, though scattered thinly, held the interior from Canada (between the Great Lakes and the mouth of the St Lawrence, except for the snowy wastes around Hudson's Bay and half of Newfoundland island) to the warm waters of the Gulf of Mexico at New Orleans. By the end of the war, the sandwich filling had expanded to the Mississippi in the west and to Hudson's Bay in the north, and the French outer layer had dwindled to the rump of Louisiana. That relic apart, there was no European power in continental North America but Britain. It was a glorious western empire for the new king, George III, who had come to the throne in 1760.

It had to be paid for, however. And it was the attempts to make the American colonists pay their share, without the fig leaf of parliamentary representation in London, that precipitated the Revolutionary War. 'Spin' would invest the revolution with a certain nobility, but as Dr Johnson wrote in *Taxation No Tyranny* (1775), 'How is it that we hear the loudest *yelps* for liberty among the drivers of negroes?'

Noble or not, the war ended with a complete defeat of British arms and an independent United States of America. But how did a handful of 'regular' American regiments and few thousand militiamen defeat a nation which had only just – and so spectacularly and comprehensively – helped defeat two of the greatest continental powers in a worldwide war? In military terms the answer was generalship, but an equal cause was parsimony: for Defoe's warning, that it took three years and 30,000 lives to make a new army, was as apposite now as it had been after the Peace of Ryswick.

For the cuts following the Seven Years War were savage. In 1763 the land forces in His Majesty's pay (including German mercenaries) stood at 203,000. They were straight away reduced to just seventy regiments, each of a single battalion and at lower strength. Yet they had to garrison a vastly expanded empire. By 1764 the army had been further cut to 45,000, of which 12,000 were to police Ireland. And so when in 1763 the capable General Amherst was succeeded as commander-in-chief in North America by General Thomas Gage, the army was already stretched.

Gage himself was a difficult man. He knew the colonies well but his record in the field was at best mixed; his faulty tactics had led to disaster at the battle of Monongahela, and Amherst himself had questioned his judgement. He was confident of his own ability, however, and his brother was a government minister. And to his credit he had not been silent in the ten years before 'the shot heard round the world' at Concord, Massachusetts (April 1775), warning repeatedly of the need for more troops to counter the increasing truculence of the American militia. But he mishandled the opening skirmishes of the American Revolutionary War, and General William Howe, his deputy, whose brother had been killed at Ticonderoga, took over command in October.

Howe was a capable soldier, and he knew the country. He had led the first troops up the Heights of Abraham at Quebec and had been with Amherst at the capture of Montreal. But the early bruising battle at Bunker Hill in June, where the Americans held their ground like seasoned troops, had shaken his confidence severely: 'A dear-bought victory,' wrote Major-General Henry Clinton, one of his commanders; 'another such would have ruined us.' After Bunker Hill, indeed, Howe became excessively cautious and reluctant to seek any direct confrontation. But he saw clearly that the rebellion was much more than London supposed. It was not the work of a minority of colonists who could be faced down easily by a local display of force (even if he had had the force). Begging for 20,000 more troops, he wrote to

London that 'with a less force . . . this war may be spun out until England will be heartily sick of it'.

With one of those exquisite twists of eighteenth-century irony, Howe's letter landed on the desk of a man outstandingly ill-equipped to respond to it. For the Secretary of State for the America Department, the minister responsible for the higher direction of the war – the determination of strategy, the appointing of generals and the allocation of resources – was none other than Lord George Sackville, reinvented as Lord George Germain. Despite the Minden court martial's verdict that he was 'unfit to serve His Majesty in any military capacity whatever', the new King had taken him to his bosom, never having got on with his grandfather and so delighting in reversing his decisions.

Sackville's views of the insurrection across the Atlantic, where he had never set foot, were straightforward, and set the tone for the 'counter-insurgency' campaign: 'the rabble . . . ought not trouble themselves with politics and government, which they do not understand . . . these country clowns cannot whip us'.

With this underestimation of the threat, a commander-in-chief rapidly losing his nerve, and six to eight weeks' sailing between the issue and receipt of reports and orders, the stage was set for tragedy. Sackville and the prime minister, Lord North, would proceed on the assumption that the American militias could not stand against British regulars, and that the war would be the same as that which they had recently fought in Europe. Marlborough would have wept.

10

The Country Clowns Whip Us

Revolutionary America, 1777–83

THERE HAD NEVER BEEN A SIGHT LIKE IT: A BRITISH ARMY UTTERLY defeated, grounding arms and surrendering. And in the rotunda of the Capitol in Washington, reminding every United States legislator and visitor alike that independence was won by force of arms, not merely by declaration, there is a huge painting of the scene. Lieutenant-General John Burgoyne in full dress uniform gives up his sword to General Horatio Gates, who also wears full dress, but dark blue, the colour of the American regulars. Gates, however, is refusing to accept the sword, indicating that he will treat with Burgoyne as a gentleman, and inviting him into his marquee from which flies the Stars and Stripes. Gates's officers, also clad in regular uniform, with not a racoon hat to be seen, observe the ceremonies with due gravity, while those in red coats appear bewildered and dejected. Burgoyne's force had left Canada 10,000 strong; they had lost 4,000 in the following six months, and now the remaining 6,000 were prisoners of war – small numbers by continental European standards, but

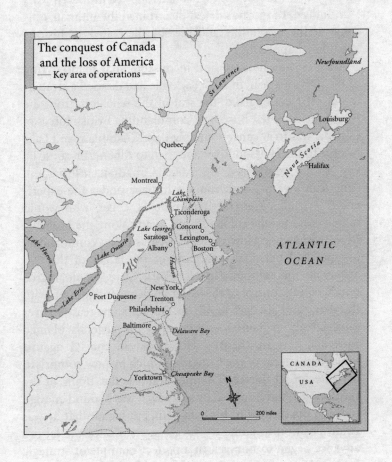

The conquest of Canada
and the loss of America
— Key area of operations —

St Lawrence

Newfoundland

Louisburg

Quebec

Nova Scotia

Halifax

Montreal

Lake Champlain

Ticonderoga

Lake George   Concord

Saratoga   Lexington

Albany   Boston

Lake Ontario

Lake Huron

Lake Erie

Hudson

Fort Duquesne

New York

Trenton

Philadelphia

Baltimore

Delaware Bay

ATLANTIC OCEAN

Yorktown   *Chesapeake Bay*

N

0   200 miles

CANADA

USA

the shock of surrender was felt round the world, for Saratoga had brutally revealed the hopelessness of Sackville's strategic assessment, as well as the dangerous weakness of the British military after the economies that had followed the Seven Years War. Indeed, Burgoyne's defeat determined the ultimate outcome of the revolution, and thereby the course of history. What had happened?

After Howe's hesitant and inconsequential manœuvring of the previous year, in 1777 Sackville had devised a strategy to cut off New England, the seat of the rebellion, from the rest of the colonies. Burgoyne, commanding British forces in Canada, was to advance south via Lake Champlain and Lake George to the Hudson River and then to Albany. En route he was to be reinforced by 2,000 Canadian militiamen, American loyalists and Iroquois Indians under the experienced Brigadier-General Barrimore St Leger, who would march east into the colony from Canada along the Mohawk River. Meanwhile Howe, who as well as being commander-in-chief had personal command of the larger of the two armies in the colonies (in New York City), would march north, taking control of the lower Hudson, and join up with Burgoyne and St Leger in Albany. But for some reason never satisfactorily explained – afterwards the generals blamed Sackville, and he them – Howe instead moved against Philadelphia, seat of the Congress (which by now comprised delegates of all thirteen colonies and was the 'brain' of the revolution). And he did so by sea, which was a sound enough move in the context of his own plan, but meant that he was in no position to come to Burgoyne's aid once the latter's advance began to unravel. In a perfect example of strategic futility, Howe succeeded in taking Philadelphia – but not before the Congress had slipped away.

Burgoyne's army, consisting of 3,000 British regulars, 4,000 Brunswick mercenaries, and 1,000 Canadians and Indians, had been tramping south for a month when Howe left New York. They retook Fort Ticonderoga, the main American base along the chain of lakes, on 6 July (it had been captured by Benedict Arnold in a brilliant action in May 1775), but unfortunately not with the defenders inside. The next day at Hubbardton they caught up with the Americans in a bloody clash from which the rebels finally broke and ran – at which point Burgoyne made a bad decision. Why is not entirely clear, but he had a deal of self-regard that could tip over into swaggering over-confidence. Instead of taking the most direct route to Albany, by boat via Lake George and then, after the relatively short portage to Fort Edward, down the Hudson River, he struck out over-land east of the lake via Fort Anne to harry the fugitives and bring them to battle. But the forests of Vermont and upper New York were not the plains of Germany or Flanders: the terrain hindered his own movement and greatly favoured the defenders.

Destroying rebel forces en route was not essential to Burgoyne's mission of linking up with Howe to cut off New England. It would be a bonus, but only if achieved without detriment to his primary object. In any case, he had secured Fort Ticonderoga, and once he had joined up with Howe the rebels would be hopelessly outnumbered and restricted in their options. In other words, there was no need to fight. Some fifty years later the Prussian strategist Carl von Clausewitz would enumerate the principles of war, the first of which is selection and maintenance of the aim: that is, commanders correctly identifying the object of military action and then directing all effort towards achieving it, a concept the ancients understood well enough. Howe's extra-ordinary diversion to Philadelphia may have been the more spectacular breach of the principle, but Burgoyne's was just as culpable, and just as calamitous.

His army now made slow progress in the wilderness, allowing the Americans to recover. The rebels chopped down trees to block what tracks there were – heavy work, but heavier still for the pursuers, who had to clear each obstacle to get the guns and waggons past. It took until 1 August to get to the Hudson at Fort Edward, by which time Burgoyne was running out of supplies. And here he received a letter from Howe telling him of the change of plan.

A further week passed while Burgoyne took stock of the situation, his supplies critically low. In the end he decided to send 1,000 of his Germans and Indians to forage east towards Bennington, Vermont. This proved disastrous, both militarily and in propaganda terms. The Vermont militia had already been alerted, and all but a handful of Burgoyne's detachment were killed or captured. Yet even if his men had been able to requisition supplies in a regular fashion, in exchange for cash or promissory notes, commandeering food and animal fodder at this time of year was a depredation which the colonists could only have resented. In fact, so heavy-handed – brutal, even – was the foraging, that loyalists across New England began changing allegiance, while Congress exploited the 'atrocities' throughout the colonies.

Burgoyne had already had to leave behind almost a fifth of his force to secure Fort Ticonderoga. Now, after Bennington, his Indian scouts, on whom he relied heavily in the forests, began deserting him. And to make matters worse, although he would not hear of it until the end of the month, St Leger's force, his only source of reinforcements, turned back at Fort Stanwix 80 miles west on the Mohawk.

In contrast with both Howe's and Burgoyne's mis-conceived moves, the American commander-in-chief General George Washington had acted with a sure strategic touch. He knew little of Burgoyne's object and progress, and even less about what Howe intended to do after embarking his army at New York, for Washington had no navy to shadow the fleet. His crucial decision, therefore – whether to send reinforcements north to counter Burgoyne or south to meet Howe (if indeed Howe was headed south) – rested on military logic rather than on definite intelligence. A weaker general would have delayed until Howe's object became clear, by which time it would have been too late to intervene decisively either way, but Washington recognized that the greatest danger was a junction between Howe's and Burgoyne's armies, and decided therefore to march against Burgoyne to prevent it. For good measure, too, he sent ahead his most pugnacious colonel, the legendary (and later infamous) Benedict Arnold, with 500 sharpshooters of the 11th Virginia Regiment, some of his best troops, and called out every man of the New England militia.

But who would lead this northern army? Congress, on hearing that Fort Ticonderoga had been abandoned without a fight, had sacked General Philip Schuyler from his command of the Northern Department – one congressman echoing Voltaire: 'we shall never hold a post until we shoot a general' – and now appointed the gentlemanly General Gates in his stead. It was risky: Gates was a former British officer (although so were many), and his experience was limited. Nevertheless he quickly recognized the strength of the position that Schuyler had been preparing at Bemis Heights on the Hudson, 20 miles downstream from Fort Edward, and sent every man he could muster there – 15,000 regulars and militia – to block the British advance.

At Fort Edward, learning of the strengthening American defences, Burgoyne calculated that he had three options. He could try to take Bemis Heights, although the Americans outnumbered him two to one. He could cross to the east bank of the Hudson, under fire, and try to bypass the position, although this would expose his flank. Or he could acknowledge that the strategic situation had changed radically, and return to Canada. But turning back was not in Burgoyne's character. And crossing the Hudson River – risking a fluid battle against superior numbers in country he did not know – would be hazardous in the extreme. The course that most suited him (he had made his name leading a bold attack on a Spanish fortress during the Seven Years War), would be to attack Bemis Heights; there, at least, he knew the enemy's position and intention. And so he chose to attack.

Having moved up the bulk of his force by the middle of September, and after a preliminary reconnaissance, on the nineteenth Burgoyne began the assault with a right-flanking approach through land belonging to a loyalist, John Freeman, whose farm the Americans had failed to picket. By mischance, however, that very morning Benedict Arnold had finally persuaded Gates, despite their mutual loathing, to allow him to send troops to picket the farm, denying Burgoyne therefore his only real chance of success.

The British advance in three columns was by no means inflexible, for each regiment now had a light company (thanks in no small part to Wolfe's experiments twenty years before). These were hand-picked men, chosen for their field-craft and intelligence, and though their equipment was weighty enough (they carried the same musket, the 'Brown Bess', as the other companies) their role was one of scouting and skirmishing, and they were trained to fight in open (loose) order, as well as in close. For the attack that morning the light companies had been mustered as a single battalion, and with the 24th Foot they formed a brigade under Brigadier-General Simon Fraser, their object to turn the American left flank. Mist delayed the advance for several hours, but in the early afternoon the leading troops of Fraser's brigade emerged from the woods into the fields of Freeman's farm. Here they met with spectacularly accurate rifle fire from Arnold's Virginians. Every officer was hit, and the rest fled back into the forest.

The Americans followed up over-hastily, however, running on to the steady bayonets of the grenadier companies which had been mustered as a battalion under Fraser's fellow Scots brigadier, James Hamilton. In turn, they fled for the trees south of the farm.

There was now a 'race for the ground' – the fields and buildings of Freeman's farm – with Burgoyne trying to get both Fraser's and Hamilton's brigades forward, while Arnold tried to steady his nervous militia by getting the regulars on the heights behind to extend their line towards the farm.

Fraser's brigade tried again to move round the American left flank, but once more the accurate rifle fire of the Virginia sharpshooters drove them back, this time with even heavier loss. Again the Americans counter-charged, Arnold himself leading, and again they were repulsed. But despite repeated requests from Arnold, Gates would not reinforce the left, in the end becoming so frustrated with his pugnacious subordinate that he put him under arrest.

Meanwhile in the centre the Brunswickers under Major-General Friedrich Riedesel had managed to forge their way through a ravine which the Americans had thought impassable, and with their support Burgoyne was at last able to take the farm and its surrounding land. But to do so had cost him 600 dead and wounded, a tenth of his force, and still he had been unable to make any impression on the main American defences. On the other hand the Americans had suffered about half that figure – and more militiamen were arriving by the hour. Burgoyne therefore decided to dig in at the farm, leaving Gates to consolidate his position on the heights.

Burgoyne now had to take a long, hard look at his situation, for supplies were getting desperately low again. A letter from Major-General Henry Clinton, commanding the remaining troops in New York City, promised reinforcements by the end of the month, and so, in spite of the worsening odds, Burgoyne decided to hang on. In the event Clinton did not set out until 3 October and turned back after meeting strong resistance in the Catskills, but by then Burgoyne had realized he would either have to withdraw or assault unreinforced. On 7 October he attacked with 2,000 men – a third of his remaining strength – while the Americans, further reinforced in the three weeks since the first battle, had placed their best troops – 4,000 provincial regulars, with 1,200 militia in support – on the left flank, where they expected the main attack.

Burgoyne's plan of attack, at least in theory, was sound: what it lacked was the troop numbers to prevail. Brigadier-General Simon Fraser, with the light infantry, some rangers, Canadian militia and the remaining Indians (around 700 men) were to make a wide right hook to secure positions for the artillery. Major-General Riedesel's 1,100 Brunswickers, with support from Fraser's artillery, would then make the main attack on the American centre. Meanwhile, Major-General William Phillips, a veteran artilleryman and Minden hero, would lead just 400 men of the grenadier battalion, with the lighter artillery, against the American right to try to prevent their reinforcing the extended left flank.

Burgoyne knew the odds were heavily against him, and he timed the attack for early afternoon so that he could use the cover of darkness to withdraw if necessary. Even so, things began badly. On the left, Phillips' men were thrown back with heavy casualties, the general himself taken prisoner along with the grenadiers' commanding officer. On the right, the Americans – the Virginia riflemen to the fore again – stopped Fraser's brigade moving west by deadly marksmanship and aggressive counter-attacks.

And now came a gift to the revolutionary cause, and in time to Hollywood film-makers. Chafing at his confinement in Gates's headquarters, Benedict Arnold suddenly rushed from his tent, leaped astride a horse and galloped towards the firing on the far left of the American line. Once there it did not take him long to sum up the situation: despite the

British losses, Fraser was rallying his troops with astonishing nerve. Turning to Daniel Morgan, the roughneck colonel of the 11th Virginia, Arnold said, 'That man on the grey horse is a host in himself and must be disposed of.'

Morgan, who had never forgotten the flogging he had received while serving under Burgoyne as a teamster in the Seven Years War, detailed his best marksman, Timothy Murphy of Schoharie County, New York, to bring down the brigadier. Forty years on, at Waterloo, an artillery officer asked the duke of Wellington's permission to open fire on Napoleon, whom he had just spotted on the ridge opposite. The duke flatly refused: 'It is not the business of a commander-in-chief to fire on another,' he said stiffly. Neither Arnold nor Fraser was as elevated in rank as the duke, but the affair revealed the nature of war with irregular troops, and would show many a British officer what the ruthless pursuit of victory required. Murphy loaded his long double-barrelled rifle, climbed a tree, and at a range of 300 yards fired one round. Fraser fell mortally wounded. The attack at once began to stall, and soon afterwards the light battalion and Canadian militia were streaming to the rear.*

In the centre, however, Riedesel's Brunswickers were making surprising progress, over-running several strongpoints. Arnold now galloped back to the centre of the heights and managed (not without difficulty) to persuade the shaken American regulars and militia to counter-attack. Leading the charge in person, he was struck by a musket ball in the leg, which was then broken when his horse fell on him. But his counter-attack drove Riedesel's men back down the hill to their original starting line, and after an hour of

---

* Timothy Murphy is Schoharie County's Revolutionary hero. Owner of one of the first double-barrelled rifles, he was known to the Indians as 'the magic man whose gun shoots without reloading'.

heavy fighting all Burgoyne's men were either casualties or back where they had begun.

Nightfall was their saving, as Burgoyne had foreseen it might have to be, but only for the moment. Over the next two days they tried to get back to Fort Ticonderoga, but Gates's men moved rapidly to block their escape. Finally the survivors rallied at Saratoga, 40 miles south of the fort, surrounded, hopelessly outnumbered and in no condition to fight. Burgoyne began to parley, and on 17 October 5,791 men grounded arms, more than a third of them unfit for duty. It was not, strictly speaking, an unconditional surrender, but it might as well have been, for the consequences would be the same.

The same and immediate. American confidence at once surged. From here on, although there would be many ups and downs, the Congress could hold out for the supreme goal of complete national independence, knowing that their troops could deliver victory and that the colonists from whom the militias and regulars were raised knew it too. Conversely, loyalist support, which in any case was always overestimated in London, began leaching away. Critically, across the Atlantic the French were quick to reach the same conclusion, and not three months later they signed a treaty of mutual assistance with the Congress.

Britain was now at war with its old and reinvigorated adversary, and without allies to pin down French troops on the Continent. Indeed, Spain and even Holland threw their lot in with the Franco-American alliance. The outcome was now almost inevitable. Britain was soon no longer master of the oceans, nor, critically, of American coastal waters. British troops could no longer be switched easily from one place to another – from Canada to Boston, from New York to Chesapeake Bay, from Philadelphia to the Carolinas, or to any number of vulnerable points – because the Royal Navy, neglected almost as much after the Peace of Paris as the army had been, found itself drawn south into the Caribbean to protect Britain's sugar islands, the major engines of wealth.

Four years after Saratoga almost to the day, on 19 October 1781 at Yorktown, Virginia, Major-General the Lord Cornwallis surrendered another army, this time to General George Washington in person. Cornwallis was outnumbered three to one (almost half the opposing troops were French), and reinforcements could not get through from New York because the French navy controlled Hampton Roads. And although bitter skirmishing would continue for another eighteen months, after Yorktown the war was truly lost. In September 1783 Britain recognized the reality and sought to mitigate wider damage by conceding independence to the Congress and negotiating a peace with France and her allies.

The consequences of the American war for the army were profound. Never again would Britannia *not* rule the waves: in Whitehall all thoughts of the red coat were eclipsed by those of the blue. The Royal Navy would be the prime strategic instrument; the army would furnish a few essential garrisons, the flotsam and jetsam, almost, of global sea power. So thoroughgoing was this strategic vision that within two decades the navy would be consuming two-thirds of Britain's gross domestic product, while the army would wither and almost die. Soon indeed, as the nineteenth-century historian Macaulay would put it,

The English Army, under Pitt [the Younger, prime minister from 1783], was the laughing stock of all Europe. It could not boast of one single brilliant exploit. It had never shown itself on the continent but to be beaten, chased, forced to re-embark, or forced to capitulate. To take some sugar island in the West Indies, to scatter some mob of half-naked Irish peasants, such were the most splendid victories won by British troops under Pitt's auspices.

And yet the defeat in America had not been wholly, or even primarily, military. Although two British armies (there is no better term, for Burgoyne and Cornwallis were operating independently) had surrendered, this had come about through want of neither courage nor fighting technique, nor even of generalship at the tactical level, for the Americans had frequently been worsted by shrewd and adaptable British commanders. There had, however, been abject failure at the operational (or 'theatre') level – the level at which campaigns are planned and conducted, decisions are made which battles to fight (and where and when), at which intelligence is managed and resources are allocated. At the operational level, Burgoyne failed to maintain the strategic object of his march south, namely to meet up with Howe's army in order to cut off New England from the other colonies. And he failed in the commander's first duty – the preservation of his force. Similarly, Cornwallis had had the opportunity to slip from the trap by quitting Yorktown and marching south into Carolina, but he held on too long through faulty intelligence and inadequate cooperation with the Royal Navy.

There were reasons for these failures (and Burgoyne had cause to be bitter about the way he had been abandoned) but they were failures none the less. They were also failures with strategic consequences beyond the loss of the American colonies and the switch of resources to the navy. Saratoga and Yorktown would loom large in the consciousness of the next generation of army officers – most notably Sir John Moore and the duke of Wellington – who in turn influenced not only the generation that followed them but to a significant degree every generation since. When for example in 1989 the army published (effectively for the first time) its concept for the design of campaigns – *The British Military Doctrine* – the illustration on the booklet's jacket was of the duke of Wellington at Waterloo; and it remains so in its most recent edition. The impact of any surrender is profound, but the British experience in America has shaped the way the country's generals have thought ever since, if not always consciously. Invariably short of men and materiel (the hardware of war), they have instinctively sought an indirect approach to winning wars (with one exception – the First World War, although even here the politicians hankered after the easier option). Sometimes they have rescued their armies in such dramatic fashion that deliverance has been presented as victory, as at Corunna in 1809 and Dunkirk in 1940; at other times the imperative of avoiding casualties has led to accusations of timidity, as in East Africa in 1941 or Singapore in 1942.

And yet there was another conclusion to be reached about Saratoga and Yorktown. Britain's strategic and economic position was not in fact gravely weakened by the loss of America. Indeed, if the retention of the American colonies had been truly in the nation's vital interest, Lord North's government would have been obliged to saturate them with troops as soon as the reality of the conflict was revealed – which it never did. To George III the colonies were something almost sacred, while to the majority of Parliament (the Commons certainly) they were really no more than a means of increasing prosperity. The war had in fact been prosecuted half-heartedly; it was therefore not surprising if generals in the field, and in turn their subordinate commanders, began to make their own calculations about how much a battle was worth fighting, not just in terms of the duty to preserve the force, but in terms also of the object of the battle. The Americans – regulars and militia alike – had been fighting for the independence of their country; put crudely, the British had been fighting for a commercial policy. Indeed, in due course the constitutional purpose of the United States Army would be expressed simply as the defence of the Union: whenever it has fought, it has done so in the knowledge of that high principle (albeit 'defence' has sometimes been elastically defined – most fatally, of course, in Vietnam). But the British army has not always fought on so fundamental a principle. As a result, the US and British armies have developed differently in character and doctrine – at least until recent years – and their generals have frequently thought differently too. Even in the Second World War, when Britain was fighting for national survival (as well as in the Far East to defend her commercial wealth), there was often a tendency to avoid direct engagement with the enemy's main force, seeking instead an indirect approach – what others, the Americans especially, derided as 'stratagems of evasion'.* Only twice in that war – at El Alamein and on D-Day – was there an acceptance by commanders that no price was too great for the attainment of the goal.

The battle at Bemis Heights is therefore worth pondering, for on this tactical action the war turned. And the battle turned on the actions of individuals – Benedict Arnold and Timothy Murphy in particular. And if Murphy had not been so fine a shot it might have turned the other way. The lesson of this episode, that the most modest tactical success can transform the strategic situation when neither side significantly outnumbers the other, is one the United States has carried with it in its doctrine of overwhelming force. The British, meanwhile, always short of men and materiel, have frequently had to resort instead to unpalatable 'stratagems of evasion'.

* * *

* Though it had been the CIGS, Sir Alan Brooke, who used the term to describe Churchill's obsession with 'sideshows' in the Mediterranean: see pp. 139, 473.

After a second Peace of Paris brought the American war to an end in 1783 the British army began a rapid slide into dejection and incapability. The cuts did not look too savage at first: only those additional regiments raised for the war were disbanded, for India now replaced America as the focus of the maritime strategy, the source of the commercial wealth that would sustain and enrich the nation. But the naval building programme was voracious in consuming that wealth, and there was not even enough money to keep the 'garrison army' efficient. By the end of the decade recruiting had fallen to so dire a pitch that the army could not even reach its establishment of 52,000 (which included 6,400 to stiffen the native regiments raised and paid for by the East India Company). And little wonder, for the soldier was paid just 6d a day, the same as his predecessor in Charles I's 'guards and garrisons', and even that was eroded by the stoppages of pay for food, clothing and countless 'reckonables'. To make matters worse, no commander-in-chief was appointed, so that the army's interests were looked after by successive civilian secretaries at war who proved uncommonly idle and corrupt. The army into which the eighteen-year-old ensign the Honourable Arthur Wellesley,* later the duke of Wellington, was commissioned in 1787 stood in as low a state of efficiency as it ever had, or ever would again.

And the army *had* been efficient, no matter what the drubbings in America suggested. It had been a make-do-and-mend sort of efficiency, however: it did not compare with, say, the extraordinary military machine that the Prussian army had become (though a Prussian army of the same numbers would have fared no better than the British against the Americans). But the regiments that had battled in the forests, hills and swamps had adapted quite remarkably

---

* At that time his name was actually spelt *Wesley*.

well, in the end fighting more like light infantry than musketeers of the line. And all of this sank in, taking root in the recesses of the army's mind, so that when the moment came, two decades later, for Sir John Moore and his associates to begin retraining a part of the army as light infantry, it was more a dusting off of something put away than a buying of something new.

But while America had shown the army a type of warfare it would see increasingly in its imperial role – in the 'small wars' – campaigning on the Continent of Europe would remain an affair of mass and of ever-increasing firepower. When Britain merely dabbled in this, her fingers got badly burned. And she dabbled too often: it would be a decade and a half into the war with Revolutionary and Napoleonic France before the British army could begin dictating terms. For first its leaders, both military and political, needed to learn – or rather, relearn – the old Marlburian trade.

11

Regeneration

War with Revolutionary France, 1792–1801

LIKE BENJAMIN FRANKLIN'S TWO IMMOVABLES OF HUMAN existence – death and taxes – in the eighteenth century there were two certainties of British life, at least in retrospect: war with France, and more red coats. More red coats, that is, after first a drastic reduction in their numbers. But as the century entered its final decade, the nature of war with the old enemy was changing. Britain was fighting a very different sort of France and a very different sort of French army, and the war – or more accurately wars – would see the ranks of red swell to a size that would be unmatched for another hundred years.

In February 1793 the 'Committee of Public Safety', the executive government of Revolutionary France, had declared war on every power in Europe except Russia. This time the contest was neither dynastic nor commercial, however, for the ends of the Revolution were ends of principle and ideology: the fight against it was a fight for the survival of Britain's constitutional monarchy (and to many, therefore, for the survival of Britishness). And at first it looked as if the Revolution's resources would be as infinite as its goals, for the Committee of Public Safety passed a measure authorizing a *levée en masse*, and France would soon have 850,000 men in its army. But because the Royal Navy, 'the wooden walls', had been so determinedly strengthened in the decade before, that vast army of Frenchmen would not be able to come to England by sea. On the other hand, because the British army had been so comprehensively run down, Britain would not actually be able to take Paris, the only certain way of defeating France. The strategy would have to be indirect, therefore: the Royal Navy would squeeze the enemy commercially, seize its colonial wealth – principally the sugar islands of the West Indies – and use those gains to subsidize the continental powers traditionally able to field large armies. It would be Russia, Prussia and Austria that would do the 'heavy lifting' in battle with the massive, vigorous new armies of *La République*.

So for ten years the British army had little to do except, as Macaulay put it, take some sugar island, courtesy of the Royal Navy, or 'scatter some mob of half naked Irish peasants' who made trouble for absentee landlords or who, in 1798, rose more ominously in support of a half-cock French landing. Occasionally the army might try some 'descent and alarm' in a diversionary attack, usually in the Low Countries, but the legatees of the great victories of Blenheim, Dettingen and Minden would soon be, in Macaulay's words again, 'beaten, chased, forced to re-embark, or forced to capitulate'. It was breaking windows with guineas once more, and a double humiliation after retreat from America.

To make matters worse, yellow fever killed so many men in the West Indies that military recruiting was desperately blighted. Britain, unlike France, had no system of conscription – nor much likelihood of being able to enforce one. The navy had the press gang, but the army's only means of coercion (the ballot) was employed in filling the ranks of the militia,* the nation's unlikely safeguard in the event that a few transports should evade His Majesty's ships in the Channel.†

Nor were some of the measures taken to make up for lost time very helpful. The decision in 1793 to appoint the first commander-in-chief since the American war was a good one, but the decision to appoint the 76-year-old Lord Amherst, the hero of Ticonderoga in the *annus mirabilis* of 1759, whose health and vigour was even feebler than his age suggested, was certainly not. When he was replaced after two years by the duke of York, Henry Dundas, the secretary of state for war – at last a cabinet post in its own right – complained that 'Amherst was a worthy and respectable old man . . . but the mischief he did in a few years will not be repaired but by the unremitting attention of many'. In 1795, then, the army's stock stood as low as the life expectancy of its troops.

Remarkably, however, the army did not entirely lose heart. Or rather, enough of its capable officers did not. Indeed, a sort of grim determination now took hold which would be a useful example during the frequent periods of military doldrums in the following two centuries, perhaps best exemplified by the remarks of the young Lieutenant-Colonel Arthur Wellesley after the ill-fated expedition to Flanders in 1793–4: 'At least I learned what *not* to do, and that is always a valuable lesson.'

---

* The 1757 Militia Act re-established militia regiments in England and Wales under the control of the county lieutenancy, with a form of conscription in which each parish made a list of adult males and held a ballot for compulsory service. If any of the chosen men was unwilling to serve in person he was required to find another man to serve in his stead.
† Other than the abortive landings in Ireland, the French were able to get only the briefest of footings on the soil of the realm – in February 1797 at Fishguard in Pembrokeshire, of all places. The landing was seen off by the Pembroke yeomanry – and, so legend has it, by the women of Fishguard, who in their traditional red cloaks and high black hats looked like so many infantrymen on the headland above.

A valuable lesson indeed, if scant comfort to countless widows and orphans. There were, though, some less bloody lessons to be learned during the lean years of the 1790s. The 'variety of steps in our infantry and the feebleness and disorderly floating of our lines', of which the young James Wolfe had complained after Dettingen, were replaced, if with some resistance, by a standard drill. It had been devised by Colonel Sir David Dundas, known (and not always affectionately) as 'Old Pivot', for the drill was based on Frederick the Great's instructions, with their emphasis on the lines of infantry changing direction, and thus the direction of their fire, by pivoting on the left, right or centre points of the line – a slower method than hitherto, but one that cured the problem of Wolfe's 'floating'. Still, 'made in Germany' was not a recommendation to every officer, and the 1792 *Rules and Regulations for the Movement of His Majesty's Army* took a deal of enforcing – which the duke of York's appointment went some way to doing, convinced as he was that it was better to have an imperfect drill understood by all than a hundred and one different variants all claiming to be the perfect system.

Indeed, the painfully slow reform of the army really began with York's outrageous promotion to field marshal and commander-in-chief at the age of 32 – and without the distinction of success in the field. There was just something in the character of George III's younger son that recommended him to generals far more experienced than himself. It was, of course, prudent for any ambitious officer to be on good terms with the King's son, but this alone could not have explained the near-universal respect for York's energy, devotion to the army, administrative ability – and, though it had been comparatively little tested, physical courage. Not for nothing was he known to all ranks, rather like the marquess of Granby, as 'the soldier's friend'. If only he had won even a minor victory in the field and had possessed the

'The grand old duke of York': commander-in-chief at 32.

figure of a Wellesley or a Moore rather than the corpulence of his brother the Prince Regent, there might today be a worthy statue of him in Whitehall.

One of York's most pressing tasks was to reform the purchase system. In the decade or so of stagnation after the American war the worst practices had returned, and there were as many new ones. Other than in the artillery and engineers, where commissioning and promotion were still conducted on merit and tightly controlled by the Board of Ordnance, in many regiments the officers were neither capable nor even at duty. The system of a hundred years before, whereby a man of modest means could enrich himself (or at least make a living) through foreign service had developed into a system by which a rich man could buy himself an agreeable, prestigious life. The rough and ready country gentry of Marlborough's day did at least have to be present in order to make their money, whereas now the 'gentleman' could use his family means to buy himself a fine uniform and the admiration of London or Bath, and as long as his regiment was not required for duty overseas his presence on parade was scarcely desired. When his regiment *was* posted abroad he simply exchanged into another that was returning home: there were always impecunious officers keen enough for the cheaper life in India or, while life actually lasted, the West Indies. Sometimes it did not even have to be the tropics to put off a blade: Beau Brummell famously sold out of the 10th Light Dragoons when they were posted to Manchester on the grounds that he had 'not enlisted for foreign service'.

York managed to curb some of the worst excesses of purchase but in truth it needed the active support of the entire officer corps to end the abuse, and in a sense every officer, good or bad, had a stake in the old system. Arthur Wellesley himself had risen to command of the 33rd Foot (which by then he had taken to India) entirely by purchase and without doing a day's actual duty in several of the regiments through which he had advanced in rank. What York did manage to do was impose minimum qualification times, so that an officer at least had to spend a decent period in each lower rank before buying his way into the next one up, and establish a reporting system by which he was able to advance outstanding officers on merit without purchase. Frederick the Great had abolished purchase in the Prussian army and substituted a rigorous system of selection and training of cadets, and Revolutionary France had swept away purchase along with much else of the *ancien régime*; but both countries had universal conscription and therefore a better class of soldier. Perversely, perhaps, Britain's 'scum of the earth' responded quite extraordinarily well to being led by the sort of men who would pay for the privilege of doing so – backed up, of course, by the lash. The job of these 'gentlemen's sons' was by and large straightforward: 'the NCOs showed us how to fight,' wrote one old soldier after Waterloo, 'and the officers how to die'.

There continued to be a stiffening of professionally minded officers, however, many of them from the 'marches' and beyond – Scotland and Ireland: men of good, ancient family whose fathers' estates could not run to supporting a fashionable life either out of or in uniform, or who found the prospect of life in their native counties somewhat too confined. Arthur Wellesley was one of these, although his family was a shade grander than the average laird's son or rack-renting Irish landlord.

The regiments in which they served were changing, too, both outwardly and in substance. Gone were the proprietor-colonels, for the financial arrangements had become much tighter. Colonelcies were now largely honorific, and increasingly the domain of senior officers or royalty. Command in all its executive functions was in the hands of the lieutenant-colonel, who nine times out of ten would have purchased the rank, although as the war rolled on and new regiments were raised (and casualties mounted), there was a marked increase in 'field promotion' – promotion without purchase as reward for merit, or in strict seniority to fill dead men's boots.

And the regiments now went by different titles. Originally known by its colonel's name – Barrell's Regiment, for example, or Howard's* – and from 1751 by its number in 'the line', in 1782 each regiment was identified with a particular part of the country in an attempt to stimulate recruitment. So the 34th Foot now became, for instance, the 34th (Cumberland) Regiment of Foot, and the 9th Foot became the 9th (East Norfolk) Regiment. Some of the most senior regiments were allowed a grander title, such as the 4th (King's Own) Regiment, whose colonel was General Sir John Burgoyne, rehabilitated after his Saratoga humiliation. Some were distinguished as 'Fusiliers' in addition to their territorial affiliation, such as the 21st (Royal North British Fusilier) Regiment, or (later) as 'Light Infantry' rather than mere pedestrian 'Foot', such as the 43rd (Monmouthshire) Light Infantry. As in other reorganizations of the British infantry in the past two centuries there were territorial anomalies, such as the renaming of the very Scottish 25th Foot as the 25th (Sussex) Regiment – until 1805, when its recruiting area was transferred to the Scottish Borders, whose name it then took. These county affiliations served the regiments well in more ways than the original recruiting intention, but their greatest test – the rapid expansion of the army in the First World War – proved that the system was indeed an act of genius, if accidental. Would, for example,

* There were in fact two regiments whose colonels were called Howard, and to distinguish them they added the facing colour of their uniform to the name – hence the Buff Howards (known as the Buffs) and the Green Howards, names which lasted over 200 years.

the county of Durham have been able to raise fourteen extra infantry battalions in 1914–15 quite so easily, and the War Office have had them battle-ready so quickly, had not the battalions been able to take on the instant identity of the Durham Light Infantry? It is doubtful.

The 'county' system continued until the twenty-first century, when the cull and reorganization of 2006 left only one county name in the whole of the army as the regiments disappeared into 'regional' – or, in the case of Scotland, Northern Ireland and Wales, *national* – multi-battalion groupings. The exception was the three-battalion Yorkshire Regiment, which says much for the consciousness of that independent-minded county, as well as the shrewdness of the regiment's elders in holding fast to a firm local identity when the strong recruiting power of its former regiments (the Green Howards, the Prince of Wales's Own Regiment of Yorkshire, and the Duke of Wellington's) was lost.*

Changes in the infantry were by no means the only developments during the 1790s, which in many ways was the decade of waiting. While the duke of York was gathering up the reins at the Horse Guards as best he could, the army had the great good fortune to find there was now an able man in charge of the Board of Ordnance – Charles Lennox, third duke of Richmond. Besides thoroughly reorganizing the Ordnance's administration and the supply of gunpowder, Richmond raised the Royal Corps of Artificers to supervise military construction, the Royal Corps of Artillery Drivers to replace the civilian hireling gun teams, and the Royal Horse Artillery to support the cavalry. Without these reforms the

---

* Though there is anecdotal evidence that the more 'cosmopolitan' officer cadets are not so attracted as once they were now that the regiment's colours are nailed so firmly to the Yorkshire mast, the 'Dukes' and the Green Howards in particular always having had a wider appeal. The law of unintended consequences is never more evident than in changes to the regimental system.

army would simply not have been able to take the field against the rapidly modernizing forces of Revolutionary France.

But when would the opportunity come to put these innovations to the test? In 1799 there was another ignominious expedition to Flanders, led gallantly but ill-advisedly by the duke of York in person, and there were more costly and futile descents on the enemy coast, urged by the prime minister, the younger Pitt, just as his father had championed the equally ineffectual tip and run raids forty years earlier. Although these showed an offensive spirit of sorts, they showed no instinct for, and arguably no understanding of, the *decisive* use of force. The breakthrough – in Egypt – was to come almost by accident.

On 8 March 1801 at Aboukir, at the mouth of the Nile, the British army carried out its first successful opposed landing from the sea. The ships were under the command of Admiral Alexander Cochrane, uncle of the famous Sir Thomas – the 'sea wolf', the only captain who could rival Nelson in daring – and himself a bold and resourceful officer, fruit of the two decades of spending on 'the safeguard of the sea'. The landing force was commanded by a fellow Scot, Sir Ralph Abercromby, with yet another, the 39-year-old Major-General John Moore, as his deputy.

Into the boats at the ships' side clambered 5,000 red-coated marines and infantry – 5,000 muskets with little expectation of keeping their powder dry, knowing the work would have to be done with the bayonet. Once the boats were full, redcoats packed tight as the cargo on a slaver, the 'bluejackets' began pulling for the shore.

Half a mile distant, in the sand dunes of the Nile Delta, 2,000 Frenchmen lay waiting for them. An onshore breeze made the sea choppy, and as the boats ploughed through the swell several of them were overwhelmed, sinking with the loss of all but the strongest and least encumbered swimmers. Cannon shot now smashed into others, the French guns hardly needing to take aim against so numerous a flotilla. And then, as the boats began running in through the surf, into the breakers charged the French cavalry to cut at the assault troops before they could scramble out and gain a footing.

But the tide of redcoats was as unstoppable as the waves themselves. Men landing on a beach find little cover and, after a galling fire on the run-in, have little thought but getting at those shooting at them. And there were none more determined to go to it with the bayonet that morning than the 42nd Highlanders, the Black Watch, the only regiment in the army as yet allowed to wear the kilt (and with the red hackle in their bonnets, a distinction whose origins are long forgotten but which even today evokes a fierce sense of specialness). The fighting, hand to hand and with little quarter, did not last long. The cold steel of the Forty-second and half a dozen other regiments – English, Welsh and Irish (truly a microcosm of the newly constituted United Kingdom of Great Britain and Ireland) – at last silenced the guns and chased the rest from the dunes and back towards Alexandria. The assault landing, ever a precarious business, had succeeded: could Abercromby now complete his task of destroying the French army in Egypt?

It was a daunting enough task. The French had been in Egypt for three years. It had been General Napoleon Bonaparte's great egotistic mission, a speculative strategic adventure which vaguely threatened British interests in India by menacing the Levant. Bonaparte had come by sea, but a month later, in August 1798, at the battle of the Nile Nelson had made sure he could not leave that way, famously sailing into the shallows on the enemy's blind side and destroying the French fleet as it lay at anchor. With his army washed up amid the pyramids it was not long before Bonaparte was back in Paris on the pretext of fulfilling his destiny. But a stranded army abandoned by its general was still an army, and it was Abercromby's own destiny to deal with it.

He had some 14,000 men against 20,000 – unpromising odds, but the French officer opposing him was not of the first water. The brilliant General Kléber, whom Bonaparte had left in charge, had been assassinated by a Syrian student the year before, and in his place now stood Jacques-François de Menou, or Abdullah as he preferred after his 'conversion' to Islam.* Menou was to prove as sedentary as Abercromby was active, though at 48 he was sixteen years Abercromby's junior.

The route to Alexandria, the French army's base, offered neither room for manœuvre nor opportunity for surprise. Abercromby's lodgement was at the point of a narrow spit of land in the Nile Delta, with the sea on one side and Lake Aboukir on the other. But he lost no time in enlarging the perimeter to gain depth in case of a counter-attack, and to make space for his regiments to recover after their passage from Sicily. On 13 March, five days after the initial landing, he struck out from the bridgehead, skirmishing briskly with Menou's outposts at Mandora, and by the twentieth had managed to extend his front across the 3 miles of the isthmus, with his right flank resting on the ruins of the ancient Roman city of Nicopolis and his left on the lake and the Alexandria canal. Expecting a counter-attack, he placed John Moore's division on the right of the line, the Guards brigade in the centre and three more brigades on the left towards the lake, backing them with a second line of two brigades and dismounted cavalry. And this would prove prudent, for, having progressively lost the initiative over the past fortnight,

---

* Menou's change of faith was apparently no more sincere than his erstwhile leader's conversion, in which Bonaparte took the name Ali. Egypt was evidently worth a mosque.

Menou now managed to stir himself into a night attack.

He struck in the early hours of 21 March, but Abercromby, anticipating the move, already had his men stood to arms. The main weight of the attack fell on Moore's division, in particular the 28th (North Gloucestershire) Regiment. As day broke, the Glosters (the spelling is one of those cherished peculiarities of every regiment) found themselves so hard pressed that their colonel ordered the rear rank to turn about to fight off an envelopment – a drill not found in the Dundas regulations. With both front and rear ranks simultaneously engaged and therefore unable to support each other, all the Glosters could rely on was their speed of reloading, the vigour of their bayoneting and the toughness belied by their pastoral recruiting area. They literally fought back to back, the French dead piling up by the minute to both front and rear, until the attack was utterly spent.

In recognition of this singular feat of arms the Glosters were awarded the distinction of wearing Sphinx badges both at the front and at the back of their head-dress – another of those cherished peculiarities which 200 years later would still somehow convince a Glosters soldier that he was a part of something special, still make a man fight just a little harder because he knows that others wearing the same badge have managed to fight hard in the past.*

The battle of Alexandria was as hard fought as any one of Marlborough's great 'quadrilateral'. The casualties were heavy on both sides, but the French could make no impression on the British line, and towards late morning, with their cavalry blown, their guns running out of powder and their infantry exhausted, they turned tail for the security of the ancient port of Egypt after which the battle was

---

* And, indeed, the distinction was taken into the new regiment of which the Glosters became a part in 2007, The Rifles.

named. A few months later Menou and his besieged, demoralized army surrendered.

Aboukir (Abu Qir) and Alexandria were magnificent turning points in the British army's reputation – the culmination of, in the words of one of Bonaparte's officers (later a Russian general), Henri Jomini, 'l'époque de sa régénération'. The infantry especially had somehow regained that doggedness and defiance which seemed to have reached its high-water mark at Minden and been on the ebb ever since, particularly after Yorktown. In Egypt British generals had planned things well, and executed them even better; and superior French numbers had been overcome by sheer fighting spirit. The army of Revolutionary France which would soon under 'Emperor' Napoleon become the Grande Armée was not invincible after all. Although it took a little time for the truth to sink in (the cabinet had agreed to unfavourable peace terms in October before even knowing the outcome of the fighting*), the King's ministers could at last see that a well-found and well-generalled British army supported by the Royal Navy could gain decisive results. The question that they now had to answer was *where* that decision should be sought.

But fate played a cruel, if ultimately obliging, trick at Alexandria. Armies are much the better for being led by generals who take a good share of the shot rather than just the major share of the prize money. Sir Ralph Abercromby, who despite his years was in the thick of the fighting at Alexandria (at one point grappling hand-to-hand with two French dragoons), had been fatally wounded. Moore, his deputy, who was also wounded (as well as three other generals), took command in the hour of the dying hero's

---

* The Peace of Amiens, described by the playwright and politician Sheridan as 'a peace which all men are glad of, but which no man can be proud of'.

triumph – with eerie portent of his own later fate. And Moore, thrust into the prominence of command, would thereby gain in due course the authority to work his own reforms, and take them to the enemy on the mainland of Europe.

In Egypt, then, the British army had served notice that it was to enter the fight on land and as a force to be reckoned with – exactly as Marlborough had done by his march into Bavaria a hundred years before. And the army's elastic order of battle would stretch to its greatest length to date. Once again, officers were learning how to rebuild an army – and in the face of their most dangerous enemy yet.

12

The Hero's Grave

The Iberian Peninsula: first steps, 1808

WITH THE VICTORIES IN EGYPT IT WAS AS IF THE ARMY HAD suddenly found a military version of the philosopher's stone: everything it now touched seemed to turn to gold. Even in India, where French money and advisers were sent to gnaw at the growing British commercial and territorial strength, there were impressive victories against dramatic odds. In 1799 the young Colonel Arthur Wellesley, still in his twenties and commanding a division, had drawn one of the most troublesome thorns in the East India Company's side – Tippoo, Sultan of Mysore. Wellesley's march on Tippoo's capital, Seringapatam, and its siege and storming, were as superbly organized as the 1793 campaign in Flanders – in which he 'learned what *not* to do' – had been disorganized.*

Wellesley's eldest brother, Richard, was governor-general in Calcutta, principal of the governors who ruled the East

---

* The taking of Seringapatam formed the background to Britain's first detective novel, Wilkie Collins's *The Moonstone* (1868).

India Company's 'presidencies' of Madras, Bombay and Bengal. After Seringapatam he appointed Local Major-General Wellesley governor of conquered Mysore. It was a spectacular act of nepotism, but an inspired one nevertheless, enabling the future duke of Wellington to add diplomatic and administrative skills to those of the soldier – skills that would be of incalculable value when in Portugal and Spain he found himself grappling with the demands of a major expedition reliant on allies.

In 1803 Wellesley also gained the experience of planning and conducting an entire campaign, and in arid country against a vastly more numerous and elusive enemy, the Mahrattas. It was as fine a grounding for the Peninsular War as he could have wished, not least for its setbacks and reverses which gave him a healthy appreciation of 'time and space', one of the most intangible factors in warfare. And at the culminating battle of Assaye he learned just how close-run a battle could be and yet be saved by the commander's determination and cool head. In later life he would claim that of all his battles he was proudest of Assaye – perhaps not surprisingly, for he had very nearly lost it, and by his own miscalculation and uncharacteristic impetuosity. India taught the duke his trade as a commander at the tactical level, at which battles are fought, and also at the operational level, at which campaigns are planned and carried out. And thanks to his brother's patronage he learned what war meant at the strategic level too, where the political ends are translated into a scheme of action, both military and non-military. In these ways India laid the foundations for the success of the British army in Europe in the early nineteenth century, as it would again for the new generations in 1914–18 and 1939–45.

By 1801 the fight had been knocked out of Britain's continental allies, and when war broke out again within a

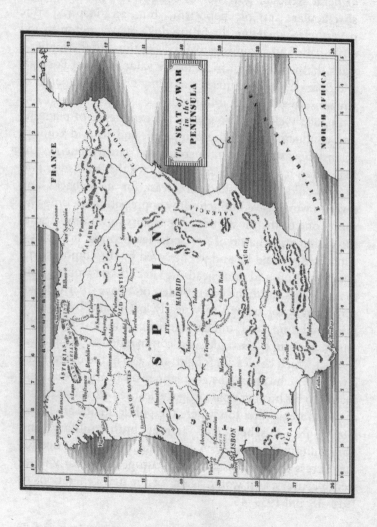

The SEAT of WAR
in the
PENINSULA

year of the delusory Peace of Amiens the whole power of France and its conquered territories was turned on Britain. The army at once mustered in Kent and along the Thames estuary to repel the expected invasion, while the Royal Navy began the remarkable campaign that was to culminate in Nelson's destruction of the Franco-Spanish fleet at Trafalgar two years later in October 1805, wiping out Bonaparte's hopes of landing on English soil.

The concentration of troops in the south-east had unintended beneficial consequences. For the first time, senior officers now had a chance to organize systematic training at brigade level and above. It was a blessing for many regimental commanders too, for since barracks were still a rarity their men were dispersed all about the countryside in billets and could rarely be mustered for training in more than company strength. Major-General John Moore, lately returned from Egypt, was given command of the key coastal defences from Dover to Dungeness. At the prompting of the duke of York, he began a training scheme for light infantry at Shorncliffe camp near Folkestone, the beginning of the sustained effort to gain agility on the battlefield as well as devastating volley fire.

It was at this time, too, that the rifle made its first appearance in the hands of regular British troops. In 1800, again on the initiative of the duke of York, an 'Experimental Corps of Riflemen' made up of detachments from several regiments had been mustered for training in Windsor Forest under Colonel Coote Manningham, who had thought out his ideas while serving in the West Indies. Once proficient with the new Baker rifle* and its associated *jaeger* tactics, these experimental riflemen were supposed to return to their units

---

* A trial had been held at Woolwich by the Board of Ordnance in 1800 to select a standard rifle pattern. The rifle designed by Ezekiel Baker was chosen (see 'Notes and Further Reading').

as sharpshooters. However, a chance raid on Ferrol in north-west Spain gave the Experimental Corps its chance to put theory into practice, and it so spectacularly proved the rifle-man's worth that afterwards the corps was reconstituted as a regiment of the line – the 95th (Rifle) Regiment of Foot – and a second battalion was raised. They were renamed the Rifle Brigade in 1816.

The 95th Rifles, popularized by Bernard Cornwell's 'Sharpe' novels, were not quite the first troops to carry the rifle and wear green. In December 1797 the 5th Battalion of the 60th Royal Americans had been raised in England by Baron Francis de Rottenburg, who armed them with rifles and dressed them in green jackets with discreet red facings. Moore used Rottenburg's treatise on riflemen and light infantry, and Manningham's regulations, to train his brigade at Shorncliffe. Its original three components, the 95th Rifles and the 43rd (Monmouthshire) and 52nd (Oxfordshire) Light Infantry, would evolve through various forms of name and county affiliations to their amalgamation in 1966 as the Royal Green Jackets (whose reputation for marksmanship, innovation and fast marching was matched only by that for the mass production of generals). Their motto was that given by Wolfe to the 60th Royal Americans at Quebec: *celer et audax* – 'swift and bold'. Historians have sometimes observed that the French light infantry – *voltigeurs* and *tirailleurs* – were a sort of military manifestation of Rousseau's concept of the natural man, in contrast to the close ranks of stiff-necked *mousquetaires* of the *ancien régime*. What animated their British counterparts is difficult to say, but the rifle regiments quickly became, if not exactly fashionable, then a *corps d'élite* and a key part of the military machine. When eventually Lieutenant-General Sir Arthur Wellesley took supreme command in the Peninsula he issued orders that the 60th and the 95th were always to form the vanguard when the army moved.

But all these innovations did not yet amount to a system as such, and the military successes – outside India – were small-scale affairs. There was no permanent organization of brigades, let alone divisions; and therefore no staff. What would today be called combat support (artillery and engineers) continued to function independently, much as in Marlborough's day – a separate 'army' of the Board of Ordnance – giving rise to the same problems. Likewise, what is now called combat service support (transport and supply) remained an extension of the Treasury. And although the uncooperativeness of these departments has sometimes been exaggerated, it was still too often the case that while the commander of the fighting troops looked towards the enemy, the commanders of the artillery, engineers and commissariat looked back towards London. Commissariat officials were rewarded for their frugality, whereas a field commander wanted all the supplies he could get. The old quartermaster's joke that if he had been meant to give away kit then his offices would have been called 'issues' and not 'stores' was writ large in the commissariat.

For all the duke of York's reforms, therefore, and for all the innovations and small-scale successes, the army was in no fit state yet to mount a major, sustained campaign on the Continent. British musketry may have been better than French in the firefight, but Bonaparte's army was simply too good an instrument of war to meddle with. Once wrong-footed by its superior powers of manœuvre and logistics, no army seemed able to resist the heavy firepower it could bring to bear, or the advance of its columns of infantry – as the Austrians, the Russians and then the Prussians learned in 1805 and 1806 at Ulm, Austerlitz and Jena. Like the cobra, the army of Bonaparte could be avoided, and it could be put off by great agility. But let it strike and it was deadly.

Britain had to come to grips with the Grande Armée somewhere, however, and soon. Indeed, the second great period in the making of the British army would be one almost of an army in search of a battlefield. And as with the first a century before, although there would be several men with a right to share in the laurels, one man would stand way above the rest – so far above the rest, in fact, that, just as with Marlborough before Blenheim, had he fallen before the definitive battle there would have been no one to fill his boots. So towering a figure did the duke of Wellington become that his exasperated 'By God, I don't think it would have served had I not been there!' after Waterloo might have applied equally at any time in the preceding seven years.

And he might easily *not* have been there. Arthur Wellesley's exploits in India had come to the attention of Whitehall – not surprisingly, for his brother wrote the dispatches – and the Tory administration was doubly approving of a capable general who was also 'one of us' (he was soon to be an MP and privy counsellor). Wellesley would be especially favoured by the support of his fellow Anglo-Irishman Robert Stewart, Lord Castlereagh, secretary for war between 1805 and 1809, and foreign secretary from 1812. Elder brother Richard, now Marquess Wellesley and back from India, was able with Castlereagh to provide a certain seamlessness at the strategic level, a sort of 'top cover' without which younger brother Arthur would have found the politics even harder going during the later years of the Peninsular War. But now, at the beginning of 1808, Lieutenant-General Sir Arthur Wellesley (he was made Knight of the Bath in 1804 and promoted after the expedition to capture the Danish fleet at Copenhagen in 1807) was not only the youngest lieutenant-general in the army, he was the most junior. While he had every right to expect some independent command, he understood full well that Sir John Moore and others senior to him would have command of any really great undertaking.

That great undertaking was now, at last, at hand. For

although Bonaparte had dealt with the major continental powers, Britannia ruled the waves, and as well as blockading French naval ports the Royal Navy continued to range freely among French colonial possessions, intercepting her trading ships and those of French conquests and neutrals alike. In turn, the only way Bonaparte could conceive of defeating Britain was by destroying her commerce, for without treasure from exports she could neither pay for the navy nor offer subsidies to continental allies. He therefore declared a Europe-wide embargo on trade with Britain, the so-called 'continental system'. It was soon failing on two counts, however. First, the New World increasingly took up the trade which the continental ports refused; and second, in desperately trying to close off every point of trade Bonaparte over-reached himself by invading Portugal through Spain, and then replacing the Spanish Bourbon king with his own brother, Joseph Bonaparte.

El Tres de Mayo, Goya's masterpiece, was according to the art critic Sir Kenneth Clark 'the first great picture which can be called revolutionary in every sense of the word, in style, in subject, and in intention'. In it a French firing squad executes Spanish captives in the early hours of the morning: the day before, 2 May 1808, in Madrid and throughout Castile the Spanish had risen up against the French. For the Spaniards' former ally – alongside whose ships they had fought against Nelson at Trafalgar – was now an occupying power.

With Spain in rebellion against Bonaparte, Britain at last had her chance to enter the war on land in strategically favourable conditions. For in the Iberian Peninsula the French were operating on long, exterior lines of communication, while the British would be campaigning on interior lines and in a part of Europe whose lengthy coastline seemed tailor-made for the exercise of her naval strength, with Gibraltar the sure and steady rock on which that naval power could pivot between Atlantic and Mediterranean. And the difficult country of the interior, its scale and its climate, somehow seemed to favour the temperament and experience of Britain's soldiers, especially her commanders: it was like India, but nearer home. The uprising and the British intervention would be the beginning of Bonaparte's 'Spanish ulcer'.

That intervention began wonderfully well, and with luck, too, for had the Spanish left it another few weeks Britain would not have been able to move in so fast and with such rapid success. Arthur Wellesley had been about to embark at Cork with 9,000 men to sail to the assistance of the Venezuelan revolutionary General Miranda when suddenly Spain was no longer the enemy and it was therefore no longer expedient to help the Venezuelans seize their independence. Instead, Wellesley's force received orders to proceed to Spain.

Though he now faced an entirely different kind of undertaking, Britain's youngest lieutenant-general was not in the least dismayed. Before sailing he attended a dinner in Dublin, remarking on his future task in a way that showed both his extraordinary percipience and his confident grasp of where the strength of the British army lay compared with Bonaparte's vast military machine:

> Why, to say the truth, I am thinking of the French that I am going to fight: I have not seen them since the campaign in Flanders, when they were capital soldiers, and a dozen years of victory under Buonaparte must have made them better still. They have besides, it seems, a new system of strategy which has out-manoeuvred and overwhelmed all the armies of Europe. 'Tis enough to make one thoughtful; but no matter. My die is cast, they may overwhelm me, but I don't think they will out-manoeuvre me. First because I am not afraid of them, as everybody else seems to be; and secondly, because if what I hear

of their system of manoeuvre is true, I think it is a false one against steady troops. I suspect all the continental armies were more than half beaten before the battle was begun – I, at least, will not be frightened beforehand.

In the summer of 1808 the undaunted Wellesley landed at Corunna in north-west Spain to consult with the Galician junta, but the Spaniards gave him a cool reception, not at all happy to see British troops, the old enemy, on Spanish soil. He therefore sailed on to Portugal, landing at Mondego Bay well out of range of the French Marshal Junot's army in Lisbon 100 miles to the south. Junot was one of Bonaparte's favourites. He had risen from the ranks and distinguished himself in the Italian campaign, but had received a head wound in Italy which made his judgement erratic – or so it was said. Since Bonaparte had been pouring troops into the Peninsula for a year, however, even a marshal with a sore head was not going to be easy to oust. In any case, Wellesley's force could only be a small part of the army that would eventually be needed to cooperate, however remotely, with the Spanish armies (and the Portuguese would be in no shape for months). More troops were already on their way: 4,000 under Major-General Brent Spencer and 10,000 under Lieutenant-General Sir John Moore – to whom, as senior general, overall command would pass when he arrived.

But therein lay a problem. Moore was a Whig, and an outspoken one at that; it would not do that a general who had been so critical of the Tory government's military policies should now be in command of its principal campaign on land. More senior lieutenant-generals were therefore ordered to Portugal: first, Sir Harry Burrard, whose nickname was 'Betty'; second, Sir Hew Dalrymple, who was known as 'the Dowager'. The nicknames were not auspicious. 'Betty', although he had done his share of fighting, was more at home on Horse Guards Parade; 'the Dowager' had not seen service for fourteen years. How could such inapt appointments be made after so many years of learning the hard way – in Wellesley's words, what not to do? In truth, the duke of York was still not master of the system: it was the civilian secretaries – the cabinet, even – who controlled appointments at this level. And in truth too a commander-in-chief in a distant theatre of war, as Moore would be, did require a sound political sense as well as sound military ability – but not as a substitute for it. It was not the first time that a general possessing one quality but not the other would be tested in both; and it certainly would not be the last.

Before Betty or the Dowager could do much damage, however, Wellesley had won two quick victories. Setting off without delay towards Lisbon, on 17 August he bumped into Junot's advance detachment of 4,000 men at Roliça, whom he forced from their positions by a combination of manœuvring on the flanks and direct assault, his two rifle regiments leading. And then four days later the main body of the French in turn attacked him – some 13,000 men under Junot's personal command. To meet them, Wellesley deployed his main firing line in what was to become his trademark defensive tactic – the reverse-slope position, the regiments concealed behind a ridge, lying down, muskets loaded and bayonets fixed.

For the first time since the battle of Alexandria the French were about to feel the full effect of British musketry allied with the best of generalship.* Two dozen regiments of the line stood or lay concealed behind the crest of the ridge

---

* There was an earlier example of British musketry – at Maida on 27 June 1806, when a British force of 5,000 men commanded by Major-General John Stuart, which had sailed from Sicily and landed in the Gulf of Sant'Eufemia in the Kingdom of Naples, defeated a French force of 6,000 after fifteen minutes' volleying – but Stuart's subsequent generalcy proved not to be the equal of his infantry's firepower (see 'Notes and Further Reading').

south of the little Portuguese village of Vimeiro, 100 miles north of Lisbon, as 2,000 veteran French infantrymen in their habitual close-order columns, supported by hand-wheeled guns and screened by the customary cloud of *tirailleurs*, marched confidently towards the centre of the position.

In command of the brigade in their path was Henry Fane, cornet at thirteen, captain the following year, veteran of many an affair in Flanders and many a raid up and down the French coast – and not yet thirty. To counter the *tirailleurs* he had detached four companies of the green-jacketed 60th (Rifles) and ranged them at the foot of the slope. Most of the riflemen lay down to take a steady aim, some on their backs supporting the barrel of the rifle in the 'V' of their crossed feet. At 600 yards, well beyond smooth-bore musketry range, they began sniping, taking a steady toll of the *tirailleurs* so that those who were not hit were soon falling back on the brigade columns, against which the guns of the Royal Artillery were already firing the new 'shrapnel' ammunition, peppering the densely packed masses of infantry with musket balls from overhead like lethal hailstones. The French artillery replied, though with limited effect on men lying in 'dead ground'.

As the French neared the bottom of the rise the riflemen withdrew. Without *tirailleurs* in front of them to flush out the main body concealed behind the crest, as the French reached the top of the slope they blundered into the 50th (West Kent) Regiment. At 100 yards the 'Dirty Half Hundred',* drawn up in two ranks not the regulation three, opened fire. The French, still in column, wavered but continued. The 50th's flank companies now wheeled inwards to

---

* The nickname they would acquire at the battle of Salamanca when the dye of their distinctive black facings stained their faces, it being a sweltering hot day, and the sleeve a useful sweat wipe.

add a deadly enfilade fire (exactly as the 52nd would do at the climax of Waterloo seven years later). The French broke and bolted, abandoning their guns and throwing away their muskets.

Soon the columns to left and right were wavering too as seventeen regiments of redcoats now rose up and began pouring fire into the mass of Junot's infantry. The French tried to deploy and return fire, but Wellesley ordered his line to charge, driving them headlong back down the slopes. Then he loosed his cavalry at them – two regiments of light dragoons – and sealed a famous victory. It had taken him just two hours.

But next 'Dowager' Dalrymple, who had just arrived, failed at the one thing he was meant to be good at: judging the political–strategic situation. Junot sued for peace. Wellesley was all for pressing on for Lisbon and finishing the job with powder, but Dalrymple agreed to a conditional surrender: Junot's troops would evacuate Portugal and return to France, but they would leave with all their arms and equipment – and in British ships.

When the details of this 'Convention of Cintra' reached London, the three British signatories – Sir Hew Dalrymple, Sir Harry Burrard *and* Sir Arthur Wellesley – were peremptorily recalled and arraigned before a court of inquiry. Though all three were eventually acquitted, only Wellesley returned to active duty, and even then under a cloud, so appalled was the public reaction to Cintra (the press, then as now with a penchant for the pun, took to calling it a *Hewmiliation*). Even Wordsworth railed against it, and Byron, ever the soldier manqué, fumed in *Childe Harold*:

> And ever since that martial synod met,
> Britannia sickens, Cintra! at thy name;
> And folks in office at the mention fret,
> And fain would blush, if blush they could, for shame.

How will posterity the deed proclaim!
Will not our own and fellow-nations sneer,
To view these champions cheated of their fame,
By foes in fight o'erthrown, yet victors here,
Where Scorn her finger points, through many a coming year?

The immediate beneficiary of the episode was Sir John Moore, who now took command of the army in Portugal, soon to grow to 42,000 men. Although this was now only about a fifth of Britain's army, on paper at least, it was all that London could spare, the rest being committed to overseas garrisons, home defence and what would later be known as internal security. Indeed, the foreign secretary, the brilliant but exasperating George Canning, warned Moore that his army was 'not merely a considerable part of the dispensable force of this country. It is in fact *the* British army . . . Another army it has not to send.'

Preservation of the force is the first duty of a commander, but such a warning wrote it in capitals. And yet Canning's arch-rival Castlereagh, the secretary for war, had high strategic expectations of the 42,000 – as well he might, for it was the largest army that Britain had sent abroad since Marlborough's day. His instructions to Moore were 'to co-operate with the Spanish armies in the expulsion of the French from that Kingdom'. Indeed, any less ambitious an aim would have seemed unreasonable for so large an investment, and at what was now to be the army's main point of effort.

Moore's instructions, however, presumed the cooperativeness of Spain, a former enemy of Britain. And although Marshal Junot had been sent back to square one – in his case Rochefort – the French had plenty more counters on the board. These now began to come into play. In October the emperor Napoleon, as he had crowned himself, free of the distractions of the big continental players after victories at Ulm, Austerlitz and Jena, marched into Spain at the head of 200,000 men – as many as Britain had under arms world-wide. His plan of campaign was, as ever, direct (like, reputedly, his lovemaking): he would thrust to Madrid in crushing strength, and then on to Lisbon, flinging aside contemptuously all in his wake, detaching a corps here and there to deal with whichever of the several rabbles that passed for Spanish armies was dolt enough to stand in his way. He had neither the inclination nor the patience for long campaigns.

Implicit in Castlereagh's instructions to Moore was the willingness not only of the Spanish troops to fight, but, crucially, of the Spanish authorities to meet his logistic needs. The first would soon prove a dubious assumption, for factionalism in the Spanish army added another dimension of unreliability. As for the second, logistic support, as Moore quickly learned, was but a vain hope. A century and a half later, when NATO was putting together the plan for the defence of Western Europe against the Soviet tank armies, one of its fundamental principles was that logistics were to be an entirely national responsibility; where any reliance was to be placed on 'host-nation support' it had to be with the most meticulous prior agreements drawn up in legal form – much as Marlborough had done before his march into Bavaria. It was a doctrine born of hard experience. 'Interoperability', which holds out the promise of logistic partnership, has always been regarded as something of an illusory goal by seasoned warriors, and Sir John Moore's experiences would ever stand as a warning to those who thought otherwise.

No more cooperative than towards Wellesley earlier in the year, the Spaniards would not let Sir David Baird and his 15,000 reinforcements land at Corunna; so they had to sail round the north-west point of the peninsula to Vigo instead, which took time. And since Moore had to leave 12,000 men to protect Lisbon, it meant that half of the force with which he intended to manœuvre against the French – including the bulk of the cavalry, of which he was especially short – would be joining him late. Indeed, in the increasingly uncertain situation that Moore faced, cavalry would have been his greatest asset, for had they been allowed to range and reconnoitre deep, the perilous position into which he now began to march (and would persist in marching) could have been recognized earlier.

Knowing that it would likely be a full month before Lord Paget (later Lord 'One Leg' Uxbridge, of Waterloo fame) and his cavalry division would land with Baird's reinforcements, and knowing even less about the country into which he was to advance without maps or pre-positioned 'combat supplies' (ammunition, food and forage), but knowing that he must move fast if he was to be of any assistance whatever to the Spanish, Moore set off on 16 October for the frontier, intending to rendezvous with Baird when and where he could.

Advancing on three widely separated axes, in a month he was at Salamanca, 100 miles north-west of Madrid.* Here he learned that the situation had indeed taken a turn for the worse: Napoleon had already smashed the Spanish army in the centre of the country, and Valladolid, not 70 miles to his north-east, was in French hands. Without his heavy artillery, which was travelling by a different route – none too quickly, because of faulty intelligence on the going – and with no sign of Baird, Moore faced rapid annihilation. But astonishingly, Napoleon now obliged him by halting his westward advance and directing his attention on Madrid. Given a fortnight's breathing space, Moore was able to gather in his artillery and begin planning an orderly withdrawal on

---

* Part of his force had actually reached the Escorial, only 20 miles or so short of Madrid, before turning back north-west to rendezvous with the others.

Lisbon. But then he learned that Madrid was fighting, and decided that he must support the Spanish defenders – for this had been the import of Castlereagh's instructions. He decided therefore to mount a diversionary attack on Marshal Soult's corps of 20,000 covering Napoleon's flank and line of communications at Valladolid.

It proved a major error. How far Castlereagh's instructions obliged Moore to make so desperate a feint have been debated ever since, but there is no doubt that 'the Whig general's' unbending sense of honour propelled him to keep fighting for as long as there was hope of rousing the Spanish to resist. The stakes were high, after all, for Moore believed that if Spain fell, so again would Portugal, leaving Britain with no toehold on the mainland of Europe except Gibraltar, and France with all the strategic advantage of owning the entire Atlantic coast. And yet Castlereagh's instructions did not oblige him to fight except in cooperation with the Spanish *armies*. The doughty civilians of Madrid scarcely constituted that. Moore's own estimation of their capacity to resist the French, and the capacity of the Spanish forces outside the capital to rouse themselves, was anyway not great: 'I much fear that they will not move,' he wrote in his diary on 11 December, 'but will leave me to fight; in which case I must keep my communications open with Astorga and Galicia.'

The strategic situation had changed, too, for the cabinet had not made its plans on the assumption that Napoleon would flood the country with troops. There were now possibly three times the number originally supposed, though even Moore underestimated the total, such was the lack of good intelligence from the Spanish. He was wary, none the less, writing to Baird that 'I mean to proceed bridle in hand, for, if the bubble bursts, we shall have a run for it.'

And so he proceeded north-east instead of south-west, 'bridle in hand', brilliantly screened by his two cavalry regiments – 1,200 sabres under Castlereagh's younger brother,

Lord Charles Stewart. Scouting deep, Stewart's men learned that Valladolid had been evacuated, and that Soult, believing the British to be withdrawing west, had moved to Burgos, 60 or so miles north-east. The British cavalry had not yet gained half the reputation of the infantry – even Granby's bald head could not quite displace the image of Sackville's motionless wig and Prince Rupert's uncontrollable Cavaliers – but in the month's campaign in Galicia, especially once united under Paget, the regiments of hussars and light dragoons showed what they could do when handled firmly and boldly. Without them – without Paget, perhaps – when Moore had to run for Corunna the army could scarcely have made it.

One of Stewart's regiments was the 3rd Hussars of the King's German Legion. The KGL had been formed in 1804 after Napoleon over-ran the electorate of Hanover and disbanded its army, many of whose officers and soldiers fled to the Elector George's British realm.* Equipped in just the same way as British troops, they had a high reputation for efficiency and reliability (at Waterloo, Wellington deployed them exactly as if they had been British), and their cavalry – five regiments of hussars and light dragoons, the two types differing in name and dress only – were generally reckoned superior in horsemastership to the British. The diary of Captain James Hughes, in temporary command of the 18th Hussars, describes how after an inspection in 1813 in the Pyrenees the brigade commander invited him to visit the 1st Hussars of the Legion to see how well the horses looked in comparison with the Eighteenth's. There is an affectionate, if not always accurate, picture of them in Thomas Hardy's *The Trumpet Major*.†

---

* The Legion grew to around 14,000; it was disbanded in 1816.
† When the miller's party visit the cavalry camp on the downs, 'They passed on to the tents of the German Legion, a well-grown and rather dandy set of men, with a poetical look about their faces which rendered them interesting to feminine eyes'.

On 19 December Moore and Baird met up at last, at Mayorga 80 miles west of Burgos. Baird's troops were led in through deep snow by the splendid 15th Light Dragoons (now styled 'hussars') who had done so much to recover the cavalry's reputation in the Seven Years War by their *élan* at Emsdorff. Awaiting them were the pickets of the 4th (King's Own) Foot, yet again in the cannon's mouth, along with the 28th (North Gloucestershire) who had fought back-to-back with Moore at Alexandria; beyond the pickets waited also many more of the best infantry regiments of the line. And in with Baird marched two battalions of the 1st Foot Guards, their field uniform no different from that of the line regiments, yet their bearing such that Moore remarked, 'Ah, they must be the Guards.'

Moore had decided to regroup this army of 30,000 to spread the experience evenly through the brigades, for some of the reinforcements had not lately seen action. He therefore formed four infantry divisions, each under an unusually able commander, three of them fellow Scots (Baird, Hope and Alexander Fraser) and the other, Edward Paget, the younger brother of the cavalry commander. And for the first time he formed two light brigades as flank guards, one consisting of two light infantry battalions of the KGL under the Hanoverian general Karl Alten, the other of two battalions of the 43rd and the 52nd Light Infantry together with the second battalion of the 95th Rifles under yet another Scot, the formidable Robert 'Black Bob' Craufurd.

In Madrid a day or two later Napoleon learned what was happening. He at once flew into an energetic rage, gathered up men from all directions – 80,000 of them – and marched north to deal with the impertinent British. In his path lay the snow-covered Guadarramas, mountains that even in summer would have tested an army. But mere mountains, even storm-bound mountains, could not stop Napoleon in his pride. Down from the saddle he clambered time after time, waist-deep in snow, leading his columns as much as driving them. An aide-de-camp wrote of the ordeal:

> I found the whole of the Imperial Guard at San Rafael . . . The storm had been so terrible on the mountain that many men and horses had been swept over precipices, where they had perished. The grenadiers, exhausted with fatigue, were sleeping on the frozen ground covered with masses of snow and ice beside their fires, which were all but extinguished by the rain and hail which were still falling . . . There was not a square foot of shelter . . . not already invaded by sleepers piled one on top of the other.

On Christmas Eve, all unknowing, Moore's troops, in great high spirits despite the equally wintry conditions on their side of the Guadarramas, set out to cross the Carrion River to attack Soult in his quarters. Halfway there, Moore learned that Napoleon was making for him, and in strength. With Canning's words ringing in his ears – 'Another army it has not to send' – and with the knowledge of what Burgoyne's surrender at Saratoga and Cornwallis's at Yorktown had brought in their wake, he saw no alternative but to abandon the attack on Soult's corps and run for safety. This would, however, require an even greater exertion than the French had made to force the Guadarramas, for the route that 'the British army' would have to take would be through the mountains of Galicia.

But had not Moore and his men won a victory in drawing off so many French troops from Madrid? In fact Spanish resistance there had already collapsed; and elsewhere, if there was resistance at all, it was ill-coordinated, self-serving and sporadic. Moore's diversion had been to no avail and would earn him no laurels. Indeed, even if he could save his force Moore knew he could expect opprobrium. He was, after all, 'a Whig general'.

And *could* he save his force? If he had been withdrawing

from Salamanca, before his bold – some might say foolhardy – thrust on Valladolid, he would have been falling back on decent lines of communication into Portugal covered by strong fortresses en route, with evacuation if necessary from Lisbon. The path he had now 'chosen' lay due west, over arduous country he did not know at all, and it led only to the trickier ports of Corunna and Vigo.

The mood of Moore's army at once swung from high spirits to low. No one seemed to understand that they were trying to evade the greatest peril; and when they did learn that Napoleon was coming, they could not understand why they ran without any sort of fight. But one part of the army would be very much in action from the outset. Indeed, the retreat to Corunna stands as probably the finest service the British cavalry ever performed, for as soon as Soult learned of Moore's movements and the approach of his emperor he loosed his own vastly superior numbers of cavalry west to harry his would-be attackers. So superior were they, indeed, that they ought to have been able to overwhelm Moore's cavalry rearguard and in no time turn retreat into rout. But Paget's five regiments of hussars, tightly handled by their commander with what would later be recognized as 'cavalry genius', went at their work with a relish and skill that utterly unnerved their opponents. Paget and his 2,000 sabres bought vital time for the infantry to get across the swollen rivers of northern Castile and into the relative safety of the Cantabrian Mountains, and had they not been able to do so the game would have been up for Moore by New Year's Day.

But it was still 200 miles to Corunna. It was deep winter, there was no supply system in place, and the terrain was as inhospitable as the local people through whose villages, few and far between, the army now trudged, hampered all the while by stragglers from the disintegrating Spanish army of Galicia, and harried by Soult's cavalry. With every mile too the grumbling got worse – not so much at the conditions (for could not a redcoat take it?) as at the thought of running without a fight. Some regiments maintained their discipline; others did not. Legitimate foraging by starving troops turned into murderous pillaging. Wine and brandy reduced whole regiments to supine rabbles. At one point Moore halted the retreat, put up the gallows and made a few examples of the worst offenders, but even this did little to check the leaching of discipline and morale. If only they could stand and fight, the men complained – as did many of their officers, some of them generals. But Moore, outraged and wholly astonished at the rapid collapse of so large a part of his army, insisted he would fight only at a time of his free choosing. The retreat continued.

Finally, after three weeks of despondent marching, with stragglers and camp-followers being cut down every mile of the way by Soult's harriers, the army reached Corunna. Here the Royal Navy, as many a time past and in the future, was waiting to take them off. And as if they had reached some sort of Promised Land, when the regiments staggered down to the coastal plain the sun came out at last. With the mountain snow behind them, the plain lush in its midwinter greenery and the orange trees full with fruit, spirits started to lift of their own accord. But they positively soared at the news that now Sir John Moore would fight. When the order went out to take post, regiments which had barely passed muster during the march took up their arms as if on parade in Hyde Park. And when the enemy did appear, Moore's depleted, sickness-ridden, ragged, half-starved army fought as well as any since Marlborough's time, beating off the French so determinedly that, with the help of the Spanish garrison to cover their embarkation, the whole force was got aboard the ships and away to England in just four days – and without a single Frenchman getting into the town to shoot them off.

Corunna was a true 'set-piece' defensive battle, one of the

few to that date in the army's history. It played well to the strengths of the disciplined volleying that the regiments had been perfecting over the years, to the defiant character of the men under arms, and to the sangfroid of British officers once brought to battle in front of their men. But the departing army left much behind: guns, baggage, horses – slaughtered horses, not a pretty sight, and an even uglier job. And they left behind their commander, for at his hour of victory Moore, like Wolfe and Nelson, fell to a bullet, mortally wounded. But unlike Nelson's, Moore's body was not brought home: he was buried in the field, in the early morning of 17 January, wrapped in his cloak and blanket, and with little ceremony – though with much reverence. Eight years later the Irish parson Charles Wolfe wrote 'The Burial of Sir John Moore after Corunna', beginning, in the beat of the slow funeral march:

Not a drum was heard, not a funeral note,
As his corse to the rampart we hurried;
Not a soldier discharged his farewell shot
O'er the grave where our hero we buried.

The lines reflect the sense of loss which the regiments felt, for although the retreat had been infamous, and they had cursed their general in their incomprehension, the final battle – and the escape of '*the* British army' to fight another day – had restored their self-respect, and with it the respect and affection even for their commander-in-chief. To some extent the same would be true after Dunkirk nearly a century and a half later, although it would take longer for the army to understand Lord Gort's achievement there than it did to grasp Moore's at Corunna.

But in a later stanza Wolfe hints at what might have befallen Moore's reputation had he not died in battle:

Lightly they'll talk of the spirit that's gone,
And o'er his cold ashes upbraid him—
But little he'll reck, if they let him sleep on
In the grave where a Briton has laid him.

The upbraiding had already begun, indeed, before the fate of the army – or even its predicament – was known in England. 'The truth is that we have retreated before a rumour – an uncertain speculation – and Moore knows it,' Canning complained: 'O that we had an enterprising general with a reputation to make instead of one to save!'* Most believed this to be unfair, especially since there was more than a suspicion of prejudice against Moore, the 'Whig general'. Even the arch-Tory Castlereagh spoke up for him.

But the army took note of the criticism. The fickleness of political support was at once evident to its senior officers, as were the hazards of using military force for political rather than strictly military purposes. The Spanish, until lately allies of France, had not proved themselves trustworthy allies of His Majesty, though their loathing of the invader Napoleon was evident enough. The perils of operating over difficult terrain and long distances without proper lines of communication had been hammered home, though the potential of the Royal Navy as a 'strategic multiplier' was again plain to see. Equally plain was the fragility of discipline in the instrument with which Moore's successor would have to finish the job. The army's organization was hopelessly ad hoc, its generals were dangerously recalcitrant, its rank and file ('the scum of the earth') reverted to type as soon as the wind blew unfavourably, and the regimental officers in too many cases were unable or unwilling to do their duty – or perhaps simply ignorant of what it was. These were

* Canning to Lord Bathurst, 9 January 1809.

the cautionary lessons of Moore's campaign; and Sir Arthur Wellesley was taking careful note.

And yet, when the order 'take post' was given, no matter what the circumstances, there were brigadiers and major-generals accomplished enough in mustering their brigades and divisions, and regimental officers who set a standard of courage that no army in Europe could better. And, of course, the 'scum of the earth' could fight with a faithful tenacity that moved men to tears. This, then, was the tricky, cranky, erratic but potentially magnificent instrument that was now placed in the hands of the only man in Britain with the qualities of character and the experience of war to replace Sir John Moore. Wellesley was now recalled from Ireland once more and told to take the army back to Portugal. Wellesley had admired Moore; he had offered his service to him the year before: 'you are the man, and I shall with great willingness act under you'. He had studied the Corunna campaign, and in doing so had developed decided views – views that would in time, slowly but very surely, take the army over the Pyrenees and into France. And so in its way the Corunna campaign was a prelude to eventual victory over Napoleon as significant as Dunkirk to VE Day. The stage was now set for perhaps the greatest trial in the making of the British army, a trial in which the steel would be further tempered and the weapon forged in a shape that would alter little in the best part of a century – and which would indeed be recognizable two whole centuries later.

13

The Spanish Ulcer

The Peninsula, 1809–14

THE STATE OF SIR JOHN MOORE'S MEN AS THEY LANDED PIECEMEAL along the south coast of England shocked civilians and soldiers alike. Storms had scattered the transports the length of the Channel; as a result, for days on end the able-bodied and sick alike tramped the roads for miles to the regimental depots. Edward Costello, just enlisted in the 95th, watched the arrival of his new regiment at their barracks in Hythe:

> The appearance of the men was squalid and miserable in the extreme. There was scarcely a man amongst them who had not lost some of his appointments, and many, owing to the horrors of that celebrated retreat, were even without rifles. Their clothing too was in tatters, and in such an absolute state of filth as to swarm with vermin. New clothing was immediately served out and the old ordered to be burnt.

With reports of the returning troops in such a condition, and of the indifferent support from their supposed allies the

Spanish, His Majesty's ministers might have recoiled from sending more troops to the Peninsula. But most of them, whatever their private misgivings, still maintained a show of public confidence: 'The British army', wrote Canning to his representative in Madrid, 'will decline no difficulty, it will shrink from no danger, when through that difficulty and danger the commander is enabled to see his way to some definite purpose.' And there were still 7,000 British troops in the Lisbon garrison, along with a few thousand Portuguese who might pass for an army: they too would have to be evacuated if they were not reinforced. Besides, Britain could hardly abandon Portugal, having convoyed its royal family to Brazil and appropriated its navy.

But with the French ranging in Spain, could Portugal be defended? Sir John Moore had thought not. Sir Arthur Wellesley believed otherwise. He now laid out the situation in a memorandum to his old friend Castlereagh, the secretary for war, concluding that Portugal could indeed be defended 'whatever might be the result of the contest in Spain' *if* Britain put an army there of no fewer than 30,000, with plenty of cavalry and artillery 'because the Portuguese establishment must necessarily be deficient in these two branches'; and *if* the Portuguese army were reorganized under British command; and *if* its every piece of uniform and equipment was sent from Britain. He took it for granted that the Royal Navy would retain command of the sea.

The memorandum, a model of strategic appreciation, did the trick: what remained of Moore's army was reconstituted and reinforced, and Sir Arthur Wellesley, despite his lack of seniority and his recent if unmerited Cintra tarnish, was appointed to command. His brief was limited and circumscribed, however. The defence of Portugal was his primary mission; going on to the offensive in Spain was to be a matter for his judgement, but it was not to be done without the express consent of His Majesty's government. He would in fact lose no time in going on to the offensive, however, for at this stage, before the French had time to consolidate their hold on the country, he was convinced that attack was the best means of defence.

Within three months of the evacuation of Corunna, Wellesley arrived in Lisbon and began reorganizing his new army to his own liking, forming divisions which could operate semi-independently (still something of an innovation), appointing British officers to commands at all levels in the Portuguese army and incorporating a Portuguese battalion into every British brigade. A fortnight later he was able to march north with 18,000 men to eject Moore's oppressor Marshal Soult and his army from Oporto, Portugal's second city and a strong centre of national resistance. He thought it prudent, however, to leave a strong garrison at Lisbon in case of a counter-stroke by a second French army under Marshal Victor in Castile, though this meant that his potentially winning move – sending General Beresford, now a marshal of the Portuguese army, with 6,000 men to try to block Soult's retreat – might fail for want of adequate strength.

On 12 May 1809, with a brilliant improvised crossing of the Douro led by the 3rd (East Kent) Regiment – 'The Buffs' – Wellesley retook Oporto. By the end of the month he had chased Soult out of Portugal, and although Beresford was not able to close the trap, the French had little option but to abandon most of their heavy equipment and run into wild Galicia. Here the Spanish partisans, the *guerrilleros* – the makers of 'little war' – received them most uncivilly. Goya's paintbrush has recorded some of the tortures meted out to captured French troops – of which barn-door crucifixions were but the most symbolic – for the guerrillas practised simple terrorism. And such became the terror of the guerrillas, indeed, that thousands of French troops would have to be used to escort couriers and convoys throughout Spain.

By the middle of July Wellesley had cleared Galicia and made good progress towards Madrid, winning a famous if costly defensive victory at Talavera towards the end of the month, for which he was ennobled as Viscount Wellington. But he was finding it all but impossible to concert his actions with the Spanish, for their armies operated under no central direction. And when he learned that Napoleon was sending even more reinforcements (though the 'emperor' himself never set foot in Spain again) the new Lord Wellington withdrew with the bulk of his Anglo-Portuguese army back across the border, while the Spanish armies in Old Castile fell back on Cadiz, their provisional capital – each separately to await the renewed French offensive.

Leading the invasion of Portugal the following year would be the marshal whose military prowess Napoleon admired the most, André Masséna. But Wellington had been thinking ahead, and Masséna's way to Lisbon would be barred by the biggest engineering effort of the entire war in Europe: the Lines of Torres Vedras, constructed at Wellington's command and in large part laid out by him. And they were built at such speed and in such secrecy that when Masséna came upon them in October 1810 he angrily demanded why he had been told that the way to Lisbon lay wide open. 'Wellington has had this made,' was the reply, at which Masséna exploded: 'Que diable! Il n'a pas construit ces montagnes!'

Wellington had not needed to throw up mountains: Estremadura, the region to the north of Lisbon, is hilly, mountainous even, as the French knew well enough. What he had done was link the hills between the Atlantic and the estuary of the Tagus in a chain of blockhouses, redoubts and ravelins, using all the natural defensive features of the country. The first of his three formidable lines, some 35 miles north-east of Lisbon and almost the same distance in length, ran through the town of Torres Vedras itself and covered every road and track to the capital, with a semaphore telegraph system operated by the Royal Navy which could send a message from one end to the other in eleven minutes. A second line, even stronger, was being built 6 miles to the south but was not complete until 1812, with a further, much shorter line, on the coast west of Lisbon to cover the embarkation beach if, like Sir John Moore, Wellington was forced to evacuate the army. The French only ever saw the first, however. For not only had the fortifications every impression of impregnability, Wellington withdrew the army into his fastness through a deep belt of 'scorched earth'.* As winter came on, Masséna's army simply sickened and starved. In March 1811 they gave up altogether and slunk away to the border once more, leaving behind most of their transport and hundreds of hamstrung mules.

The Lines of Torres Vedras, built by Portuguese labourers under command of Lieutenant-Colonel Richard Fletcher of the Royal Engineers, cost £100,000. At £80 million in today's prices (only half as much again as a single Eurofighter), they must rank as one of the most cost-effective items of defence expenditure in history – certainly since David's sling.† And they set something of a benchmark for what the Royal Engineers might do on operations. But it was not just the economy of the undertaking, nor perhaps the secrecy nor even the stopping power of the Lines which imprinted itself on the minds of officers for generations to come: it was Wellington's anticipation of the need for them. Indeed, when the army defined its doctrine of 'battle procedure' before the Second World War (the process by which a commander receives his orders, makes his reconnaissance and plan, issues

---

* He had fought an active defence astride the border throughout the spring and summer in concert with the Spanish fortresses, delaying the French invasion longer than he had expected.
† HMS *Victory*, launched in 1765, had cost £63,000 – in excess of £104 million at today's prices.

his orders, prepares and deploys his troops and executes his mission), one of the four principles upon which the procedure was based was 'intelligent anticipation of future tasks'.

The notion of a reserve position, a bastion, a firm base into which a force could withdraw when the situation beyond turned unfavourable, was hardly new: every castle in Europe owed its construction in part to that principle. But the Lines of Torres Vedras, including the protected embarkation beach, were something more: they gave Wellington 'balance', allowing him a greater opportunity for boldness than he might otherwise have had. Nothing quite like it was ever achieved again, although the idea was certainly strong in the minds of the desert generals of the Second World War: El Alamein was first an 'impassable' defensive line, dug, mined and wired as a *ne plus ultra* to protect Cairo, before it became the springboard for the British army's greatest offensive battle honour of all time.

Torres Vedras showed above all what a good 'eye for ground' Wellington had, and therefore exalted the virtue in the minds of his successors. The skill was associated with Wellington's love of fox- and hare-hunting in the Peninsula, his only recreation, and the conflation of the two has had numerous echoes since. Siegfried Sassoon, in the first volume of his chronicle of the First World War, *Memoirs of a Foxhunting Man*, writes lyrically of the eye for ground which riding to hounds gave him. And the late Lord (Bill) Deedes, reflecting on his time in Normandy in 1944 as a young officer in the 60th Rifles, observed of the officers of their supporting armoured regiment, the 13th/18th Royal Hussars, that those who had spent a good deal of time in the hunting field seemed able to read the ground best.*

* When the author was senior subaltern of his infantry battalion in Germany in the mid-70s, the new commanding officer, a seasoned soldier who had been a master of Cumbrian foxhounds, exhorted him to get the rest of the subalterns out hunting to improve their eye for ground. And this was the nuclear age!

The eighteenth-century notion of battle, which Napoleon had developed rather than revolutionized, treated the ground almost as incidental. In the Peninsula, however, Wellington took every opportunity to exploit the French preference for the attack, and chose the ground on which to receive them in such a way as to maximize his advantages and minimize his weaknesses. Before withdrawing into the Lines of Torres Vedras, for example, he fought a linear, defensive battle on the long ridge at Bussaco, which his veterans would recall as they stood on the ridge of Mont St Jean five years later at Waterloo. And although Wellington cannot be pigeon-holed simply as a defensive-minded general (his time in India and his early offensives in Portugal showed otherwise), he certainly had a fondness for advancing into the enemy's country, choosing a good piece of ground and then waiting for the attack. It was said a hundred years later that whereas the French army of 1914, with their doctrine of *attaque à outrance* (attack to excess) would prefer to abandon a position and then retake it with *panache*, the British infantry would cling on like limpets.* Writing of his experience of the desert war in 1944, Rommel too spoke ruefully of the British army's 'well thought-out guileful methods of defence'. It went back to Wellington: not for nothing was the principal defence exercise of the platoon commanders' battle course at the School of Infantry called until recently 'Bussaco Ridge'.†

Yet the will to go to it with the bayonet, as the infantry had

---

* After the battle of Albuera (May 1811) Beresford wrote: 'Every individual nobly did his duty; and it is observed that our dead . . . were lying, as they fought, in ranks, and every wound was in the front.'
† After the school's move from Salisbury Plain to the Brecon Beacons, it was thought appropriate to update the exercise name – but only as far as 1942. It is now called 'Sittang Bridge' after the battle in Burma (at which, coincidentally, the 2nd Battalion of the Duke of Wellington's Regiment played a prominent part).

at Steenkirk and would again in the Crimea – and on the Western Front in the First World War, and at El Alamein, and on many more recent occasions – was always an asset that Wellington sought to exploit, if under the strictest control. And not just to exploit, but when necessary to drive to the very limit. Two years after Masséna abandoned his 'siege' of Torres Vedras, the sanguinary ruthlessness of which Wellington was capable when circumstances demanded it was made dramatically evident when he in turn laid siege to Badajoz.

The years between Torres Vedras and Badajoz had been a crab-like business of advance and withdrawal. Wellington himself said, and not merely in self-justification, that the best test of greatness in a general was to know when to retreat, and to dare to do it. He had certainly dared to retreat well enough, and it had cost him a deal of trouble in London, though he never lost the support of the most important of the King's ministers. Fortunately, too, the duke of York was back at the Horse Guards. He had been forced to resign in 1808 when his mistress was caught selling commissions (and his complicity therefore presumed), but his innocence had been painstakingly established and the Prince Regent, whose hands held the reins during George III's madness, was willing to overlook the questionable judgement of keeping a mistress who was prepared to trade beyond her competence. So with York's backing re-established, in 1812 Wellington began another offensive, besieging the great border fortresses of Ciudad Rodrigo and Badajoz. The first, and smaller, of these fell in January with a deal of blood – including that of 'Black Bob' Craufurd who, as ever leading from the front, fell mortally wounded in one of the breaches, in which he was afterwards ceremoniously buried.

Once inside Ciudad Rodrigo, the army's discipline had

faltered, as Lieutenant John Kincaid of the 95th Rifles described:

> A town taken by storm presents a frightful scene of outrage. The soldiers no sooner obtain possession of it than they think themselves at liberty to do what they please . . . without considering that the poor inhabitants may nevertheless be friends and allies . . . and nothing but the most extraordinary exertions on the part of the officers can bring them back to a sense of duty.*

But at least in Ciudad Rodrigo no Spanish civilians had been killed gratuitously. At Badajoz it was a different matter. The city sits on a slight rise overlooking the Guadiana River. A massive wall and bastions, a broad moat and outworks, and forts on the surrounding heights made it one of the strongest fortresses in Europe. Yet it had been taken by the French in March 1811 without assault – the Spanish commander succumbing (or so it was believed) to gold rather than lead.

Wellington began his siege in March the following year, and after a month there were three breaches in the walls which he deemed 'practicable' – wide enough to assault, though in truth they were too narrow. He still had no heavy artillery worth the name to batter away with, nor enough engineers for mining, and the breaches were therefore the best he could hope for. So on the night of 4 April two divisions battled for five hours to get through two of the breaches while a third feinted against the other to draw off the defending troops. But without success.

When Wellington received the news that the divisions could make no progress he turned white. 'I shall never forget

---

* Sir John Kincaid, a faithful chronicler of the war, wrote two volumes of reminiscences, *Adventures in the Rifle Brigade* and *Random Shots from a Rifleman*.

it till the last of my existence,' wrote James McGrigor, his chief medical officer: 'The jaw had fallen, the face was of unusual length, while the torchlight gave to his countenance a lurid aspect; but still the expression of the face was firm.' And so was his resolution. Wellington could not afford to be repulsed that night: his deputy, Beresford, had twice failed to take the fortress the year before, and the 'croakers', as Wellington called them – officers on leave in London who complained of his excessive caution – were not serving his cause. There was intelligence, too, of a French army marching to the relief of the city. And perhaps a rebuff at Badajoz, at the very gateway to Spain, might have shaken the confidence of the army – in itself and in him. His reputation, after all, was founded on not losing battles. So it must have been in some measure of desperation as well as firm resolve that Wellington ordered his two reserve divisions to take the walls by 'escalade' (a pretty term for using ladders). A reserve is used in defence to avert a crisis, and in offence to reinforce success; but here Wellington was reversing the formula – and crudely. Quite evidently it is easier to enter a fortress through a breach in the walls than by trying to scale them; yet here he was ordering his two reserve divisions to do more than the first two divisions had just failed to achieve.

The 'human factor' now intervened fortuitously – and Wellington was always a lucky general – as the magnificent 5th Foot* managed to get a lodgement atop the walls; and how they did so says much about the army's fighting spirit then and since. For the first man into the fortress that night was not a thrusting ensign fresh from the playing fields of Eton, nor a hardened serjeant who knew how to fight, nor a corporal keen for his extra stripe, nor even one of

---

* They became the 5th (Northumberland Fusiliers) Regiment of Foot in 1836. The 'Fighting Fifth' was the regiment in which John McManners, whose thoughts on war conclude the introduction to this book, served during the Second World War. They were long known as one of the toughest in the army.

Wellington's 'scum of the earth', a rum-fuelled private soldier from the coaly hovels of the Tyne or wherever the Fifth managed to find their men. It was in fact their commanding officer, Lieutenant-Colonel Henry Ridge. To protect himself against the missiles raining down from the battlements, he told a handful of his men to hold their muskets above his head in an umbrella of bayonets, and to advance up the ladder behind him. When they reached the end of the ladder, a good 15 feet short of the top of the wall, he told one of them to stand on the top rung, and another to climb on to his shoulders, and then he, Ridge, climbed on to the second man's shoulders and hauled himself over the parapet. He was cut down and killed soon afterwards, but not before his men were swarming over the battlements, following his example.

There were 5,000 French defending Badajoz that night, and Wellington lost 5,000 men taking it. But the losses were disproportionately high among the officers: of the twenty-three infantry battalions that took part, in fifteen of them three out of four officers were killed or wounded, and in five of them every single one. By the time Wellington's men got into the fortress, their blood was boiling from the fight, and they were without their officers. And, as at Ciudad Rodrigo, discipline faltered – or rather, this time broke down altogether. Although the French were largely spared, the Spanish civilians suffered very badly indeed, for there was a strong though generally unfounded suspicion that they were collaborators. Murder, rape and robbery followed for a full forty-eight hours before Wellington was able to regain control of his men – helped in no small measure by the exhaustion of the fortress's wine cellars.*

---

* Wellington had countermanded the old rule of sieges that gave 'no quarter' when the defenders had held out after a breach had been made in a fortress wall. This had led to a more protracted defence of Ciudad Rodrigo, and Wellington's men resented the order, but at Badajoz they followed it almost to the letter as far as the French were concerned. He had

Badajoz stands as one of the greatest fighting feats of the British army, and one of its worst instances of indiscipline. And it demonstrated the fragility of the system once more. Wellington himself had been all too aware of it: he had seen for himself the climactic aftermath of countless skirmishes and battles – not least at the sack of Seringapatam – and he was clear enough about the remedy. But there was only so much that flogging could achieve: he insisted too that the officers do their duty in every detail, not just lead bravely, and he harried them mercilessly – so mercilessly, indeed, that one of them, Lieutenant-Colonel Charles Bevan, blew his brains out after such a rebuke. Bevan commanded the admirable 4th (King's Own) Regiment, which had scarcely put a foot wrong in its 150 years' existence and had lost more men in the storming of Badajoz than any other regiment. It is tempting to suppose that Bevan wanted to show Wellington and the rest of the army that the rebuke had been unwarranted.*

Quite what Wellington expected of his officers was not always clear-cut. Reminiscing many years later with Earl Stanhope – his 'Boswell' – the duke was very decided that

> The Guards are superior to the Line – not as being picked men like the French – for Napoleon gave peculiar privileges to his guardsmen and governed the army with them – but from the

Footnote continued from p 227
issued no specific instructions about the Spanish (and it is suggested that some commanders such as General Sir Thomas Picton actually encouraged the bloodlust), for manifestly they were allies. Badajoz is a case where commanders perhaps understood the uncertainty of the extent of their disciplinary grip, and thought it best not to test it.

* In Bevan's case, however, it was more on account of tardiness in carrying out orders to move rather than dereliction of disciplinary duty, and there was widespread sympathy for him. There is a fine account of the affair in *Wellington's Scapegoat* by Archie Hunter (2003).

goodness of the non-commissioned officers. They do in fact all that the commissioned officers in the Line are expected to do – and don't do . . . It is true that they regularly get drunk and go to bed soon after, but then they always took care to do first whatever they were bid. When I had given an officer in the Guards an order, I felt sure of its being executed; but with an officer of the Line, it was, I will venture to say, a hundred to one against its being done at all.

Perhaps over the port and nuts before a good fire at Walmer Castle, the duke's residence as Lord Warden of the Cinque Ports from 1829 until his death there in 1852, and in the company of a much younger man who hung on his every word, the Great Man made his point with a certain expansiveness – exaggeration even. But what he recalled was essentially true: the Guards had established an ascendancy. That, after all, was what was expected of Guards regiments in any army. And although, as Wellington said, they were not picked men, nor given privileges as in the French army, nor used in any sense to govern the rest of the troops, they came increasingly to set an example of how things should be done. Sandhurst, which began to set the tone for the formation of officers in the Victorian era, and has continued to do so ever since, was and is run by the Guards. Although the commandant may be from any regiment, the adjutant and the Academy Serjeant-Major are always from the Foot Guards.* And when additional officer training units were established during the two world wars and the period of National Service that followed, the Guards ran these too.

But it was not only in administration, training and

---

* Sandhurst's Academy Serjeant-Major ranks above every serjeant-major in the army. As a warrant officer, the appointment of 'Conductor' in the Royal Logistic Corps ranks above the Academy Serjeant-Major; but a conductor is not a serjeant-major.

discipline that the Guards became pre-eminent. They had by Wellington's time established a formidable fighting reputation. Other regiments often fought as well as they, but none ever fought better. After Waterloo, writing of the crucial defence of Hougoumont on the right flank by a composite force from the three regiments of Guards, Wellington said simply: 'No troops but the British could have held Hougoumont, and only the best of them at that.' And they have never truly faltered since – a remarkable record, for every regiment has had its setbacks, a time which might be whispered as 'not their finest hour', when an attack has failed, or a position has been over-run because of some dereliction, mishap, lack of skill or plain lack of resolve. But not the Guards. Indeed, when news that 'the Guards are in the line' ran up and down the trenches of the Western Front, it was always with a frisson.*

Badajoz was a turning point, but it would be a full year before increasing Anglo-Spanish pressure would force the French into irrecoverable retreat. In the summer of 1812 Wellington gained a spectacular victory at Salamanca, occupying Madrid soon afterwards and pressing on east; but he was checked at Burgos, and had to withdraw once more before a renewed French offensive all the way back to Portugal. The following year he and the Spanish armies came out of winter quarters renewed, however, whereas the French were much weakened by the continuing guerrilla operations and by the destruction of their eastern army in the Russian campaign. And so at last Wellington could begin his sustained advance to the Pyrenees and beyond (and his own advance to both duke and field marshal). He fought

* Sir Arthur Conan Doyle's poem about an attack on the Western Front, 'The Guards Came Through', captures the frisson perfectly (see 'Notes and Further Reading').

his final battle of the Peninsular War not in Spain but in France, at Toulouse in April 1814 – before the news could reach him that Napoleon had abdicated.

By the end of the war the army had become as efficient as any on the Continent, in its own very British way. The French colonel Thomas Bugeaud, who would win laurels as a marshal of France in Algeria in the 1830s, had been a regimental commander in Spain, and observed simply, 'The British infantry is the most dangerous in Europe. Fortunately it is not numerous.' As for Wellington, who later in life admitted to one fault in command – 'I should have given more praise' – he said of his Peninsular army simply, 'I could have taken it anywhere and done anything with it.'

And the army's success was of his making – his personal making. Sir Arthur Wellesley, in turn viscount, marquess and finally duke of Wellington, had not had a day's leave since landing in Portugal in 1809, and his daily rate of work – at his desk or in the saddle – had been prodigious. It was necessary, he said, that a general be able to trace a biscuit from Lisbon all the way to the army in the field. This was the way that war was made, and he had learned how to do it in India. Marlborough had learned it through a longer process of experience in Flanders, and through a practical intuition, a doctrine of common sense which Wellington shared. Both commanders, in their own fashion, had evolved a way of making war that perfectly suited the nation's character and the realities of its politics and economy.

The army's foundations had been laid out, if erratically, in the Civil War and at the Restoration; they had then been dug deep and well in Marlborough's time; and the walls had been securely raised in the Peninsula. There was much building work still to do, and there would be earthquakes, fire and flood that would shake the edifice to its foundations. Neglect would bring dilapidation, so that the roof would fall in from time to time; but the structure would remain fundamentally sound, and no assault on it would bring about complete collapse. No army of any sovereign nation in 1814, except that of the United States, would be able to look back a century and a half later and make that claim.

14

A Damned Nice Thing

Waterloo, 1815

AND SO CAME THE ARMY'S (AND WELLINGTON'S) GREATEST TEST: the battle that defined the rest of the nineteenth century and whose name is deeply etched in the minds of soldiers today, even if they know little of its details. 'Waterloo' stands for something timeless and fundamentally unshakable in the way the British army conducts itself in defence: endurance to the end by the man with the rifle; self-sacrificing offensive action by the cavalry; limpet devotion by the artillery to its guns; officers standing side by side with their men; the triumph of simple duty; fortitude in the face of seemingly overwhelming odds – and, of course, some brilliance. That the battle was, in Wellington's words next day, 'a damned nice thing – the nearest run thing you ever saw in your life' makes it all the more powerful an exemplar. Victor Hugo, ever anxious to rewrite any fact of Waterloo to explain that it was either in truth a French victory or at worst a *diabolical* defeat, calls it in *Les Misérables* 'a battle of the first rank won by a captain of the second'. But he makes this extraordinary claim ostensibly to make a much greater point: 'What is truly admirable in the battle of Waterloo is England, English firmness, English resolution, English blood. The superb thing which England had there – may it not displease her – is herself; it is not her captain, it is her army.'

Hugo has never found himself in strong company with that view of Wellington. Nevertheless the victory was the army's too, certainly; and the manner of victory did indeed show what the army had become. For however vaunted the opponent, however inadequate the British appeared by comparison, there was some ingredient in Wellington's army, some sort of leaven, that could make it rise to any occasion. Wellington himself likened his way of campaigning to the practical business of horse furniture: 'The French plans are like a splendid leather harness which is perfect when it works, but if it breaks it cannot be mended. I make my harness of ropes, it is never as good looking as the French, but if it breaks I can tie a knot and carry on.' And this make-do-and-mend approach seemed to reach right down to the ranks of red: no matter what the setback, every man would simply 'tie a knot' and move on. In 1914 the Kaiser is supposed to have referred to Britain's 'contemptible little' army (see 'Notes and Further Reading'); the soldiers of that army immediately started calling themselves 'the Old Contemptibles' and proceeded to deal with the myth of Prussian military invincibility with contempt. A recurring feature of history has been the enemy's underrating of the British army, failing to see beyond its sometimes ramshackle appearance to what lies deeper. An army that could tie knots, and use them as it did at Waterloo, could be expected to do so again.

The 1815 campaign was an affair of a few days only before its climax at Waterloo. Forced to abdicate the previous year as the armies of Britain, Austria, Prussia and Russia began

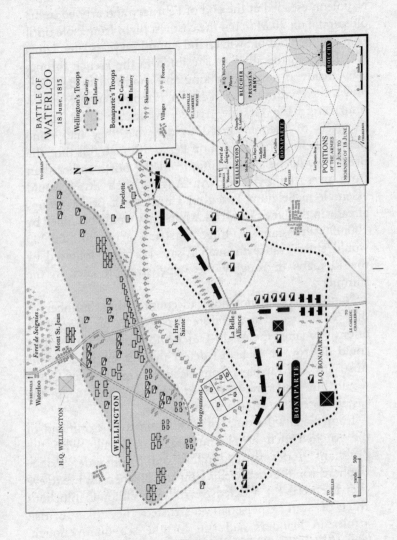

BATTLE OF
WATERLOO

18 June, 1815

Wellington's Troops
Cavalry
Infantry

Bonaparte's Troops
Cavalry
Infantry

Skirmishers

Villages   Forests

N

TO OHAIN

Papelotte

Forêt de Soignes

Mont St. Jean

TO BRUSSELS
Waterloo

H.Q. WELLINGTON

WELLINGTON

La Haye Sainte

La Belle
Alliance

Hougoumont

BONAPARTE

H.Q. BONAPARTE

TO
LE CAILLOU
CHARLEROI

TO
NIVELLES

yards   500
0

TO
CHAPELLE
ST. LAMBERT,
WAVRE

POSITIONS OF THE ARMIES
17 JUNE TO
MORNING OF 18 JUNE

H.Q. BLÜCHER
Wavre

PRUSSIAN
ARMY

GROTS END

Chapelle
St. Lambert

BONAPARTE

WELLINGTON

Forêt de
Soignes
Waterloo

Mont St. Jean
La Haye Sainte
La Belle
Alliance
Le Caillou

Les Quatre Bras

TO
BRUSSELS

TO
NIVELLES

TO
CHARLEROI

miles   5

closing on Paris and consigned to exile on the island of Elba, Napoleon escaped at the end of February and arrived in his old capital on 20 March – the costliest flight from exile until Lenin slipped out of Switzerland a hundred years later. Louis XVIII, whom the allies had restored to the French throne, was already an unpopular king and soon fled to Belgium, now united with the Dutch crown. Napoleon at once began resurrecting the Grande Armée, while the allies at the Congress of Vienna agreed to do what they had done the year before and close in on Paris, although this time the British would be approaching from the north rather than from over the Pyrenees. An Anglo-Dutch army would assemble in Belgium, based on a nucleus of British troops already in the Low Countries, where they would be joined by 100,000 Prussians. A huge Austrian army of 200,000 would enter France through Alsace-Lorraine, followed by the Russians in roughly the same strength later in the summer.

Napoleon could not hope to counter these numbers if they combined. He therefore decided on an immediate offensive against the Anglo-Dutch and Prussian forces: if he could destroy, or even just defeat them it might deter the Austrians and Russians from risking a similar fate. And it would buy him time to raise more troops and to put the eastern fortresses on to a strong footing.

Immediately, the duke of Wellington took command of the Anglo-Dutch army and began staff talks with the Prussians under their doughty old commander Prince Gephard von Blücher. Wellington and Blücher saw eye to eye over the need for a defensive strategy, although Blücher's chief of staff, Gneisenau, distrusted the duke: British armies in Flanders had run for the sea often enough. The allies were, in fact, too weak to risk an offensive, especially since the reliability of the Dutch–Belgian element was questionable and half the British regiments were inexperienced.* Besides, a defensive strategy best suited Wellington's instincts; certainly it played to the British army's strengths.

In turn, Napoleon knew he was not strong enough to fight both Wellington and Blücher in the same battle. He would therefore have to deal with them one before the other, making sure they could not combine. This – dealing with two armies – he had done often enough before. Success consisted in hitting one of them so hard that it had to withdraw to recover, turning to the other army and destroying it utterly, and then if necessary returning to the first to destroy it in detail. The trick was never allowing one of them to render support to the other.

Wellington and Blücher understood this perfectly, and were resolved to remain in close mutual support. The flaw in allied cooperation lay, however, in their respective lines of communications: if they gave way in the face of a huge French offensive the two armies would naturally begin to diverge, for the British supply lines ran north-west to the Belgian ports, while the Prussians' ran east to the Rhineland. And it was this fault-line, the boundary between the two armies, that Napoleon intended exploiting.

Wellington had another factor with which to grapple in his calculations: the French axis of advance on Brussels (their presumed objective) might be from the south through Charleroi, or from the south-west through Mons or even Tournai. An offensive from the south-west would threaten his lines of communication, and he would have to dispose a considerable force to guard them. He could not deploy the army forward on the border until he knew Napoleon's real

* For the most part Wellington did not have his old Peninsula battalions: a good many of his veteran troops had been sent to North America for the 'War of 1812', an eruption of festering sores, a mixture of US grievances and unfinished business. Its outbreak defies all good sense, and its conduct is unworthy of study. It was poor, nasty, brutish and, mercifully, short.

axis of advance, for there would not be time to switch positions. And besides, the problems of billeting and supply, particularly fodder for the cavalry and artillery, obliged him to have his troops more dispersed than he would have liked. But he was confident that he would have good and early intelligence of French moves: he had his trusted spies and observing officers in Paris and elsewhere. If he had had greater confidence in his cavalry he might have arranged some reconnaissance with the Prussians across the French border: it would at least have spared him tactical surprise when Napoleon appeared with all his men – to the complete amazement of both him and Blücher – before Charleroi on 15 June. Wellington was not exaggerating when he exclaimed, 'Napoleon has humbugged me, by God!'

On the sixteenth the French inflicted a sharp reverse on the unsupported Prussians at Ligny, Blücher himself being unhorsed and almost killed. Racing south from Brussels, gathering up the army as he went, Wellington managed to fight an action on the extreme right (west) of the battle area at Quatre Bras. Had Marshal Ney – *le brave des braves*, as Napoleon had dubbed him – whose job it was to deal with any appearance of the Anglo-Dutch, pressed his attack with more vigour, Wellington might have been forced from the ground. By nightfall, however, he was still in possession of the crossroads which gives its name to the village and the battle. There is probably some truth in the idea that Ney did not press the attacks because he knew Wellington's trick of keeping the main part of his force concealed, and that he was wary of being wrong-footed (he had seen the British army at Bussaco and Torres Vedras). Such is the 'virtuous circle' of good tactics.

Whatever the reason, however, Ney's apparent timidity allowed some semblance of cooperation between the allies at a crucial moment. Despite Gneisenau's urging that the Prussians withdraw east, in the belief that Wellington had failed to show sufficient resolve in supporting them, Blücher promised to fall back north instead, and to continue the fight. For his part, Wellington would withdraw to the ridge of Mont St Jean before the village of Waterloo on the southern edge of the Forest of Soignes and astride the Charleroi–Brussels high road, and there give battle when Napoleon switched his main effort – as both he and Blücher knew he must.

Next day, the seventeenth, in heavy rain, the Prussians began to withdraw as agreed. Emmanuel de Grouchy – who had commanded Napoleon's escort on the retreat from Moscow, but who had only now been made a marshal – at once advanced to harry them, and by one of those mis-readings of the situation usually ascribed to 'the fog of war' sent back a report to Napoleon that Blücher was withdraw-ing east. He had, in fact, mistaken the supply train for elements of the army itself. Thus confirmed in his own good opinion of his strategic skill, Napoleon turned with a vengeance on the Anglo-Dutch, believing the Prussians to have been knocked out of the contest. Wellington's with-drawal to Mont St Jean went well, however: just as Lord Paget and his cavalry had covered Sir John Moore's escape from east of the Esla River, allowing the infantry to gain a head start for Corunna, so now would the Marquess of Uxbridge – as Paget had become – cover the infantry back from Quatre Bras.* The wet and weary regiments of foot which tramped to their watery bivouacs on the ridge that night, joining those whom Wellington had ordered up directly, had much to thank Uxbridge's cavalry for, though they had not actually seen anything of them – appreciating

---

* After Corunna, Sir Stapleton Cotton had replaced Uxbridge in command of the cavalry. Uxbridge had earlier run off with Wellington's sister-in-law, and the Horse Guards thought it best to keep the two men apart. By 1815, however, the hatchet had been buried sufficiently deeply to allow his return.

only the absence of French *cuirassiers* harassing them on the march back.

Waterloo today – despite the huge memorial mound of which the duke complained in later years – is remarkably as it was that day in June 1815, Sunday the eighteenth. The ridge is not high, but there is a pronounced slope, which was rather more pronounced 200 years ago.* It was not so much an obstacle to the French advance as an elevation which gave the defenders a slight psychological advantage and, more importantly, Wellington the opportunity to work his favoured tactical trick of concealment behind the crest. The ridge itself is just short of 3 miles long, about two-thirds of which Wellington was able to occupy with infantry. He posted two brigades of cavalry on the left (east) flank, and another on the inner right, with his heavy cavalry (two brigades) in the centre, and his artillery interspersed with the infantry. Indeed, in essence the deployment was little different from that at Edgehill and the other battles of the Civil War – except that it was a far longer line. But there were three man-made features which gave the position its real strength. First of these were the villages of Papelotte, Frichermont and La Haye, all clustered at the foot of the slope on the left flank, and Braine l'Alleud on the extended right. Garrisoned by the Dutch and Belgians, these villages gave a solid pinning to both ends of the position, and since Wellington had no idea how steady his allies might be in the open, the solid cover of masonry gave him some assurance that they would stand and fight. Second, at the foot of the slope on the right inner flank was the little chateau of Hougoumont which might act as a rock breaking up the waves of an attack on that flank. It was exposed, however,

* Established definitively by recent topographical survey: see Andrew Roberts, *Waterloo, Napoleon's Last Gamble.*

and once under attack would be difficult to support, and so into Hougoumont Wellington put several companies of the Guards, with some German riflemen and a few Dutch Nassauers – 1,500 men in all. Finally, in the centre, halfway down the slope on the high road, was the little farm of La Haye Sainte. This might have the same effect as Hougoumont in breaking up an attack on the centre, and it would certainly be useful cover to the French if they could take it. Wellington therefore placed a battalion of the King's German Legion (about 400 men) inside the farm courtyard, with companies of the 95th Rifles in some sandpits across the road. His red-coated infantry and the rest of the KGL he placed in the centre and right of the line, where he expected the main weight of the attack to fall, and his untried Dutch-Belgians and Hanoverians on the left.*

The arguments continue to this day about what or who won Waterloo: the earth was too wet for the French artillery, the French marshals were faint-hearted, the Prussians arrived in the nick of time; Napoleon's piles, even. The fact is – as a visit to the battlefield shows at once – that the masterly choice and use of ground was everything.

Thomas Hardy, in that rather strange epic verse-drama *The Dynasts*, pictures the ridge of Mont St Jean, and the ridge opposite on which the French were camped, the night before the battle:

CHORUS OF THE PITIES
 . . . And what of these who to-night have come?

---

* Hanover was only lately delivered from occupation by the French for the better part of fifteen years. The electorate's troops (sometimes known as Brunswickers) were to Wellington's mind untried, although their allegiance was not as suspect as that of the Dutch and Belgians. The KGL were Hanoverians, too, of course, but they had been British to all intents and purposes for a decade and more.

CHORUS OF THE YEARS
The young sleep sound; but the weather awakes
In the veterans, pains from the past that numb;

Old stabs of Ind, old Peninsular aches,
Old Friedland chills, haunt their moist mud bed,
Cramps from Austerlitz; till their slumber breaks.

CHORUS OF SINISTER SPIRITS
And each soul shivers as sinks his head
On the loam he's to lease with the other dead
From to-morrow's mist-fall till Time be sped!

(The fires of the English go out, and silence prevails, save for the soft hiss of the rain that falls impartially on both the sleeping armies.)

Hardy uses the term 'English' as it was used loosely in his time: there were Scots, Irish and Welsh regiments the length of the ridge, and half of the allied army had German or Dutch or French (Walloon) as their first and perhaps only language.

There was no patrolling during the night, and little movement at Mont St Jean other than ammunition waggons and the odd commissary's sparse load of rations, nor any forward movement by the French, exhausted as they were by the battle at Ligny and the following march. Only sentries keeping sodden watch.

As the trumpets and bugles blow reveille, Hardy describes the scene in his 'stage directions':

THE FIELD OF WATERLOO
An aerial view of the battlefield at the time of sunrise is disclosed
. . .

The sky is still overcast, and rain still falls. A green expanse, almost unbroken, of rye, wheat, and clover, in oblong and irregular patches

242

undivided by fences, covers the undulating ground, which sinks into a shallow valley between the French and English positions. The road from Brussels to Charleroi runs like a spit through both positions, passing at the back of the English into the leafy forest of Soignes.

The latter are turning out from their bivouacs. They move stiffly from their wet rest, and hurry to and fro like ants in an ant-hill. The tens of thousands of moving specks are largely of a brick-red colour, but the foreign contingent is darker.

Breakfasts are cooked over smoky fires of green wood. Innumerable groups, many in their shirt-sleeves, clean their rusty firelocks, drawing or exploding the charges, scrape the mud from themselves, and pipeclay from their cross-belts the red dye washed off their jackets by the rain.

At six o'clock, they parade, spread out, and take up their positions in the line of battle, the front of which extends in a wavy riband three miles long, with three projecting bunches at Hougoumont, La Haye Sainte, and La Haye.

Looking across to the French positions we observe that after advancing in dark streams from where they have passed the night they, too, deploy and wheel into their fighting places – figures with red epaulettes and hairy knapsacks, their arms glittering like a display of cutlery at a hill-side fair.

They assume three concentric lines of crescent shape, that converge on the English midst, with great blocks of the Imperial Guard at the back of them. The rattle of their drums, their fanfarades, and their bands playing 'Veillons au salut de l'Empire' contrast with the quiet reigning on the English side.

A knot of figures, comprising WELLINGTON with a suite of general and other staff-officers, ride backwards and forwards in front of the English lines, where each regimental colour floats in the hands of the junior ensign. The DUKE himself, now a man of forty-six, is on his bay charger Copenhagen, in light pantaloons, a small plumeless hat, and a blue cloak, which shows its white lining when blown back.

On the French side, too, a detached group creeps along the front in preliminary survey. BONAPARTE – also forty-six – in a grey overcoat, is mounted on his white arab Marengo, and accompanied by SOULT, NEY, JEROME, DROUOT, and other marshals. The figures of aides move to and fro like shuttle-cocks between the group and distant points in the field. The sun has begun to gleam.

The rain had stopped, and the sun had indeed begun to dry out the armies, if not the ground. The French seemed in no hurry to attack, troops forming up on the ridge opposite Mont St Jean as if for a review. In fact the bands played and Napoleon rode the length of the line raising cheers. There has always been speculation that he delayed the opening of the battle so that his artillery could have greater effect: the drier the ground, the greater the ricochet of the solid shot. But this seems unlikely: time was the one commodity that Napoleon had always said was not his to dispose – 'Ask of me anything but time,' he told his marshals – and therefore it was not to be squandered. The truth is probably that his officers could not muster the army into its battle positions any more quickly (they had been dog-tired), and that Napoleon and his gunners made the best of this with the consolation that 'at least the ground is drying out with every minute'.

The battle is usually divided by historians into five phases. It began just before midday with a colossal bombardment by Napoleon's massed batteries directed at the centre-left of the allied line, together with a diversionary attack on Hougoumont to tempt Wellington to reinforce that flank at the expense of his centre and reserve. The French ploy was a shade too obvious, however, especially since the attack on Hougoumont was practically unsupported by artillery (which any serious attempt to turn a flank would have needed), and so Wellington simply stood his ground. Indeed, the attack soon began to serve him, for the Guards held out so resolutely, and were supported so deftly by the brigades on the right flank, that they drew in and tied down more and more French – all day, in fact, a sort of mini Spanish ulcer. There were many heroes of that day-long action, but none more praiseworthy than the admirable Corporal Joseph Brewer of the Royal Waggon Train who, when powder was running low, volunteered to gallop his ammunition tumbrel from the ridge, under fire, and into the chateau – which he did, to the cheers of the defenders.

Just before launching his main attack on the allied centre (at about 1.30 p.m.), the second phase, Napoleon learned that the Prussians had withdrawn not east and out of supporting range, but north towards Wavre. He therefore formed a defensive right flank, and this would draw increasing numbers of troops from his reserve for the rest of the day – just as Hougoumont drew in more and more of Ney's left wing. Meanwhile, the great juggernaut that was the assaulting corps – 16,000 men, a quarter of Napoleon's whole force, under the comte d'Erlon – began its march in column towards the ridge and to the east of La Haye Sainte.

They had 1,300 yards to march, over wet, loamy earth, through corn 6 feet high,* and in full view of the allied guns now being rapidly run forward along the chemin d'Ohain. The preliminary bombardment by the massed battery had done some damage to Wellington's troops, but not nearly as much as Napoleon hoped or supposed (for he could not see the defenders behind the crest of the ridge), and the allied infantry now came forward in line two deep and began volleying into the densely packed columns which were already under a galling fire from the skirmishers – especially from the 95th Rifles in the sandpits. Unsupported by

---

* 'Corn 6 feet high' is sometimes disputed, but selective cultivation has reduced the height of wheat by almost a half since Waterloo.

artillery, and with the cavalry on their flanks unable to influence the fighting to their front, the French columns began to waver and then turn tail – but not before the French cavalry had cut up a battalion of Hanoverians sent to support La Haye Sainte, forcing the nearby battalions on the ridge to form square (a salutary reminder that it did not do to make a mistake in front of French cavalry) and allowing the assaulting columns almost to gain the crest. Seeing this, Uxbridge, without waiting for orders, launched his two heavy cavalry brigades – the Household and the Union (so-called because it consisted of English, Scots and Irish dragoon regiments) – at the French right flank, driving them off and turning the repulse into rout. But Uxbridge's cavalry galloped on too far, cutting up some of the grand battery, and were in turn cut up by French lancers. They were saved from complete disaster only by a counter-charge from the light dragoon brigade on the far left of the allied line.

Skirmishing continued throughout the afternoon along the whole length of the ridge and in front of the three anchor points of Hougoumont, La Haye Sainte, and the villages on the left of the allied line. But the next real challenge came from the massed French cavalry. In this third phase of the battle some 7,000 assorted horsemen – *cuirassiers*, dragoons, hussars and lancers – but with little horse artillery and no supporting infantry, came on in a great host towards the allied centre-right between La Haye Sainte and Hougoumont. The allied infantry formed squares to meet them. In fact the squares were largely oblong-shaped, with the defenders forming the sides in two ranks, the front kneeling – a bristling hedge of bayonets and a fair length of musketry. Outside the squares, the gunners fired until the last safe moment and then ran for the cover of the nearest square, racing back out to their guns again as soon as the French had swept past.

A sort of stalemate thus developed: the French horsemen

galloped – and trotted, and even walked – about the ridge with impunity, but in turn could do no harm. The infantry, secure in their squares save for the occasional plaguing artillery (though some did suffer badly from the French guns), could do nothing to send them back. Uxbridge's heavies were in no state to mount much of a counter-attack, and Wellington (who with his staff had to keep taking refuge in a handy square as he rode about the field) had given express orders to his cavalry brigadiers on the flank not to leave their places. The attacks continued for an hour and more, until at about half-past four Wellington was heard to say, 'The battle is mine, and if the Prussians arrive soon there will be an end to the war.' This was optimistic, and no doubt deliberately so, but soon afterwards cannon fire to the east signalled that the Prussians were indeed making progress.

Things were looking distinctly shaky for the allies on the ridge, however, where casualties were mounting terribly, and Wellington had to do a deal of realigning and reinforcing. Then, at about six-thirty, disaster threatened as La Haye Sainte fell to a coordinated attack by French infantry, cavalry and artillery – the fourth phase, and the only decent piece of French coordination in the entire battle – which made the allied centre look distinctly vulnerable just as the duke was managing to stabilize the battered inner flank above Hougoumont. The infantry casualties in the centre, just behind the crossroads, had also been rising alarmingly, for the ground here was not so favourable to the defenders, offering less shelter. The 27th (Inniskilling) Regiment alone had lost 400 men to the grand battery before firing a single musket, and their brigadier now sent a note to Wellington asking if his brigade, by this time down to a third of its starting strength, could be relieved for a while. He received one of the 'backs to the wall' orders that in desperate moments have often screwed the British army's courage to the sticking post: 'Tell him', said Wellington to an aide-de-camp, 'that what he asks is impossible: he and I, and every Englishman [*sic*] on the field, must die on the spot we now occupy.'

And for many on the ridge that day, it looked as if that indeed would be the outcome: a battle in which all were killed. Yet Wellington himself, riding to wherever the action was most intense, remained not only alive but untouched, even as others of his staff were maimed or killed at his side. His senior ADC Lieutenant-Colonel Sir Alexander Gordon and his quartermaster-general (chief of staff) Sir William de Lancey were mortally wounded; another of his ADCs, Lord FitzRoy Somerset, lost an arm;* and Lord Uxbridge famously lost a leg. 'By God, sir, I think I have lost my leg,' he said quietly to Wellington as grapeshot smashed his knee. The duke's reply, more solicitous than it perhaps sounds, was simply, 'By God, sir, I believe you have!'

Imperturbable throughout, Wellington now ordered what few reserves he had to the centre. A little later, on learning that Prussian cavalry had reached the furthest end of the allied line, he ordered Sir Hussey Vivian's hussar brigade to the centre from the left flank, where it had stood inactive for most of the day with the light dragoon brigade.

Coming up, Vivian was shocked by what he found: 'the ground actually covered with dead and dying, cannon shots and shells flying thicker than ever I heard musketry before, and our troops – some of them – giving way'. Wellington himself had to gallop to where some Brunswickers, un-characteristically, were recoiling. If ever there was a general with a perfect grasp of how in war the tactical was some-times short-wired to the strategic it was he, which is one of the reasons he was everywhere that day. By about seven o'clock that evening he had stabilized the allied line.

The Prussians now began arriving on the left flank in large

---

* FitzRoy Somerset was to become Lord Raglan and begin a fashion for shoulderless coats – the 'Raglan sleeve'.

numbers, and with them, as Wellington had foreseen two hours or so before, came the prospect not just of winning the battle but of strategic victory.

Napoleon knew this perfectly well too. He had two options, therefore. The first was to form a covering force and retire: the British were exhausted, after all, and the Prussians, though scarcely engaged that day, had force-marched a fair distance; it might have been possible under cover of darkness to get away to see to the defence of Paris. Or else he could throw the dice once more – one last time – in an attempt to break through the allied line and get into the Forest of Soignes before nightfall and thence move towards Brussels. What good that might bring ultimately, when the Austrians and the Russians came to the borders of France, and with a Prussian army bruised but intact, could only be conjectured. His whole campaign, how- ever, which would become known as 'The Hundred Days', was more visceral than cerebral. And what good would come of defending Paris? Without *La Gloire* Napoleon was nothing. So he threw the dice one more time: indeed, he threw in the Garde – the Garde Impériale, which had never been defeated. In fair- ness it can be argued that he perhaps genuinely believed the allied line to have been so weakened that the moral superiority of an attack by the Garde would be irresistible. He was, how- ever, wrong.

The Garde – five battalions of the 'Middle Guard' backed by three of the 'Old', in all some 5,000 men – marched for the ridge between La Haye Sainte, which was now in French hands, and Hougoumont, dividing into two distinct masses as they came up the slope. And by one of those quirks of war the force advancing on the left marched directly for the place where the British Guards lay prone behind the crest of the ridge.*

* The numbers taking part in the attack, and their precise formation, have never been firmly established. By this time the smoke on the battlefield was thick and widespread, and there is even some doubt as to whether all the attacking troops were from the Garde. Some of the battalions advanced in column, and some appear to have done so in square.

Again, Wellington was at the decisive place: 'Now, Maitland! Now is your time!' he called to the brigade commander as the French broached the ridge. And knowing, almost certainly, that this was not just the decisive place but, as Clausewitz would later term it, the culminating point, he could not resist giving the order direct to the brigade: 'Stand up, Guards! Make ready! Fire!'

At 50 yards the effect was devastating. And as two allied brigades from the right flank moved to join the musketry, Colonel Sir John Colborne commanding the 52nd Oxfordshire Light Infantry took his men forward to open an enfilading fire on the Garde's right flank.

It was enough. '*La Garde recule!*' was the most astonished cry the Grande Armée had ever heard. But even as they fell back, their supporting artillery checked the ardour of the allied troops following up.

'Go on, Colborne! Go on! They won't stand! Don't give them time to rally!' called the duke. And then, making sure they would not, he took off his hat and began waving the whole line forward.

It was all over but the pursuit – and that he would leave to the Prussians.

What, then, was the legacy of the Napoleonic period – Wellington's legacy? Above all, there was now a confidence that the British army could face anything. At Waterloo, Napoleon in person had thrown everything he had at the allied line, and it had been the steadfastness of the British infantry that had stopped him in his tracks. Battalions did not turn and run, as did some of the Dutch and Belgians (and even Hanoverians – including, infamously, the Duke of Cumberland's Hussars); squares did not break. These were not all Peninsula-hardened veterans; yet somehow the notion of what was expected of them – nothing less than what would have been expected of the battalions that had fought in Spain – had taken deep root in the ranks. The regiments themselves, through a dozen years' continuous experience under arms (and with surprising continuity among the officers, despite purchase), had developed distinct identities, with an additional sense of pride and discipline.* And it was the discipline of volleying, by the Guards and the infantry of the line, that was the foundation of success in battle. Light infantry work was an accessory, artillery a supplementary, cavalry – if it could be trusted at all – an auxiliary. Wellington's insistence on carefully arranging all aspects of his logistics before a bold advance, and then fighting a defensive battle on well-chosen ground, was the formula for success on campaign. The infantry, when required to, could attack with all the dash of the best in the world, and the cavalry never needed restraining from the charge, but on the whole Wellington and his army preferred to let the enemy come on. And this was the sense of how war should in future be made: 'They came on in the old way,' said Wellington of the Grande Armée at Waterloo; 'and we saw them off in the old way.' It did not seem to occur to anyone that armies in the future might look for a new way.

So a sort of self-satisfaction entered into the army's soul – at least into that of its spiritual directors at the Horse Guards. The musket had been the army's mainstay; why change it for the rifle? Artillery had been useful against densely packed columns at relatively short ranges; why have

---

* Regimental pride is usually thought of as a later, Victorian development – the officers at Waterloo owning more to individual honour than to the corporate, and the soldier having no especial sense of pride in the number he bore on his shako and large pack. But when, for example, Uxbridge sent one of his ADCs to try to stop the Cumberland Hussars (Brunswickers) from leaving the field, the officer appealed to their commanding officer to 'consider the regiment's good name'. And when Sir John Colborne saw how some of his men ducked as cannon shot came over he cried 'For shame! That must be the second battalion!' (who were recruits).

heavier guns to fire at a greater distance when they had to be realigned manually with great effort? The cavalry experimented with the lance, but their regiments were increasingly becoming rich and exclusive clubs given excessively to show, the lance prized more for its fluttering pennant than as an innovative weapon. It is particularly ironic, indeed, that at a time when British industry, transport and commerce were leading the world in their inventive approach, the army sought almost consciously to revert to a pre-industrial age.

In his role as Master General of the Ordnance, and later longstanding commander-in-chief, the duke of Wellington himself has been blamed for much of this military reaction. There is no doubt that his intensely conservative view of military (and public) order did not make for innovation. He famously opposed the abolition of both purchase and flogging – though with sound enough reasoning, given his perspective. He had no cause to suppose that France would be a resurgent threat in the short to medium term (unlike in 1814, after 1815 the restored Louis XVIII disbanded the entire army to remake it in a thoroughly non-Revolutionary image), and like other statesmen he had confidence in the peace made at the resumed Congress of Vienna (at which he had been one of the delegates). And he was realist enough to acknowledge that great economies had to be made after the exertions of the 'never-ending war'. As Master General of the Ordnance he may have been a member of the cabinet, but he knew that – then as now – there were no votes in defence.

But what of the duke's legacy in methods of command, and organization? In one sense it was fundamental to the British way of war ever since – the meticulous attention to planning, but with the 'harness of ropes' approach, the personal reconnaissance and eye for ground, the supervision of the execution of detail, the sheer hard work ('attending to the business of the day *in* the day'), the exemplary courage, the understatedness. And along with the personal qualities and practices went a pragmatic yet robust political sense: he was careful to secure his political as well as his tactical flanks. Professor Richard Holmes, biographer of both Wellington and Marlborough, has compared them in just these terms: 'It is no accident that both based their success on mastery of logistics, and both were principally commanders within coalitions, always obliged to blend the military with the political, as much strategists as tacticians.'

And yet these very qualities had their downside. When, after Waterloo, Wellington said, 'By God, I don't think it would have served had I not been there!' he was almost certainly right. He had some capable and experienced generals in the field with him that day, but none who – on past form – was equal to his own task. It is curious, therefore, that he took no great efforts to make sure that if he did become a casualty all would not be lost. Uxbridge, his nominal second-in-command, asked the night before what were his chief's plans, to which Wellington replied, and with some asperity, that since Bonaparte would be attacking and had not vouchsafed his plans to him, how might he know what he would do? But then he added, more emolliently, 'There is one thing certain, however, Uxbridge, that is, that whatever happens you and I will do our duty.' It was hardly helpful.

But Wellington certainly *had* had some idea of the possi-bilities open to Napoleon, which was of course why he made his dispositions on the ridge of Mont St Jean as he did. And he had made them in detail: it was Wellington himself who had decided exactly which troops to place in Hougoumont, La Haye Sainte and the villages to left and right of the line. He himself had decided which brigades were to go where, and it is not impossible to believe also that he had directed the divisional and brigade commanders just where and how to dispose their regiments. It had been a faultless disposition, of course: Wellington had played to a T the different national characteristics and regimental capabilities. Indeed, when Baron Müffling, Blücher's liaison officer, had asked if 1,500 men were really enough to hold Hougoumont, Wellington replied, 'Ah, you don't know Macdonnell. I've thrown Macdonnell into it.' He meant Lieutenant-Colonel James Macdonnell, commanding officer of the Coldstream Guards. This was the level of decision-making to which Wellington applied himself.

He was criticized both at the time and later for his tight-gripped 'top-down' tendency. In his commentary on the diaries of Sir John Moore, for example, Major-General Sir J. F. Maurice, the official historian of the late Victorian army, observes sharply of Wellington's one criticism of the Corunna campaign (that Moore ought in anticipation of the retreat 'to have sent officers to the rear to mark and prepare the halting-places for every brigade') that 'If it really was the practice in Wellington's own army towards the end of the Peninsular War for headquarters to interfere in such a matter, then all that can be said is that it is an extraordinary illustration of the extent to which Wellington, in his utter contempt for his subordinate generals, had reduced the whole army to the condition of a mechanical instrument in his own hand.'

In the duke's hands, of course, the instrument worked. In the hands of others it did not work nearly so well. Indeed, as the years went by after Waterloo, the instrument, like a well-sprung clock, lost time (and in the Crimea it eventually stopped for a while) because it was never wound or serviced properly. Fortunately there was another instrument-maker at work – India. But India was fashioning a piece that was not to the liking of many at home, and so the army, half of which was in Britain and Ireland, was to continue its long decline throughout the nineteenth century, its true condition concealed by the fact that the Queen's enemies were far away and second-rate. The army received a bloody nose from time to time, but its heroic recovery – often accompanied by a good deal of 'spin' – only served to convince Parliament and the public that at heart all was well. Without India, however, many a bloody nose could have turned into a knock-out.

In the end, the sword that the duke of Wellington had wielded so magnificently grew rusty in his hand. But a fine sword it had been, and it could be burnished again. Writing of the army at Albuera in May 1811, Sir John Fortescue describes how the 'battalions, struck by the iron tempest, reeled and staggered like sinking ships – but suddenly and sternly recovering they closed on their terrible enemies, and then was seen with what strength and majesty the British soldier fights'.

With such fighting quality it might have been expected that the British soldier would always recover from the 'iron tempest' – and by and large he did. But he would reel and stagger a good many times in the two centuries after Waterloo before closing on his terrible enemies.

15

Eastern Questions

The Crimea, 1854–6

OF THE THREE STIRRING ACTIONS ON 25 OCTOBER 1854 IN THE Crimea, the last, the charge of the Light Brigade, is undoubtedly the best known. Five regiments of light dragoons, hussars and lancers, all considerably under strength due to the privations of the climate and the campaign – around 650 men in all – galloped behind Major-General the Lord Cardigan down a valley lined with Russian guns and riflemen to attack an artillery battery which threatened no one. As an example of incompetence and futility it is not unique, but it is perhaps the most picturesque, and certainly the most poetically lauded. 'C'est magnifique, mais ce n'est pas la guerre,' was the French Marshal Bosquet's memorable opinion. But earlier that morning there had been a sharp little action in which the double rank of bayonets of the 93rd (Sutherland) Highlanders – 'a thin red streak tipped with a line of steel' in the words of William Howard Russell, *The Times*'s war correspondent – had stopped the advance of the Russian cavalry towards the port of Balaklava, the army's

Lieutenant-General Sir Colin Campbell (later Field Marshal Lord Clyde) – one of the best commanders-in-chief the army never had.

supply base. Their brigade commander, Sir Colin Campbell, had called out from atop his horse, 'There's no retreat from here, men. You must die where you stand!' – to which possibly the most printable reply was 'Aye, Sir Colin. If needs be, we'll do that.'

Campbell, who had celebrated his sixty-second birthday only three days before, had earned his spurs in the Peninsula (he had been with Moore at Corunna – indeed, he had been at Moore's old school, Glasgow High – and then with Wellington throughout Spain), and had seen much action since then in various corners of the expanding Empire, including China and India. And on the morning of the twenty-fifth his long, broad and recent experience was translated into good judgement, for he had formed a low opinion of the Russian cavalry during the operations since the army had landed in the Crimea a month before. When, therefore, the alarm had gone up that enemy cavalry was advancing on Balaklava, and he stood-to the 93rd on outpost duty, he decided to risk sacrificing depth, and even the security of the square, in order to form a longer line and thereby gain greater firepower, drawing them up two deep rather than in the four ranks conventional in the face of cavalry.

The 93rd were one of the battalions fortunate enough to have the new Minié rifle (others still carried the Brown Bess musket that their forebears had used at Waterloo), which could be loaded quickly and was accurate enough for volley fire at half a mile.* Supported by some Turkish infantry and Royal Marines, they fired their first volley at 800 yards, a second at 500, and then again at 350 yards. As the Cossacks

* The Minié was a French design which allowed much quicker muzzle-loading. It was the slow rate of reloading that had formerly stood in the way of the rifle's introduction throughout the infantry. The Minié was a percussion lock, like the muskets it replaced, the army having belatedly modified its ancient flintlocks in the 1830s to take the percussion cap with its filling of explosive fulminate of mercury.

and hussars closed for the charge, their ranks thinned but still intact, their commander suddenly pulled up sharp, convinced that the 'thin red line' of riflemen was a decoy backed by a much stronger force waiting to counter-attack. He turned them about and began withdrawing at the walk.

Some of the Highlanders started forward for a counter-charge. 'Ninety-third, damn all that eagerness!' cried Campbell, halting them with the flat of his sword. But the eagerness was evidence enough that 100 years after Minden – and with forty years of peace in Europe – going to it with the bayonet, even against mounted men, was still the instinct of the British infantryman.

Next came evidence that the same offensive spirit that had propelled Lord Uxbridge and the Union and Household brigades into the great mass of d'Erlon's columns at Waterloo was no less alive. It was not, however, Cardigan's famous charge with the Light Brigade, but the equally astonishing and far more effective, if much lesser known, charge of the Heavy Brigade – three regiments of dragoons and dragoon guards, bigger men on bigger horses, under Brigadier-General James Yorke Scarlett. Watching it from his place at the head of his regiment was Lord Uxbridge's sixth son, Lord George Paget, deputy commander of the Light Brigade. A day or so later, after his own foray 'into the valley of death', he wrote of the Heavies:

> This noble brigade was on its way to cross the Balaclava plain from our rear (where it had been formed up in our support), with the object of giving support to the 93rd Highlanders, whose position was seriously threatened. It had just reached the level of the plain beneath us, when large masses of the enemy's cavalry appeared, rapidly advancing, and debouching, as it were, from the plain which was afterwards the scene of our charge, having crossed the ridge of redoubt hills at a point where the undulation of the ground leaves little rise from the plain itself . . .

This advancing column could not have been 300 yards from the Heavy Brigade when they first came upon their view. These last were at that moment in the act of executing their flank movement, and close on their left (between them and the enemy) were the remains of a vineyard.

Anyone who has ridden, or attempted to ride, over an old vineyard will appreciate the difficulties of moving among its tangled roots and briars, and its swampy holes. But these did not stop these noble fellows. They were caught in a position and formation quite unprepared for what was to follow, aggravated as this was by the nature of the ground. They immediately and most skilfully showed a front to their left, and advanced across the vineyard to meet a foe of many times their number.

It must be here observed that the confines of this vineyard were just on the line where the shock took place between the two advancing bodies; or rather that the Heavy Brigade had only just time to scramble over the dry ditch that usually encircles the vineyards, when they came in contact with their foes.

I am now about to describe only what I saw! — This has been called a charge! How inapt the word! The Russian cavalry certainly came at a smart pace up to the edge of the vineyard, but the pace of the Heavy Brigade never could have exceeded eight miles an hour during their short advance across the vineyard. They had the appearance (to me) of just scrambling over and picking their way through the broken ground of the vineyard. Their direct advance across the vineyard could not have exceeded eighty or one hundred yards. What a thrilling five-minutes (for it did not last longer) was the next — to us spectators!

The dense masses of Russian cavalry, animated and encouraged doubtless by the successes of the morning (for the poor slaves had been told that the Turkish redoubts had been taken from the English) — advancing at a rapid pace over ground the most favourable, and appearing as if they must annihilate and swallow up all before them; on the other hand,

the handful of red coats [the heavy cavalry wore red] flounder-
ing in the vineyard, on their way to meet them.

Suddenly within twenty yards of the dry ditch, the Russians
halt, look about, and appear bewildered, as if they were at a loss
to know what next to do! the impression of which appearance
of bewilderment is forcibly engraven in my mind on this
occasion, as well as later in the day. They stop! The Heavies
struggle — flounder over the ditch and trot into them!

Then followed anxious moments! 'red spirits and grey,' green
coats and blue! all intermingled in one confused mass!

The clatter of the swords against the helmets, the trampling
of the horses, the shouts! — in short, the din of battle (how
expressive the term) still rings in one's ears! One body must give
way. The heaving mass must be borne one way or the other.
Alas, one has but faint hopes! for how can such a handful resist,
much less make head through, such a legion? Their huge flanks
lap round that handful, and almost hide them from our view.
They are surrounded and must be annihilated! One can hardly
breathe! Our second line (half a handful) makes a dash at them!
One pants for breath! — one general shout bursts from us all!
It is over! They give way! the heaving mass rolls to the left! They
fly! Never shall I forget that moment!

There is no more show of resistance, and they soon disappear
to whence they came. It was a mighty affair, and considering the
difficulties under which the Heavy Brigade laboured, and
the disparity of numbers, a feat of arms which, if it ever had its
equal, was certainly never surpassed in the annals of cavalry
warfare, and the importance of which in its results can never be
known.*

The success of the Heavies was all the more surprising
because – like Cardigan, and even the divisional commander

---

* From *The Light Cavalry Brigade in the Crimea: Extracts from the Letters
and Journal of General Lord George Paget* (1881).

Lord Lucan – Scarlett had no experience of active service whatever.* He had been commissioned into the 18th Hussars three years after Waterloo and had risen by purchase to the rank of lieutenant-colonel in the 5th Dragoon Guards. He had then held this command for an astonishing fourteen years until on the outbreak of the Crimean War he was appointed commander of the newly formed Heavy Brigade. To have decided on an immediate attack against such a vastly superior number of cavalry – his brigade could muster only 300 sabres that morning – and against the advice of his two regimental commanders was the stuff of a Colin Campbell. Whence came his brilliant *coup d'œil*?

It was ultimately Scarlett's decision, and certainly his responsibility, but it was his two aides-de-camp, both of whom had seen service in India, who urged the attack. Faced with the same overwhelming numbers of Marathas, Pindarees or Sikhs, the trick had always been to charge at once, before the enemy decided to, gaining the moral ascendancy and also the advantage of speed (it was, after all, what had saved the young Arthur Wellesley at Assaye). 'India' had stopped the Cossacks in their tracks in front of the 93rd Highlanders, and it was 'India' again that overthrew the Russian hussars and lancers opposing the Heavy Brigade. Why, therefore, did 'India' not come to the rescue of the Light Brigade during the confusion leading to the attack on the Russian guns? Why, indeed, did it seem to speed it to its destruction?

The reason why – the title of Cecil Woodham Smith's unsurpassed account of the débâcle – is to be found in the aftermath of Waterloo. As was to be expected, there had been wholesale cuts in the army establishment (and in the navy, where almost every three-decker was decommissioned), for

---

* Lucan had, however, been an observer in the Russo-Turkish War in 1828, though little good it appears to have done him.

the usual reasons of economy and fear of militarism. But the army that remained was much the largest the country had ever supported in peacetime. Law and order at home remained the army's business, and with political, agricultural and industrial unrest rife in post-war England, the spectre of revolution was enough to frighten the authorities into keeping more cavalry regiments in being than initially intended. Even so, by 1820 the army's establishment had been reduced to 100,000, less than a quarter of its pre-Waterloo strength. Not only was it smaller, its distribution was also very different from what it had been before 'the never-ending war'. Only half the troops were in the United Kingdom, and a large proportion of these were in Ireland; there were 20,000 in India (besides the army of the East India Company) and 30,000 scattered about the world piecemeal in colonies and bases, many acquired during the war. These included 5,000 in Canada, where the bruising War of 1812 with the United States had left scars and an unresolved frontier prompting vast expenditure on military canals and fortifications. But as peace took root in Europe – international peace, for there was civil war enough – the economies increased. When the duke of York died in 1827, still in harness as commander-in-chief, his funeral arrangements were delayed by the difficulties of arranging decent ceremonial. Britain, said many a senior officer in despair, did not have enough troops to bury a field marshal.*

Nor was it just a matter of numbers. The whole system of divisional and brigade command, including the experienced staff, was dismantled. The army at home reverted to its earliest form – a loose association of regiments supervised in only the most elementary way by general officers in com-

---

* At the time Britain had sent a force the size of a small division to Portugal to stabilize the country as it slid into civil war, but this placed such a strain on the army that Wellington, as soon as he became prime minister, recalled it.

mand of military districts designed to support the local civil power.* The short-service enlistment brought in during the war was replaced by 'enlistment for life', or twenty-one years (in 1847 it was reduced to ten for cavalry and fourteen for infantry, with an option to serve to twenty-one for a pension). Barracks, usually overcrowded and insanitary, became the rule rather than the exception to the former system of billeting. And in the decades after Waterloo, four out of every ten men who lived in them were Irish.

The army became increasingly set apart – either overseas or behind barrack walls. Soldiers were seen less about the streets after Acts of Parliament founded the Metropolitan Police and the county forces. But while recruits were always hard to come by, officers were not. The sons of the aristocracy and gentry were still propelled towards the army by an aversion to trade only too often accompanied by an incapacity for the learned professions. Purchase returned with a vengeance – at vastly overinflated prices and, with peace, little free promotion. The regiments became not only increasingly autonomous once more, but increasingly closed, the officers ever smarter, the rank and file once again enlisted from the lowest order of men – Wellington's 'scum of the earth'. Indeed, it is difficult not to recall Dr Johnson's observation of an earlier age that no man would enlist who could get himself into prison, for 'A man in a jail has more room, better food, and commonly better company.' The soldier had indeed less space than that to which a prisoner was entitled, and mortality in barracks was higher than in jail.†

---

* Of the 103 infantry regiments of the line that remained on the establishment, as many as 80 would be abroad or in transit. The Guards did not serve overseas, nor the bulk of the cavalry. Artillery and engineers were penny-packeted to the colonies as the need arose.

† Dr Johnson would probably not have been surprised to learn that when during the 1970s and 1980s the demand for prison beds exceeded supply,

Equally unhealthily, as the mirror image of the generally despised soldiery in their shunned barracks, the officers, in the words of one historian, 'became ever more stiff-necked and haughty, rigid in social etiquette and distinctions, and dominated by a hierarchy of birth, wealth, kinship, connexion and fashion'.* Even if this had been only half true – and the description was probably nearer to the half truth than to the whole – it would not have been a situation conducive to efficiency. Those who could not afford the exclusivity of life in regiments at home had little option but to seek a less expensive life in a regiment abroad: the purchase system and the long peace meant that money frequently drove out capability from the home-based regiments, for it was often the more impecunious officers who studied their profession more – even if only for want of the means of expensive recreation.

And so there developed, in effect, two armies: the one at home, wearing ever more elaborate and expensive uniform and turning out from time to time at the magistrates' bidding; and the other far from home (and a long time away from home), doing the fighting. In India during the decade after Waterloo there was continuous skirmishing, culminating in one of the great set-piece sieges – at Bhurtpore, where Lord Combermere, who as Sir Stapleton Cotton had commanded Wellington's cavalry throughout the Peninsular War, took the fortress that had defeated every attempt to storm it for fifty years (and in a fashion that his old chief at Badajoz would have approved). There was a bloody and sickly war with the Burmese; there was fighting with various

---

as frequently it seemed to do, and temporary penal accommodation was sought, Rollestone (training) Camp on Salisbury Plain would regularly be requisitioned. The Home Office would have to install 'facilities' and comforts for the prospective inmates, which would then be taken out again before the camp was handed back to the MoD.

* Correlli Barnett, as ever caustic in *Britain and Her Army* (1970).

tribes throughout the new Cape Colony in southern Africa. There was war in West Africa with the Ashanti; in Kandy (Ceylon) to suppress revolt; and in China in 1839 over control of the opium trade. Out of that last war Britain gained Hong Kong; and out of war with the Gurkhas of Nepal it gained its quite extraordinary 'mercenary' regiments of tough natural soldiers who have since served alongside British regiments from the Far East to the Falklands – and on the same terms of soldierly respect that the King's German Legion enjoyed during the Napoleonic Wars. There is a story, related by Field Marshal Lord Slim in his account of the 14th Army during the Second World War – *Defeat into Victory* – of troops being served food in a rest area in India by 'memsahib' volunteers, with serving points to cater for the various diets of the different regiments – Sikhs, Hindus and Mohammedans (as they then were called). Several Gurkha soldiers had joined the British queue by mistake, and when they came to be served the volunteer pointed out that this was not their queue: they wanted the one 'over there' – to which the British soldier behind protested: 'No, Miss. Them's Gurkhas. Them's *us*!'

But if a ready respect, affection and affiliation came out of the Nepali War of 1816, a more grudging, wary respect and alienation came out of war with the Afghans between 1839 and 1842 – in the opinion of Field Marshal Lord Carver 'probably the worst conceived and executed of all Britain's politico-military ventures' (which after the disastrous reverses of the Burmese War was indeed saying something). The First Afghan War, sparked by the not irrational but grossly inflated fear of Russian designs on India, was an affair of evasions, compromises and groundless optimism that precipitated a shocking series of massacres of British and East India Company troops, including – almost unbelievably – 4,500 men and 12,000 camp followers en route to the Khyber Pass in January 1842.

At their head was the 60-year-old Major-General William

Elphinstone, who though an unquestionably brave man (he had commanded the 33rd Foot, the duke of Wellington's old regiment, at Waterloo) had not seen action since that day on the ridge at Mont St Jean. Gouty and terminally indecisive, he made the error of trusting to an assurance of safe passage from Kabul, surrendering his artillery and taking few tactical precautions. Within a week the Ghilzai tribesmen and the snow and bitter cold had done their worst. Harassed for 30 miles of treacherous gorges and passes along the Kabul River, his only King's (British) regiment, the 44th (Essex), and the few others that remained made a heroic last stand at the Gandamak Pass.* The only Briton to escape was Dr William Brydon, who staggered to the besieged garrison at Jellalabad on his starving horse, which promptly dropped dead. There could have been no more dramatic a signal that Afghans were not men to be subdued as easily as their neighbours to the south – and were cunning too, for why allow Brydon to escape when he could easily have been cut down by a Ghilzai's *tulwar*, or sniped with a *jezail*? Perhaps even then the Afghans had an understanding of propaganda and psychological warfare: better one survivor to keep the tale alive than every last man slaughtered with no witness – as Lady (Elizabeth) Butler's powerful painting of Brydon's arrival at Jellalabad, *The Remnants of an Army* (1879), showed only too painfully.

Elphinstone's disaster was an example of the extremes of 'honour' that had grown distorted in the post-Waterloo army. It was as if the army of 1815 had been reduced and then preserved in aspic in the belief that what had beaten the French could beat anyone. The duke of Wellington's resistance to change took a powerful hold on too many senior officers, and even as late as the 1850s he was still wary of innovation: when asked to approve the introduction of the

* The Afghans had required Elphinstone himself as a hostage.

new Minié rifle, he did so with the warning 'but don't call it a rifle or they'll all want to be riflemen!' But the difference was that in 1815 the army had been commanded by a general of experience and instinctive sound judgement: it is inconceivable that even in his most reactionary old age the duke would have undertaken a retreat from Kabul without precautions, and trusting in the word of anyone but a Prince Blücher.

The three generals at Balaklava in whose hands the fate of the Light Brigade rested that day in October 1854 were not in Wellington's division, or even league. Their paths to the top – Lord Raglan as commander-in-chief of the army in the Crimea, Lord Lucan as commander of the cavalry division, and Lord Cardigan as commander of the Light Brigade – are as ever instructive, and in these cases more improbable even than in the most inventive work of fiction. Raglan had not heard a shot fired since Waterloo, when as Lord FitzRoy Somerset he had been ADC to the duke – in fact, for most of the subsequent forty years he had been the duke's military secretary. His manners were perfect, his distaste for anything ungentleman-like was profound, and his unsuitability for anything but the most courtly exchange with the French was remarkable.* His approach to any problem was to try to imagine what the duke would have done – that is, what the duke would have done had the Russians been the French and had the Crimea been the Peninsula. Indeed, more than once he absently referred to the French as the enemy.

A Colonel Beatson of the East India Company's army had offered to raise a troop of *bashi bazouks* – irregular Ottoman cavalry – to supplement the Heavy and Light Brigades. Beatson had been a soldier from the age of 16, had spent

* His fluency in French and his tact had, indeed, been the reasons for his appointment, as well as his seniority.

most of his life fighting in India, and had commanded *bashi bazouks* in Bulgaria in 1854. Raglan turned him down rather sniffily and then dismissed his offer to serve on his staff. Beatson then offered his services to Lucan, who had been on half-pay (not at active duty) for so long that he did not know the new words of command for manœuvring cavalry. Lucan dismissed him with equal contempt; and Cardigan did likewise. India was simply not a place where a gentleman practised war, and it was unthinkable that anyone but a gentleman could have anything to say on the matter to another gentleman. Such was the *reductio ad absurdum* of the Wellington code. Scarlett of the Heavy Brigade, on the other hand, had seen at once the advantage of having Beatson at his side, and so took him on as an additional ADC – for which the whole of the Heavy Brigade would have cause to be grateful, since it was Beatson's advice that helped steer Scarlett to charge.

Later that day, sitting with the Light Brigade at the head of Tennyson's 'valley of death', Lucan received the written order: 'Lord Raglan wishes the cavalry to advance rapidly to the front, follow the enemy, and try to prevent the enemy carrying away the guns.' This he failed to relate to the previous order from Raglan.* Had he done so he might have put two and two together and realized that the guns at the end of the valley – the only ones he could see – were not the guns to which Raglan referred. Had he also had the presence of mind to challenge an order that made no tactical sense whatever

---

* To the *series* of Raglan's orders, indeed. The one preceding the order which precipitated the charge read: 'Cavalry to advance and take advantage of any opportunity to recover the Heights. They will be supported by the infantry which have been ordered. Advance on two fronts.' Now although this was not a model of clarity, it showed very well to what purpose the cavalry was to act – recovery of the Heights. Lucan's inability to grasp this stands as the main charge against him. And indeed against the whole command and staff system.

(the guns he could see were unlimbered ready for action: they were clearly not being carried off, nor were they any threat to Balaklava port itself or its defenders), and whose outcome could only be the destruction of one half of his command, he could have been forgiven for disdaining the services of an 'Indian' officer like Beatson. And had he conceived of some tactical plan more subtle than the frontal charge that he believed he was being ordered to execute (Raglan's order included the useful information that 'Horse artillery may accompany. French cavalry is on your left') then the laurels would rightly have been his. Afterwards, with the Light Brigade all but destroyed as a fighting force, Raglan strongly reprimanded him: 'Lord Lucan, you were a lieutenant-general and should therefore have exercised your discretion, and, not approving of the charge, should not have caused it to be made.'

Raglan's reprimand was in fact a fundamental statement of generalship. And there was a famously instructive reversal of Lucan's dereliction in this regard a century and a half later in the altercation between the Supreme Allied Commander Europe, the US General Wesley Clark, and the then Lieutenant-General Sir Mike Jackson commanding the NATO force preparing to evict the Serbs from Kosovo. The Russians had sent a flying column to seize Pristina airport, and Clark wanted Jackson to stop them. Jackson, who considered the order futile and likely to end in serious bloodshed, protested, but Clark repeated the order baldly, to which the lieutenant-general (the rank is colloquially known as 'three-star'*) replied, 'Sir, I'm a three-star general, you can't give me orders like this. I have my own judgement of the situation . . .'

Lucan had no such judgement, and his all-round in-

---

* Brigadiers have one star, major-generals two, and (full) generals four. A field marshal has five.

capability – the troops called him 'Lord Look-on' – was made worse not only by his disdaining the services of the likes of Colonel Beatson but also by his despising another 'Indian' officer, Captain Louis Nolan of the 15th Hussars. Nolan was one of Raglan's ADCs, and the man who had carried the order down from the Causeway Heights on which the commander-in-chief and his staff sat with a perfect but wholly different view of the battlefield from Lucan's.

'Lord Raglan wishes you to attack immediately!' was Nolan's verbal addition to the written order, accurately relaying the last words that Raglan had spoken to him.

'Attack? Attack what, sir? What guns?' spluttered Lucan to the despised Nolan, who was not only an Indiaman but a man who had had the impertinence and bad taste to write a book about cavalry.

But surprisingly – and as a reminder that in war only the unexpected should be expected – what should have been the Indiaman's faithful *coup d'œil* now failed Nolan. Instead of at once comprehending the differences of perspective (Raglan sat high on a hill: he could see much more, and the guns he meant were on the heights, being pulled away by Russian cavalry with lassos) he exploded in contempt for the man he saw as the antithesis of cavalry dash. Flinging out an arm without looking, in the direction of the guns that no one could see from that end of the valley, he almost yelled the fatal words, 'There, my lord! There are your guns!'

Lord Lucan, though utterly baffled by the order, turned to his brother-in-law Lord Cardigan, whom he detested even more than any Indiaman, and instructed him to charge with his Light Brigade.

Cardigan, who reciprocated Lucan's loathing, icily pointed out the lethal futility of charging guns frontally while being enfiladed in the process. To which Lucan helplessly replied that he understood him full well, 'But Lord Raglan will have it!'

Even now, a cool and experienced head could have averted disaster. Cardigan might have suggested a plan to use the Heavy Brigade, the French Chasseurs d'Afrique and the horse artillery to cover his flanks. He might have questioned Nolan to understand the purpose of the order and therefore what freedom of action he had. But both ideas eluded him. As he was later to protest, he had been given an order in front of his brigade, and he felt obliged to carry it out no matter what the consequences. Besides, speaking to Nolan would be beneath his dignity. Cardigan's chestnut charger, Ronald, might as well have been standing at the head of the Light Brigade with an empty saddle for all the use Cardigan was at that moment.

What happened next was summarized by Cardigan himself in a speech at the Mansion House the following year:

We advanced down a gradual descent of more than three-quarters of a mile, with the batteries vomiting forth upon us shells and shot, round and grape [Nolan was killed by the first salvo], with one battery on our right flank and another on the left, and all the intermediate ground covered with the Russian riflemen; so that when we came to within a distance of fifty yards from the mouths of the artillery which had been hurling destruction upon us, we were, in fact, surrounded and encircled by a blaze of fire, in addition to the fire of the riflemen upon our flanks. As we ascended the hill, the oblique fire of the artillery poured upon our rear, so that we had thus a strong fire upon our front, our flank, and our rear. We entered the battery – we went through the battery – the two leading regiments cutting down a great number of the Russian gunners in their onset. In the two regiments which I had the honour to lead, every officer, with one exception, was either killed or wounded, or had his horse shot under him or injured. Those regiments proceeded, followed by the second line, consisting of two more regiments of cavalry, which continued to perform the duty of cutting down the Russian gunners.

Then came the third line, formed of another regiment, which endeavoured to complete the duty assigned to our brigade. I believe that this was achieved with great success, and the result was that this body, composed of only about 670 men, succeeded in passing through the mass of Russian cavalry of – as we have since learned – 5,240 strong; and having broken through that mass, they went, according to our technical military expression, 'threes about,' and retired in the same manner, doing as much execution in their course as they possibly could upon the enemy's cavalry. Upon our returning up the hill which we had descended in the attack, we had to run the same gauntlet and to incur the same risk from the flank fire of the Tirailleurs as we had encountered before. Numbers of our men were shot down – men and horses were killed, and many of the soldiers who had lost their horses were also shot down while endeavouring to escape . . .

And so passed the most infamous, but by no means the worst, display of muddle and incompetence – of sheer unprofessionalism – in a war that shook the growing complacency of Victorian England. A quarter of the British (and Irish) soldiers sent to the Crimea did not return, a casualty rate worse than in the Great War – although in the Crimea four-fifths of those died from disease.

Shook the complacency, but evidently not violently enough; for although there were commissions of inquiry and a considerable reorganization of the government of the army, its worst aspects were to remain unreformed. Raglan himself died in the Crimea, but Lucan in time became a field marshal, and Cardigan, in a move that passes all understanding, was appointed Inspector-General of Cavalry.

One of the fundamental failures of the army in the Crimea had been the work of the staff – the officers responsible for administration and for the realization of commanders' decisions. Without a staff college (and arguably even with a staff college) staffwork can only be perfected by practice, and that means there have to be staffs in being, not improvised as occasions demand – in other words, permanent brigade, divisional and corps headquarters. In India there were in effect permanent establishments for the various field force headquarters, and some staff experience was therefore retained and passed on. But a permanent staff on the Prussian model, for example, would have been efficient in the writing and transmission of orders. A permanent staff on the French or even Italian model would have understood logistics better than to try to supply a whole army from the tiny harbour of Balaklava.* As soon as the Crimean War ended, however, the various hastily assembled headquarters were once more disbanded. The army reverted to the ossified regimental structure of the status quo *ante bellum*, though there were a few younger officers who, like Wellington in Flanders, had learned how *not* to do things.

Other nations took fuller note, however. The French political philosopher and historian Alexis de Tocqueville wrote to an English friend:

> The heroic courage of your soldiers was everywhere and un-reservedly praised, but I found also a general belief that the importance of England as a military power had been greatly exaggerated, that she is utterly devoid of military talent, which is shown as much in administration as in fighting, and that even in the most pressing circumstances she cannot raise a large army.

* The Piedmont–Sardinian expeditionary force of 15,000 (which arrived in the Crimea in 1855) was, as observers remarked, astonishingly well set up logistically. It also fought well, its Bersaglieri (light infantry) in particular being very fleet in the attack. Russell considered their artillery to be the best in the Crimea.

Britain quietly resolved simply not to be drawn into battle on the Continent again. Nevertheless the Crimean War had been a loud wake-up call – a very public one, thanks to the innovations of war correspondents, photography, the electric telegraph and wide-circulation newspapers. Army reform would soon follow – surely? Yet less than three years after Balaklava there would come an even greater calamity to dent the nation's prestige, one that would have much greater influence on the way the army was to be reshaped in the second half of Queen Victoria's century – mutiny in India.

16

Names, Ranks and Numbers

India and Whitehall, 1856–81

THE INDIAN MUTINY AND ITS CAUSES HAVE INSPIRED FICTIONAL and non-fictional literature in probably equal measure. For the novelist especially, the cultural, religious, racial, economic, 'nationalistic' and even simply hysterical tensions that piled up tinder-like in Bengal, the largest of the three presidencies of the Honourable East India Company, towards the middle of the nineteenth century offer boundless possibilities for the pen. And that a spark so prosaic and at the same time so symbolic – the new greased cartridges – should have ignited such a raging conflagration is gift indeed to the writer. And it must have seemed ironic to Sir Colin Campbell, not long back from the Crimea and promoted lieutenant-general, that the technological advance which had allowed him to stop the Russian cavalry at Balaklava – the rifle – had now brought about the greatest shock to British confidence in its military superiority since the American Revolutionary War.

'When will you be ready to set out?' Lord Palmerston, the

prime minister, had asked him anxiously (Lord Aberdeen had resigned early in 1855 after criticism of the muddle in the Crimea). And Palmerston had good reason to be anxious: the army of Bengal had mutinied, and its commander-in-chief was dead.

'Within twenty-four hours,' Campbell replied. And how he must have relished the prime minister's summons; for when Raglan had died two years before, Palmerston had ignored the clamour in the press for Campbell to replace him. Now, at the moment of greater danger, the Glasgow carpenter's son was recalled from virtual retirement. And, true to his word, the new Commander-in-Chief Bengal (and therefore in effect India) was aboard a Channel packet the following evening, travelling south to Marseilles by train, then east across the Mediterranean by steamship, and then overland by any means he could find to reach Calcutta, his headquarters, on 13 August 1857.

It had all begun the previous year when the sepoys – the soldiers of the East India Company's army – in Bengal were issued with the Enfield rifled musket, the same rifle that had replaced the short-lived Minié in the Crimea. Rumour had spread, incorrectly, that the cartridges (the card-wrapped powder and shot) were greased with either pig fat or beef tallow to permit easier ramming – one grease being obviously abhorrent to Mohammedan sepoys, the other equally to the Hindu, for the new drill still required the man to bite open the cartridge. When the sepoys were told they could grease the cartridges with beeswax or vegetable oil instead, this was taken merely as 'proof' that the issued cartridges were unclean. The facts were anyway soon irrelevant; and the indifference of many British commanding officers who failed to recognize the degree of alienation only exacerbated the discontent. When mutiny broke out with the murder of British officers and their families in Barrackpore – a thing which the authorities supposed wholly impossible, for the sepoys' loyalty to the sahibs and memsahibs was taken as absolute – the shock was so profound that they responded either too weakly or in so draconian a fashion as to encourage the wavering regiments to throw their lot in with the mutineers. There was no pattern to which regiments mutinied initially, nor much correlation between benign command and loyalty: even those regiments in which the officers were well liked were not immune, although there were instances of individual officers or their wives being warned of what was to happen, or helped to escape.

The army of 'John Company', as the Honourable East India Company was affectionately known by its Indian as well as British employees, was a sort of gigantic confidence trick in the best sense: for there were nearly 300,000 native troops in India, and fewer than one-tenth that number European.* The Company's officers, moreover, were a breed entirely apart from those of the Crown regiments. Recruited separately and without purchase, trained separately at Addiscombe rather than Sandhurst, and with very different (and more favourable) terms of service, a Company officer had no authority over a Queen's officer unless he was a general (a rank that could be conferred only by the Crown). But India was garrisoned by the Company's army, and the few British regiments were an insurance policy – almost literally, for they were paid for by the Company's court of directors, not the British taxpayer. And in 1857 the Company was badly underinsured.

---

* Several regiments were recruited by the Company in India from European 'settlers'. These were taken on to the British army's establishment after the mutiny and given numbers in the line, and later British territorial designations – so that, for example, two regiments of Bengal Europeans became the 101st Regiment of Foot (Royal Bengal Fusiliers) and 104th Regiment of Foot (Bengal Fusiliers), and were later linked with the Militia of Munster (Ireland) to become the Royal Munster Fusiliers (see the Cardwell–Childers reforms later in the chapter).

To the ordinary sepoy, the reverses in Afghanistan a decade and a half before, and more recently in the Crimea, while not suggesting that British power was on the point of collapse signalled nevertheless that the British were not gods. And to those more elevated Indians, like the Rani of Jhansi and Nana Sahib of Cawnpore, whose own particular grievances were turning into political ambition, the British setbacks seemed to suggest an opportunity in which the sepoys' unrest could be used to advantage.

Trouble had begun in Barrackpore just north of Calcutta in January 1857, but it was not until the end of March that the real line was crossed. There, in the afternoon of the twenty-ninth, Lieutenant Baugh, the adjutant of the 34th Bengal Native Infantry, heard that several of the regiment's sepoys were 'in an excited state', and that one of them, Mangal Pandey, was armed with a loaded musket and threatening to shoot the first European he set his eyes on (it later emerged that Pandey was high on *bhang* – cannabis).

In a British regiment, the adjutant or the orderly (duty) officer would at once have summoned an NCO, preferably a serjeant-major, to deal with a soldier who was out of his mind with drink, but the 34th had only one British NCO – Serjeant-major Hewson. Having been told that Hewson had been called, Lieutenant Baugh immediately buckled on his sword and pistol and galloped from the European canton-ments to the sepoy lines, where he found Pandey behind the signal gun in front of the quarter-guard of a dozen men. True to his doubtless braggart word, Pandey took aim with his musket, which he had somehow loaded without defiling himself with the unclean grease, and fired as Baugh rode on to the parade ground. The bullet struck Baugh's horse in the flank, bringing it and its rider down. Baugh at once sprang up and ran at Pandey firing his single-shot pistol, but missed. Before he could draw his sword, Pandey attacked him with a *tulwar* – the heavy, curved Mughal sword – felling him with slashing cuts to the shoulder and neck. As another sepoy, Havildar (Serjeant) Shaikh Paltu, tried to restrain Pandey, Serjeant-major Hewson arrived and weighed in, but he too was soon in trouble. For the quarter-guard, now under arms and joined by the orderly jemadar (sub-lieutenant), began to waver. And then some of them freed Pandey and chased off the adjutant, the serjeant-major and the loyal havildar.

At this point the garrison commander, General Hearsey, arrived with his two sons, and with both the prudence and the boldness acquired from long service in Bengal ordered the guard to do its duty and threatened to shoot the first man who disobeyed. The sepoys promptly did their duty and arrested Pandey, and a few days later he was hanged. This much was predictable and fair, and understandable to the rank and file; but the decision then to disband the 34th Native Infantry and dismiss all its sepoys with disgrace was exemplary punishment that would backfire. The regiment's adjutant had acted precipitately, an Indian officer, the jemadar, had hesitated and the sepoys had wavered, though an Indian serjeant had done his duty (Havildar Paltu would indeed be promoted to jemadar). But when Hearsey used the normal chain of command to deal with a *bhang*-crazed hot-head the situation had been defused; it was hardly the stuff of a violent or deep-seated rebellion. Disbanding the regiment, on the other hand, provoked many hundred griev-ances.

A month later in Meerut, one of the principal garrisons 40 miles or so north-east of Delhi astride the lines of communication with the Punjab, the general disquiet which the disbanding of the Bengal Native Infantry had increased became violent hostility. Hitherto the trouble had been con-fined to the infantry, but at Meerut the 3rd Bengal Cavalry were the main instigators. And with the cavalry came speed, range and galloper guns – light artillery pieces which could keep up with the sowars, as Indian cavalrymen were known. Even the presence of two British regiments – the 6th Dragoon Guards and the 60th Rifles – and heavier artillery could not check the disorder, and both the brigade and district commanders dithered. Soon the mutineers had swarmed down to Delhi where they murdered every European they could find – men, women and children – and placed the pensioned-off old king, Bahadur Shah, back on the throne that he had been only too happy to quit many years before. With the old Mughal capital in rebel hands and an Indian back on the old Mughal throne, the mutiny now had wings and apparent legitimacy. The intelligence soon reached London by galloper and the telegraph from Constantinople, and with it the news that the commander-in-chief, General George Anson, had died of cholera while assembling a force to relieve Delhi.

Sir Colin Campbell lost no time therefore in taking the fight to the mutineers, for all that his reputation was one for caution.* By March the following year he had relieved all the besieged garrisons – notably Lucknow and Cawnpore – and by the autumn, with the Mutiny broken, the last of its leaders were hunted down. Although Queen's regiments had been in the van of the counter-offensive, some of the Bengal native infantry had remained loyal, as had the Gurkhas, Sikhs and Pathans and most of the Punjab Irregular Force, together with many of the older irregular 'silladar' cavalry such as Skinner's Horse† (though all the light cavalry had mutinied) and, of course, the Company's four infantry regiments recruited from Europeans in Bengal. Critically too, the independent states – Hyderabad and Mysore much the largest – had either remained aloof or sent troops to

---

* 'Sir Crawling Camel' was one of his nicknames, more alliterative than fair.
† Silladar cavalry was not unlike British yeomanry, providing their own horses (and weapons), but in regular service.

support the British, and the Aga Khan had declared his support for the Company, which did much to neutralize calls to the Mohammedan sepoys for *jihad*, 'holy war'.

But after the Mutiny India could never be the same again. The fragility of the system had been exposed, and the danger in leaving the government of a great part of the subcontinent in the hands of a company of merchants was all too apparent. Parliament enacted several measures which effectively wound up the Company, and the government of India became the responsibility of the Crown through a viceroy, first in Calcutta and later in Delhi. The composition of the army in India was altered radically too. An 'Indian Army' was established incorporating the reformed armies of the three presidencies. It was still an army apart from that of the Crown, and its officers, both British and Indian, were still commissioned on different terms of service, but it was no longer to be treated as second rate and a thing entirely apart. Furthermore, there was to be a much higher proportion of Queen's regiments in India, with all artillery except some mountain batteries in British hands. The Indian Army establishment was to be 125,000, and that of British troops in India 62,000, in striking contrast with the ratio of nearly 10:1 before the Mutiny. This meant, of course, more than doubling the number of British troops permanently stationed in India. To achieve this there was to be a wholesale hollowing out of the colonial garrisons, together with an increase in the army's establishment of both infantry and cavalry by re-raising regiments which had been disbanded after Waterloo and by bringing the Company's European infantry regiments into the line. In consequence, half the British army would soon be either serving in India or not long returned, and its character – its language, indeed – would be increasingly 'Indian', typified by, on the one hand, even greater regimental introversion, with regiments spending up to twelve years in one station, and on the other by the habit of frequent and pragmatic campaigning. The pattern thus established would persist in India almost until Partition in 1947.

Both the Crimean War and the Indian Mutiny had exposed deep-rooted flaws in the army's system, not least in the selection of senior officers. The presumption of aristocratic merit did not disappear as a result – as the promotion of Lucan and Cardigan bore witness – but the nation had been sharply reminded of the imperative (and possibility) of competence. Whatever his first instincts had been, Palmerston had sent the Glasgow carpenter's son to quell the Mutiny, not a belted earl or the royal duke of Cambridge.

The duke, however – who had by no means disgraced himself commanding a division in the Crimea – was now made commander-in-chief at the Horse Guards, and in his first flush of youthful energy was something of an innovator. The Staff College at Camberley, previously the senior department of the Royal Military College – later 'Academy' ('Sandhurst') – was his creation, and its influence in developing military doctrine and a network of officers whose character and capabilities were known to each other would be increasingly important. For since there would be no permanent operational headquarters for half a century and more, this network was the only practicable means of mounting a military campaign with any promptness and efficiency. Long after the establishment of permanent head-quarters, in fact, the Staff College continued to produce a corps of like-minded officers, linked by the letters 'psc' ('passed staff college') after their names, who unconsciously operated a ring – albeit a large one – with a common under-standing of ends, ways and means, and in a more flexible way than, say, the Prussian 'Great General Staff' (*Grosser Generalstab*). It still does, albeit rather less exclusively (and at Shrivenham now, not Camberley) since the advent of universal staff training for regular officers replaced selection by examination.

But other reforms in the wake of the Crimean War were hardly sweeping. The appointment of secretary *at* war was incorporated in the portfolio of the secretary *for* war, the latter losing his responsibility for the colonies to a new colonial secretary. The artillery and engineers were at last brought under the control of the commander-in-chief (rather than the Master General of the Ordnance) and a permanent military transport organization was authorized, the Army Service Corps. This rapidly gained a reputation for efficiency and was in fact the only part of the army to escape criticism in the inquiries that would follow the Boer War at the turn of the century. But for all the popular Crimea outrage, military reform soon slowed to a very pedestrian pace, stifled by what has been described as 'a resuscitated complacency and conservatism'. Indeed, the duke of Cambridge was to become as obdurate an opponent of innovation as ever the duke of Wellington had been. Dukery controlled the Horse Guards for all but a handful of years in the nineteenth century; and, but for York's brief tenure, its record was a sorry one.

Perhaps the biggest effect of the Crimean War, however, lay in the change in public perception of the army. For the war was the first to be reported to the British public in 'real time', courtesy of William Howard Russell of *The Times* and the electric telegraph. Russell's reports, highly critical of both the organization of logistics and the conduct of operations (though on the whole he steered clear of directly criticizing Raglan) almost single-handedly stirred the public as well as the government into ameliorative action. Florence Nightingale recruited nurses and took them to the base hospital at Scutari in Turkey; Mary Seacole went out to the Crimea with her soldiers' comforts; Sir Morton Peto built a light railway to move supplies from Balaklava to the siege-works at Sebastopol; the great society chef Alexis

Soyer took his patent stove to the army's tented camps and personally advised on field cooking. To these and countless thousands of well-wishers at home knitting winter woollens for the troops, Russell's reports brought a sense of 'Our Boys', a powerfully sustaining moral force which has continued, if sometimes with more sentimentality than action, ever since. Indeed, it was the apparent absence of this moral force that in 2007 drove the chief of the general staff, Sir Richard Dannatt, to call for public shows of support with home-coming parades for troops returning from Iraq and Afghanistan. The success of this call, together with that of charities such as Help for Heroes, set up specifically to raise money to improve conditions at the Services' Rehabilitation Centre at Headley Court, proved a huge boost to morale in two conflicts about which public opinion was at best un-certain and at worst hostile.

Public support for the army in the mid-nineteenth century was a mixed cocktail in which compassion and social concern blended with a residual and wounded pride in the Wellingtonian period and a nascent 'jingoism' – the aggressive nationalism that was seeded by Palmerston and flowered under Disraeli. The term 'jingoism' itself was coined during the Russo-Turkish war of 1877–8 on the back of a popular music-hall song whose chorus ran:

> We don't want to fight but by Jingo if we do,
> We've got the ships, we've got the men, we've got the money
> too.
> We've fought the Bear before, and while we're Britons true,
> The Russians shall not have Constantinople!

And the cocktail was fortified by the increasingly personal interest of Queen Victoria herself, actively and substantially informed by Prince Albert. Any political tendency to insult the military – the army especially – by neglect was now checked by this watchful team. And royal watchfulness, although nowadays inevitably diminished in effect, remains a distinct check on army affairs – most strongly and obviously, but by no means exclusively, in the Household Division.* Most of the army's regiments and corps have a royal colonel-in-chief, and this can be a useful focus of loyalty when all else is going badly.

But the man who was to change the army most in the half-century after the Crimea and the Mutiny was a politician – Edward Cardwell, secretary of state for war in Gladstone's first ministry of 1868–74. In recent years scholars have rather tempered the importance of the 'Cardwell Reforms', but in three highly significant areas they were profound. First, Cardwell abolished purchase – and not without a considerable fight. He did so not entirely on the grounds that purchase in itself was detrimental to efficiency, but because it impeded more thorough reform of the organization of the infantry, and of recruiting. The substitution of merit rather than means as the primary criterion for commissioning and promotion did not suddenly open the doors to a great press of poor but capable officers, for the expenses of uniform and mess living remained an obstacle to the deserving officer who had only his pay to live on; but it laid the foundations of the meritocratic system that slowly but surely established itself in the century that followed.

The major question that Cardwell had to address was manpower, for the army was chronically under-strength as well as under-established. The demands of empire could not be ignored; nor could the growing power and belligerence of Prussia, which the congress system could not wholly check.

* The stories are legion of the present Queen's consciousness, in her role as commander-in-chief, of the observance of the proprieties in an otherwise busy age; and also at times of an almost Wellingtonian good sense in 'minor matters of great importance'. A report is sent to her twice yearly by the army on its operations and aspirations.

For now was the time of Bismarck. Brought to Berlin in September 1862 by the Prussian king to form a government during a crisis over the military estimates, the newly appointed prime minister made a speech to the budget committee of the house of deputies which set sabres rattling across Europe in a resurgent militarism that would continue for the best part of a century:

> The position of Prussia in Germany will not be determined by its liberalism but by its power ... Prussia must concentrate its strength and hold it for the favourable moment, which has already come and gone several times. Since the treaties of Vienna, our frontiers have been ill-designed for a healthy body politic. Not through speeches and majority decisions will the great questions of the day be decided – that was the great mistake of 1848 and 1849 – but by iron and blood.

Eighteen months later Prussia was at war with Denmark over possession of Schleswig-Holstein. Britain had tried to head the Prussians off diplomatically, but Crimea had exposed weakness in the army for which even the Royal Navy's strength could not compensate, so that any threats could only be hollow. Indeed, when asked by a nervous official what he would do if a British army were to land on the German coast (in support of the Danes), Bismarck famously replied, 'I shall send a policeman to arrest it!'*

The Prussian army, like those of the other continental powers, was a conscript force. It was therefore large. But its

---

* An increase in the army estimates was materially helped by *The Battle of Dorking*, an 1871 novel of the 'invasion literature' genre. It was written by Colonel (later General) George Tomkyns Chesney, and probably influenced the now much better-known *The War of the Worlds*, by H. G. Wells. Written just after the Prussian victory in the Franco-Prussian War, it describes a successful military invasion of Britain by a foreign power which the preface makes clear is Germany.

real strength lay in its vast pool of reservists – former conscripts with residual military skills who could be instantly recalled. Its mobilization system was also without peer, based on a highly developed scheme whereby the local peacetime administrative headquarters formed the basis of the mobilized field army. Reservists had only a short journey to make to their local depot, where they drew their uniforms and rifles and mustered as a formed unit before entraining for the front. In most of the other continental systems, reservists were mustered as individual reinforcements – much less efficient and much less personal. In the Prussian army, a reservist would know many of his fellow soldiers already, and probably some of his officers.

Cardwell, impressed with Prussia's mobilization for the wars with Denmark (1863–4) and France (1870–1), sought to emulate its strengths. Since there was no question of conscription in Britain he had to work through voluntary recruitment. To begin with, he reduced the length of service 'with the colours' – the time spent on duty with regiment or corps – and introduced a corresponding period on the reserve with pay of 4d a day and an annual training obligation. This would ensure a steady build-up of reservists to a total of 80,000 by the end of the century. That much was relatively easy to achieve; but the outflow of those leaving at the end of their shorter service had, of course, to be matched by intake – and the supply of Irish recruits had been drying up with the mass emigration in the wake of the potato famine. The answer, Cardwell believed, lay in a steady voluntary transfer of men from the militia (revived in 1852, though without the ballot) to the regulars. The Napoleonic Wars had shown the militia to be a significant source of recruits, for militiamen were, quite evidently, already hardened to the life – and might even have found its camaraderie appealing. It made sense, therefore, to align the regular regiments as closely as possible with the local militias or 'volunteers'. Infantry regiments had borne county titles since the last year of the American Revolutionary War, but the actual make-up of the regiments bore no relation to their nominal recruiting areas: they were rarely stationed in the county, and drummed up recruits where they chose – or, more accurately, where they could. Cardwell instituted a system of pairing regular battalions with common local connections, one serving overseas, the other training recruits to feed to the overseas battalion. These were known as 'linked battalions'. The home battalion was to be based in a depot which would also serve as both the headquarters for the local militia battalions and the receiving depot for reservists called back to the colours.

In contrast to the Prussian system, the regional administrative headquarters would not be the basis of a mobilized field army: reservists would be sent individually or collectively to where they were needed. Regular service, however, was to be in the regiment of a man's choice – and it was fully expected that reservists would serve in their former or adopted regiments too. One of Cardwell's first acts was to do away with 'general service', the provision whereby a recruit could be sent to any part of the army the War Office saw fit. As early as 1829 Palmerston, secretary at war, had recognized that

> there is a great disinclination on the part of the lower orders to enlist for general service; they like to know that they are to be in a certain regiment, connected, perhaps, with their own county, and their own friends, and with officers who have established a connection with that district. There is a preference frequently on the part of the people for one regiment as opposed to another, and I should think there would be found a great disinclination in men to enlist for general service, and to be liable to be drafted and sent to any corps or station.

On the twin foundations of efficiency and sentiment, then, Cardwell set about putting into place the third of his key reforms: 'localization', a linking and building programme which changed the face of many a British town, as well as that of the army. It was a timely idea. Local regimental pride perfectly suited the mood of mid-Victorian Britain: many a prominent architect enthusiastically drew up grand designs for the new depot-barracks in his county town, and regimental bands were applauded playing at the county fairs. Some of the linked battalions found their homes in imposing fortresses: for example, the 34th (Cumberland) Regiment and the 55th (Westmoreland) were linked, with a depot in Carlisle Castle; likewise the 91st (Argyllshire Highlanders) and the 93rd (Sutherland Highlanders) were linked, with a depot halfway between their two recruiting areas at Stirling Castle. Other pairings, such as the 62nd (Wiltshire) Regiment and the 99th Duke of Edinburgh's (Lanarkshire) were altogether less obvious; but the arithmetic – and the geography – were never going to be perfect. In this instance, the 99th being a latecomer (raised in 1824, with no county affiliation until 1836), its paired regiment would take priority in location; even so, it seems there was no suitable depot building in Wiltshire, for a brand new one was built in the county town of Devizes, with a massive neo-gothic keep. In these years many a hitherto peaceful county town suddenly found itself with a redbrick medieval citadel in its midst.

The depots were formed immediately by whichever one of the 'linked' battalions was on the home station (which included Ireland, Gibraltar and Malta) sending its 'depot company' to the new headquarters. Since Palmerston's time as secretary at war, each battalion had had one or two depot companies to train recruits; these remained at home to continue recruiting and training when the battalion was posted overseas.* The Carlisle depot, for example, was established in April 1873 with the title of the 2nd Brigade Depot, the brigade being a purely administrative title. It was commanded by a lieutenant-colonel and was home to the depot companies of the 34th (Cumberland), who were in Ireland, the 55th (Westmoreland) who had been in India since the Crimean War, and the two county regiments of militia.

Recruiting remained a persistent problem, however, and the system was soon thrown out of balance again by demands for more battalions overseas, so that by the end of the 1870s there were only fifty-nine battalions on the home establishment, with eighty-two abroad. And the sometimes incongruent linking – of which Wiltshire and Lanarkshire was but one extreme example – was not calculated to generate harmony, for the strength of the regimental system derived in part from robust independence rather than collaboration: why, for instance, should the 34th work harmoniously with the 55th just because they both recruited in the north-west?

The system probably worked best with the more senior regiments, which had retained their second battalions. It was far easier for, say, the first battalion of the 4th (King's Own) Regiment of Foot to send men to the second battalion of a regiment which had the same name and uniform. So in 1881, when Gladstone formed his second administration, the new secretary of state for war, Hugh Childers, decided to take Cardwell's system one step further by amalgamating the linked battalions into new regiments. The regimental numbers were now dropped altogether – the 4th, for example, a non-linked regiment (for it had kept two battalions throughout) became simply the King's Own – and

* The pairing also meant a saving in manpower, the overseas battalion no longer needing to maintain a depot company.

new regimental names were substituted. Some of the names were an obvious reflection of the amalgamation: for example, the 91st (Argyllshire Highlanders) and the 93rd (Sutherland Highlanders) became the Argyll and Sutherland Highlanders. But there were some other new and rather evocative titles: the 34th (Cumberland) and 55th (Westmoreland), for example, became the first and second battalions respectively of the Border Regiment; the 45th (Nottinghamshire) and the 95th (Derbyshire) regiments became the first and second battalions of the Sherwood Foresters. Regiments were now referred to locally as 'our regiment', or even '*the* regiment'. Local pride would be a powerful spur to recruiting in emergencies, and when additional battalions had to be raised, the regimental identity was an equally powerful aid to assimilation, if not quite in the way that Cardwell had expected. The Foresters, for example, would field thirty-one battalions in the First World War.

No amalgamation of 'opposites' seemed too difficult for the War Office either. The conjunction of Scottish heather and English chalk – the Lanarkshire–Wiltshire amalgamation – was in fact one of the most felicitous: the two were renamed the Duke of Edinburgh's (Wiltshire Regiment), taking the county affiliation from the 62nd Foot (which became the first battalion) and the honorific from the 99th Foot (which became the second battalion). Having then lived with the title for forty years, in 1921 the regiment switched the order round to become the Wiltshire Regiment (Duke of Edinburgh's). Long must have been the agonizing over the change; but such is the stuff of regimental identity! With these new regimental names the army would see out the Empire and fight two world wars; in large measure the names would continue right up to the wholesale amalgamations of 2006/7*.

So Gladstone, Queen Victoria's least favourite prime

minister, had significantly reshaped the army – or rather, his two war ministers had.* But the Queen was determined always to keep Gladstone up to the mark. 'If we are to *maintain* our position as a *first-rate* Power,' she now wrote to him, with her characteristic emphatic capitals and underlinings, 'we must, with our Indian empire and large Colonies, be *Prepared for attacks* and *wars, somewhere* or *other,* CONTINUALLY.'

Footnote from p292

* '"The tartan question" is one of the gravest character, far more important . . . than the maintenance of the Union with Ireland. All the thoughts of the War Office are concentrated on it, and patterns of tartans past, present, and future, fill our rooms. We are neglecting the Transvaal, and the Ashanti for the sake of well weighing the merits of a few more threads of red, green or white.' Hugh Childers' baleful remarks to the Commons in 1881 were repeated verbatim by one of his successors, Geoff Hoon, a century and a quarter later during the controversial amalgamation of all the Scottish regiments into one large regiment – the Royal Regiment of Scotland. But Hoon was no Childers, and his remarks were taken as both insensitive and dismissive at a time when the Scottish regiments were at full stretch in Iraq and Afghanistan.

* Cardwell's reforms came out of serious inquiry and committee work, but not least from the application of Cardwell's own considerable intellectual faculties, and his conscientiousness. Indeed, his health was said to have been broken by the work. Few secretaries of state since have committed such ability and integrity to the army's cause.

17

Eleven VCs before Breakfast

Rorke's Drift, 1879

THE 1964 FILM CLASSIC *ZULU*, STARRING MICHAEL CAINE AND Stanley Baker, can still raise a cheer despite its enormous toll of Chief Cetewayo's phenomenally brave warriors. It has its exaggerations – for example, of the supposed friction between the officer commanding the companyof the 24th (2nd Warwickshire) Regiment, 33-year-old Lieutenant Gonville Bromhead, who had purchased his commission, and the Royal Engineers officer, 31-year-old Lieutenant John Chard, who had not – as well as a fair sprinkling of inaccuracies, some of which were careless (such as the serjeant-major's Great War medals) and some egregious.* But the essential story is dramatically told: a small number

---

* In fact, the two officers' backgrounds were not dissimilar. Bromhead's was perhaps more obviously gentry (his father owned Thurlby Hall in Lincolnshire), though he was educated at the grammar school in Newark. Chard's father was a Devon GP, and he too was educated at the local grammar school, in Plymouth. And indeed it was Chard, not Bromhead, who was frequently entertained by Queen Victoria after his award and became a Royal favourite.

of redcoats hold off a vastly superior force of Zulu warriors, while the rest of the battalion lies dead 10 miles away at Isandhlwana, and win an unrivalled number of VCs. The year after the battle, 1880, Chard was invited to send an account of it to Queen Victoria. His report on the action that day at the little mission station on the border between Natal and Zululand reads at first like a policeman giving evidence in a magistrate's court, but the narrative soon picks up:

Early in January 1879, shortly after the arrival of the 5th Company, Royal Engineers, at Durban, an order came from Lord Chelmsford directing that an officer and a few good men of the R.E., with mining implements, etc., should join the third column as soon as possible. I was consequently sent on in advance of the company, with a light mule wagon containing the necessary tools etc., and in which the men could also ride on level ground; with a Corporal, three Sappers and one Driver, my batman, who rode one and looked after my horses ... The roads were so bad that in spite of all our exertions, our progress was slow, and ... we did not reach Rorke's Drift ['drift' = ford] until the morning of the 19th January 1879. The 3rd Column was encamped on the other side (left bank) of the River Buffalo, and the wagons were still crossing on the ponts [cable-guided raft ferries]. I pitched my two tents on the right (Natal) bank of the river, near the ponts, and close to the store accommodation there for keeping them in repair ... There were two large ponts at the river, one of which only was in working order, and my sappers were during this time working at the other, which was nearly finished, to get it also in working order. Late in the evening of the 21st January I received an order from the 3rd Column to say that the men of the R.E., who had lately arrived were to proceed to the camp at Isandhlwana at once ...

Why were Chard and his sappers needed at Rorke's Drift? Because it was the entry point to Zululand from Natal, and the Zulu king, Cetewayo, was to be on the receiving end of a punitive expedition. It was all part of being, in Victoria's words, 'Prepared for attacks and wars, somewhere or other, continually.'

War with the Zulu had been precipitated – intentionally or not – by the new high commissioner for South Africa, Sir Bartle Frere. The Transvaal had recently been annexed for the Crown by the Natal administration, and Frere believed the Zulu army of some 40,000 astonishingly well-disciplined warriors posed a grave threat to the security of the new British territory, for there had been constant border skirmishing between them and the predominantly Cape Dutch settlers in the Transvaal for years.

The Zulu were quite unlike any enemy that the army had faced. The individual warrior was formidably brave, not least because death awaited him at the hands of a superior if he once faltered in battle. Their *impis* – regiments – could move across country at a steady jog-trot for days on end. They attacked in 'horns of the buffalo' formation, fronting with a paralysing mass of spearmen – the head or breast of the buffalo – with a horde of others sweeping round both flanks like the horns of the beast to prevent escape. To defeat them, firepower was everything. To be caught off-balance by an *impi* meant certain destruction.

The military commander in South Africa was Lieutenant-General the Lord Chelmsford, a soldier of wide experience who had recently brought yet another Cape frontier war with the Xhosa to a successful end.* In January 1879 he marched into Zululand with 17,000 British and native troops in five columns. The centre column, 5,000 strong with Chelmsford at their head, began crossing the Buffalo River by the ponts at Rorke's Drift on the eleventh, and by

---

* The Xhosa were a less formidable tribe than the Zulu, lacking their leadership and discipline – but perpetually troublesome nevertheless.

the twenty-first were encamped on the plain of Isandhlwana. Fatally, they did not dig in – a precaution that no Roman legion would have failed to take, no matter how hard the ground, and even if they too had been armed with the rapid-fire Martini–Henry rifle which had replaced the Minié and the Enfield.

At Rorke's Drift, once all the columns had passed, Chard's orders were 'to select a suitable position protecting the ponts for Captain Rainsforth's Company 1/24th to entrench itself'. And having done this, Chard decided to go and seek further orders from his sapper superior at Isandhlwana. While he was there,

> An N.C.O. of the 24th Regiment lent me a field glass, which was a very good one, and I also looked with my own, and could see the enemy moving on the distant hills, and apparently in great force. Large numbers of them moving to my left, until the lion hill of Isandhlwana, on my left as I looked at them, hid them from my view. The idea struck me that they might be moving in the direction between the camp and Rorke's Drift and prevent my getting back, and also they might be going to make a dash at the ponts.

Chard returned at once to Rorke's Drift where the officer in command, Major Spalding of the 104th Foot, formerly the Bengal Fusiliers, instructed him to take command (for Chard, although a sapper whereas Bromhead was an infantryman, and older, was the senior in service by three years) while he went to bring reinforcements from nearby Helpmekaar: 'Nothing will happen, and I shall be back again early this evening,' said Spalding encouragingly. Chard wrote:

> I then went down to my tent by the river, had some lunch comfortably, and was writing a letter home when my attention

was called to two horsemen galloping towards us from the direction of Isandhlwana. From their gesticulation and their shouts, when they were near enough to be heard, we saw that something was the matter, and on taking them over the river, one of them, Lieutenant Adendorff of Lonsdale's Regiment, Natal Native Contingent, asking if I was an officer, jumped off his horse, took me on one side, and told me that the camp was in the hands of the Zulus and the army destroyed; that scarcely a man had got away to tell the tale, and that probably Lord Chelmsford and the rest of the column had shared the same fate. His companion, a Carbineer, confirmed his story. He was naturally very excited and I am afraid I did not, at first, quite believe him, and intimated that he probably had not remained to see what did occur. I had the saddle put on my horse, and while I was talking to Lieutenant Adendorff, a messenger arrived from Lieutenant Bromhead, who was with his company at his little camp near the commissariat stores, to ask me to come up at once.

Chard then gave orders for the waggons to be 'inspanned' – hitched up to the draught animals – and posted a serjeant and six men on high ground to cover the ponts, before galloping for the cluster of buildings at the mission station where the commissariat stores had been dumped. There he found a message from the Isandhlwana column

that the enemy were advancing in force against our post. Lieutenant Bromhead had, with the assistance of Mr. Dalton [commissary officer], Dr. Reynolds [medical officer] and the other officers present, commenced barricading and loopholing the store building and the missionary's house, which was used as a hospital, and connecting the defence of the two buildings by walls of mealie bags, and two wagons that were on the ground. The Native Contingent, under their officer, Captain Stephenson, were working hard at this with our own men, and the walls were rapidly progressing.

Breathless fugitives continued to arrive with news from Isandhlwana, and just after four o'clock firing was heard from beyond the Oscarberg, the heights that overlooked the drift, for some of the Zulus had Brown Bess muskets acquired from traders (they had not yet, contrary to some accounts, taken up the Martini–Henrys from the dead troops at Isandhlwana). Most of the native contingent and their white officer promptly deserted.

We seemed very few now all these people had gone, and I saw that our line of defence was too extended, and at once commenced a retrenchment of biscuit boxes, so as to get a place we could fall back upon if we could not hold the whole.

Private Hitch, 24th, was on top of the thatch roof of the commissariat store keeping a look-out. He was severely wounded early in the evening, but notwithstanding, with Corporal Allen, 24th, who was also wounded, continued to do good service, and they both when incapacitated by their wounds from using their rifles, still continued under fire serving their comrades with ammunition. We had not completed a wall two boxes high when, about 4.30 p.m., Hitch cried out that the enemy was in sight, and he saw them, apparently 500 or 600 in number, come around the hill to our south (the Oscarberg) and advance at a run against our south wall.

We opened fire on them, between five and six hundred yards, at first a little wild, but only for a short time, a chief on horse-back was dropped by Private Dunbar, 24th. The men were quite steady, and the Zulus began to fall very thick. However, it did not seem to stop them at all, although they took advantage of the cover and ran stooping with their faces near the ground. It seemed as if nothing would stop them, and they rushed on in spite of their heavy loss to within 50 yards of the wall, when they were taken in flank by the fire from the end wall of the store building, and met with such a heavy direct fire from the mealie wall, and the hospital at the same time, that they were checked as if by magic.

They occupied the cook house ovens, banks and other cover, but the greater number, without stopping, moved to their left around the hospital, and made a rush at the end of the hospital, and at our north-west line of mealie bags. There was a short but desperate struggle during which Mr. Dalton shot a Zulu who was in the act of assegaing a corporal of the Army Hospital Corps, the muzzle of whose rifle he had seized, and with Lieutenant Bromhead and many of the men behaved with great gallantry. The Zulus forced us back from that part of the wall immediately in front of the hospital, but after suffering very severely in the struggle were driven back into the bush around our position.

The main body of the enemy were close behind the first force which appeared, and had lined the ledge of rocks and caves in the Oscarberg overlooking us, and about three or four hundred yards to our south, from where they kept up a constant fire. Advancing somewhat more to their left than the first attack, they occupied the garden, hollow road, and bush in great force. The bush grew close to our wall and we had not had time to cut it down. The enemy were thus able to advance under cover close to our wall, and in this part soon held one side of the wall, while we held the other.

A series of desperate assaults were made, on the hospital, and extending from the hospital, as far as the bush reached; but each was most splendidly met and repulsed by our men, with the bayonet. Each time as the attack was repulsed by us, the Zulus close to us seemed to vanish in the bush, those some little distance off keeping up a fire all the time. Then, as if moved by a single impulse, they rose up in the bush as thick as possible rushing madly up to the wall (some of them being already close to it), seizing, where they could, the muzzles of our men's rifles, or their bayonets, and attempting to use their assegais and to get over the wall. A rapid rattle of fire from our rifles, stabs with the bayonet, and in a few moments the Zulus were driven back, disappearing in the bush as before, and keeping up their fire. A brief interval and the attack would be again made, and repulsed in the same manner. Over and over again this happened, our men behaving with the greatest coolness and gallantry . . .

For all that the duke of Wellington had not wanted the infantry's rank and file to call themselves 'riflemen' – certain that it was volley fire that decided matters – when an enemy as numerous as the Zulus closed with the firing line it was individual weapon handling and marksmanship that counted. Otherwise it would soon come to a duel between bayonet and *iklwa*, the Zulu's short stabbing spear – or rather, a contest between a single bayonet and a great many spears. Of the defenders' skill at arms Chard goes on:

Our fire at the time of these rushes of the Zulus was very rapid. Mr. Dalton dropping a man each time he fired his rifle, while Bromhead and myself used our revolvers. The fire from the rocks and caves on the hill behind us was kept up all this time and took us completely in reverse, and although very badly directed, many shots came among us and caused us some loss, and at about 6.00 p.m. the enemy extending their attack further to their left, I feared seriously would get it over our wall behind the biscuit boxes. I ran back with two or three men to this part of the wall and was immediately joined by Bromhead with two or three more. The enemy stuck to this assault most tenaciously, and on their repulse, and retiring into the bush, I called all the men inside our retrenchment and the enemy immediately occupied the wall we had abandoned and used it as a breastwork to fire over.

Mr. Byrne, acting Commissariat Officer, and who had behaved with great coolness and gallantry, was killed instantaneously shortly before this by a bullet through the head, just after he had given a drink of water to a wounded man of the NNC.

All this time the enemy had been attempting to fire the hospital and had at length set fire to its roof and got in at the far end . . .

The garrison of the hospital defended it with the greatest gallantry, room by room, bringing out all the sick that could be moved, and breaking through some of the partitions while the Zulus were in the building with them. Privates Williams, Hook, R. Jones and W. Jones being the last to leave and holding the doorway with the bayonet, their ammunition being expended. Private Williams's bayonet was wrenched off his rifle by a Zulu, but with the other men he still managed with the muzzle of his rifle to keep the enemy at bay. Surgeon Reynolds carried his arms full of ammunition to the hospital, a bullet striking his helmet as he did so. But we were too busily engaged outside to be able to do much, and with the hospital on fire, and no free communication, nothing could have saved it. Sergeant Maxfield 24th might have been saved, but he was delirious with fever, refused to move and resisted the attempts to move him. He was assegaid before our men's eyes.

Seeing the hospital burning, and the attempts of the enemy to fire the roof of the store . . . we converted two large heaps of mealie bags into a sort of redoubt which gave a second line of fire all around, in case the store building had to be abandoned, or the enemy broke through elsewhere . . .

Trooper Hunter, Natal Mounted Police, escaping from the hospital, stood still for a moment, hesitating which way to go, dazed by the glare of the burning hospital, and the firing that was going on all around. He was assegaid before our eyes, the Zulu who killed him immediately afterwards falling. While firing from behind the biscuit boxes, Dalton, who had been using his rifle with deadly effect, and by his quickness and coolness had been the means of saving many men's lives, was shot through the body. I was standing near him at the time, and he handed me his rifle so coolly that I had no idea until afterwards of how severely he was wounded. He waited quite quietly for me to take the cartridges he had left out of his pockets. We put him inside our mealie sack redoubt, building it up around him. About this time I noticed Private Dunbar 24th make some splendid shooting, seven or eight Zulus falling on the ledge of rocks in the Oscarberg to as many consecutive shots by him. I saw Corporal Lyons hit by a bullet which lodged in his spine, and fall between an opening we had left in the wall of biscuit boxes. I thought he was killed, but looking up he said, 'Oh, Sir! You are not going to leave me here like a dog?' We pulled him in and laid him down behind the boxes where he was immediately looked to by Reynolds. Corporal Scamle [Scammell] of the Natal Native Contingent, who was badly wounded through the shoulder, staggered out under fire again, from the store building where he had been put, and gave me all his cartridges, which in his wounded state he could not use. While I was intently watching to get a fair shot at a Zulu who appeared to be firing rather well, Private Jenkins 24th, saying 'Look out, Sir,' gave my head a duck down just as a bullet whizzed over it. He had noticed a Zulu who was quite near in another direction taking a deliberate aim at me. For all the man could have known, the shot might have been directed at himself. I mention these facts to show how well the men behaved and how loyally worked together . . .

Someone thought they saw redcoats in the distance coming their way, and as word spread among the defenders there was cheering, which Chard believed made the Zulus pause for a moment. In the film, the defenders begin singing 'Men of Harlech', but this is an anachronism. Although the 24th are portrayed as a Welsh regiment, they were in fact in a state of Cardwell–Childers transition: their depot had been formed five years earlier at Brecon, but it was not until 1881 that they were transformed into the South Wales Borderers, who took 'Men of Harlech' as their regimental march. Of the soldiers present, forty-nine were English, thirty-two Welsh,* sixteen Irish and twenty-two others of various nationalities. But in any case, the redcoats on the road from Helpmekaar were a mirage, and as darkness came the defenders found themselves completely surrounded. For some hours the attacks continued, though more sporadically:

Although they [the Zulus] kept their positions behind the walls we had abandoned, and kept up a heavy fire from all sides until about 12 o'clock, they did not actually charge up in a body to get over our wall after about 9 or 10 o'clock. After this time it became very dark, although the hospital roof was still burning – it was impossible from below to see what was going on, and Bromhead and myself getting up on the mealy sack redoubt, kept an anxious watch on all sides . . .

About midnight or a little after the fire slackened and after that, although they kept us constantly on the alert, by feigning, as before, to come on at different points, the fire was of a desultory character. Our men were careful, and only fired when they could see a fair chance. The flame of the burning hospital was now getting low, and as pieces of the roof fell, or hitherto unburnt parts of the thatch ignited, the flames would blaze up illuminating our helmets and faces. A few shots from the Zulus, replied to by our men – again silence, broken only by the same thing repeatedly happening. This sort of thing went on until about 4 a.m. and we were anxiously waiting for daybreak and the renewal of the attack, which their comparative, and at length complete silence, led us to expect. But at daybreak the enemy were out of sight, over the hill to our south west. One Zulu had remained in the kraal and fired a shot among us (without doing any damage) as we stood on the walls, and ran off in the direction of the river – although many shots were

* Stretching a point, for eighteen were from Monmouthshire which was not then a 'Welsh' county.

fired at him as he ran. I am glad to say the plucky fellow got off . . .

There is perhaps more in that last remark than mere admiration for individual pluck. The Zulus, for all their disembowelling of the best part of five companies of the first and second battalions of the 24th, and a good many others in the Isandhlwana column, were not demonized or written off as savages. Indeed, respect for the enemy among soldiers in the field, as opposed to underestimation of him at headquarters, was to become a feature of late Victorian colonial wars, and helped sharpen individual fighting skills. Kipling got the sense of this when he wrote:

> I fired a shot at a Afghan,
> The beggar 'e fired again,
> An' I lay on my bed with a 'ole in my 'ed;
> An' missed the next campaign!
> I up with my gun at a Burman
> Who carried a bloomin' dah,
> But the cartridge stuck and the bay'nit bruk,
> An' all I got was the scar.*

Back at Rorke's Drift, the Zulus remained menacingly on the Oscarberg for some time, until the appearance of reinforcements from Helpmekaar and the realization of their already formidable losses persuaded them to turn and lope away. Chard was 'glad to seize an opportunity to wash my face in a muddy puddle, in company with Private Bush 24th, whose face was covered with blood from a wound in the nose caused by the bullet which had passed through and killed Private Cole 24th. With the politeness of a soldier, he

---

* 'Pte Ortheris's Song', from 'The Courting of Dinah Shadd', in *Life's Handicap*.

lent me his towel, or, rather, a very dirty half of one, before using it himself, and I was very glad to accept it.'

Private Cole's towel speaks eloquently of the easier relationship between officer and soldier that had developed in the small professional army of the mid-Victorian period. It would stand in sharp contrast with the rigidity required in the mass conscript armies of the Continent, which in the worst cases would lead to a cheapening of the worth of the private soldier in the minds of senior officers.*

And then, with the understatedness typical of the Victorian officer, when their deliverance was assured Chard and Bromhead, who would each receive the VC from the Queen-Empress, allowed themselves a little relaxation: 'In wrecking the stores in my wagon, the Zulus had brought to light a forgotten bottle of beer, and Bromhead and I drank it with mutual congratulations on having come safely out of so much danger.'

Chard continues his report with brief summaries of the part played by those who received the VC and the Distinguished Conduct Medal (DCM),† and finishes with an acknowledgement of the moral support of public opinion and the active interest of the Crown, which the army had increasingly enjoyed since the Crimean War:

---

* The gulf between officer and man in the Tsar's army of the First World War is well known, but it was by no means unique. When, for example, British officers found themselves in northern Italy in 1917 after the setbacks at Caporetto, they were dismayed by the indifference shown by many Italian officers towards their men, although there was no lack of technical skill in the former or courage in the latter.

† The Victoria Cross had been created in 1856 to recognize outstanding acts of bravery by British soldiers in the face of the enemy. Eleven VCs were awarded, seven to the 2nd Battalion, 24th (2nd Warwickshire) Regiment, one to the Army Medical Department, one to the Royal Engineers (Chard's), one to the Commissariat and Transport Department (Dalton's), and one to the Natal Native Contingent (the splendid and aptly named Corporal Schiess, a Swiss aged 22).

As the Reverend George Smith [an army chaplain who had taken a very active part in the defence] said in a short account he wrote to a Natal paper – 'Whatever signs of approval may be conferred upon the defenders of Rorke's Drift, from the high quarters, they will never cease to remember the kind and heart-felt expressions of gratitude which have fallen both from the columns of the Colonial Press and from so many of the Natal Colonists themselves.'

And to this may I add that they will ever remember with heart-felt gratitude the signs of approval that have been conferred upon them by their Sovereign and by the People and the Press of England.

Chard's references to individual marksmanship show how much the army of Wellington's day was being transformed by both technology and the nature of the enemy. The Martini–Henry rifle had entered service in 1871, replacing the Snider–Enfield (a hybrid breech-loader which had itself replaced the muzzle-loading Minié). It used a fixed metal cartridge (in other words, there was no biting open the cartridge to get at the powder), could fire at the rate of ten rounds a minute and was accurate at 500 yards. When the defenders opened fire 'between five and six hundred yards, at first a little wild', it would have been volley-fire, the Zulus presenting a massed target. But as they closed, Bromhead* would have ordered 'rapid fire' (shots at a mass target as fast as the bolts could be drawn back and new cartridges inserted) so that the Zulus faced a continuous hail of bullets until they were so close as to present individual targets, when each rifleman would fire deliberate aimed shots. At this point the battle would become one of 'independent' fighting, as opposed to the Wellingtonian system which for the most

---

* Bromhead was in command of the detachment of the 24th, so he gave the direct orders.

part saw the infantry shoulder-to-shoulder, musket levelled and firing on command as part of a machine.

While the French and German armies would gain a little of this experience in their own more limited colonial ventures, this sense of individual marksmanship took deeper root in the British infantry, not least through the parallel experience in India. The Martini–Henry had not performed perfectly at Isandhlwana, to some extent because of the excessive heat – rifles had jammed because of powder-fouling and the failure of the working parts – but it had taken a terrible toll nevertheless. Before entering Zululand Lord Chelmsford had written, 'I am inclined to think that the first experience of the Martini-Henrys will be such a surprise to the Zulus that they will not be formidable after the first effort.' And although that first effort was the disaster of Isandhlwana, he had not been far wrong.

From this time, therefore, the Board of Ordnance applied itself with a new clarity of purpose: the requirement was for a robust breech-loading rifle, accurate – or at least effective – to half a mile, and capable of rapid fire. Black powder (and therefore white smoke) would for some years mean that the rifleman gave his position away when he fired, so that concealment remained academic, and the infantry's instinct for tight formation, even in open order, would sometimes nullify the advantage of range and accuracy. But from now on the infantry could expect to engage the enemy at a distance instead of having to stand their ground under artillery fire as the enemy closed in, firing a couple of volleys at less than a hundred yards and then relying on the bayonet.

Many a battle of Victoria's expanding empire had proved that the army did not want for courage, but Rorke's Drift now set the gold standard for resolute defence. No officer could quit a position in the face of seemingly overwhelming odds when Chard and Bromhead had not done so – not, at least, without inviting critical comparison. Indeed, Rorke's Drift still resonates today. When in the summer of 2006 the Taleban launched their offensive in Helmand, throwing themselves in huge numbers at the scattered platoon posts held by 3rd Battalion the Parachute Regiment, senior officers spoke of 'dozens of Rorke's Drifts' every day.* And if the Paras themselves thought first of Arnhem rather than Rorke's Drift, there were nevertheless plenty of Chards and Bromheads in those platoon posts, and plenty of equally resolute NCOs and riflemen.

What stands out from the Zulu Wars, and from the Crimea and the Mutiny, is the army's 'operational resilience' – a certain strength and tenacity that has become one of its defining characteristics. The defeat at Isandhlwana shocked the nation when the news reached London in mid-February (by ship, and then by cable from Madeira). Sir Garnet Wolseley, the model for the 'modern major-general' in *The Pirates of Penzance* (1879) and referred to in the press at the time as 'our only general', was at once despatched to relieve Chelmsford. But before he could arrive with reinforcements Chelmsford and his army had pulled themselves together, regrouped, adjusted tactics and made a methodical advance against the Zulu capital at Ulundi, where at the beginning of July they completely crushed Cetewayo's *impis*.

There is nothing to be proud of in losing the opening battle of a campaign, no matter how hard fought, if it is lost through faulty intelligence, poor tactics, bad planning, the wrong equipment, inadequate logistics or any other avoidable reason – as the army has done many times since Isandhlwana – but the ability to absorb the shock of tactical defeat, to adjust plans in the light of experience, to take the fight back to the enemy early and regain the initiative,

---

* The story of the battalion's fighting in Helmand is told superbly by Patrick Bishop in *3 PARA*.

winning the key battles of the campaign and in the end strategic victory, is the mark of maturity and true greatness in an army. The fact that Chelmsford and his forces did so from within their own resources is also significant. The distance from London to most of the seats of Victoria's small wars, despite steamships and the electric telegraph, fostered self-reliance in both commanders and field forces. And in these circumstances the regimental system came into its own, for the Victorian regiment, in which officers and other ranks served together for long enough to know each other's strengths and how to adjust after a setback, became a remarkably self-healing organization.

The British army's operational resilience would soon stand out all the more strongly by comparison with the armies of continental powers. The French, demoralized by the early miscalculations of the Franco-Prussian War (1870–1), saw Paris fall to the enemy. Again in 1914 the failure of their much-vaunted offensive plans resulted in widespread dismay (although Marshal Joffre showed a certain *élan* at the Marne), and in 1940 the complete defeat of their defensive strategy brought catastrophic collapse. And while German armies have been rightly praised for their intense resilience at the tactical level, again and again at the campaign level their generals and staffs have shown an inability to adapt to operational realities, being often too restrictively wedded to control from Berlin through their Grosser Generalstab system. Too often German armies have merely endured, and as a result have been ground down by the very elements that had inflicted the first tactical defeat.

Although initial tactical failures could have profound strategic consequences – such as those of the Crimean War in exposing the nation's military weakness – in Victoria's time they also had the benefit of 'tempering' the army, making both generals and rank and file able to withstand the sort of impact that would have destroyed many another nation's army. The tempering would be a slow process, occurring largely out of sight (and all the better for it), for as the military correspondent of *The Times*, Colonel Charles Repington, wrote in an article entitled 'Death and Resurrection' after the annihilating battles on the Somme in 1916:

> Nobody in particular noticed that between 1878 and 1902 the British Army added to the Empire an area of territory equal to that of the United States, but the British soldier naturally noticed it because he did it. In the mountains that girdle the North-West Frontier, amidst the rocks of Afghanistan, through the swamps and forests of Burma and Africa, on the Veldt, in Egypt and in the deserts of the Sudan, an Empire was being carved out by the old Army in a quiet, unostentatious but methodical sort of way ... overcoming inconceivable difficulties with small means.

Philip Mason, one of the best historians of the Raj, who had himself been an officer of the Indian Civil Service, explains part of that tempering process in *The Men Who Ruled India*: 'The routine of the Frontier accustomed men to bullets and to taking cover, to guards and sentry duty which had real purpose, to night marches and sniping ... it was infinitely superior to a field day at Aldershot.'

In exactly the same process a century later, three decades of fighting the IRA would prepare the army for its coming tests in the Balkans, Iraq and Afghanistan.*

---

* *The Men Who Ruled India* is the title of his best-known work, a two-volume history. His much shorter *A Matter of Honour* (1974) is generally considered the best history of the Indian Army (i.e. the successor to that of the East India Company). Many an officer earned his spurs on the Frontier, and some a peerage, not least among them Lord Roberts of Kandahar – of whom more in chapter 19 – who commanded during the Second Afghan War of 1878–81. That war was fought principally by Indian troops, but Roberts's methods and success were carefully noted at home.

18

The Cheaper Man

Egypt, 1882–98

A hundred years before the term was used, Kipling wrote a poem about 'asymmetric warfare', which with considerable prescience he called 'Arithmetic on the Frontier' (1886):

> A Great and glorious thing it is
> To learn, for seven years or so,
> The Lord knows what of that and this,
> Ere reckoned fit to face the foe –
> The flying bullet down the Pass,
> That whistles clear: 'All flesh is grass.'
> Three hundred pounds per annum spent
> On making brain and body meeter
> For all the murderous intent
> Comprised in 'villainous saltpetre'.
> And after? – Ask the Yusufzaies
> What comes of all our 'ologies.
> A scrimmage in a Border Station –
> A canter down some dark defile

Two thousand pounds of education
Drops to a ten-rupee jezail.
The Crammer's boast, the Squadron's pride,
Shot like a rabbit in a ride!
No proposition Euclid wrote
No formulae the text-books know,
Will turn the bullet from your coat,
Or ward the tulwar's downward blow.
Strike hard who cares – shoot straight who can
The odds are on the cheaper man.

'Asymmetric warfare' describes conflict between conventionally equipped, 'high-tech' regular troops and an enemy who takes them on, deliberately or through no choice, at a lower level of military technology. The low-tech force relies on patience, aiming at the exhaustion of the opponent's political will to continue the fight in the face of steady – if comparatively quite small – losses. The conventional force must either bring the other to battle 'in the open' – tempting them to believe that 'one big push' will finish the job, and that they have the strength to do so – or else adapt the enemy's own tactics to fight at the 'low-tech' level even better than he does. The latter requires a great deal of 'humint' (human intelligence, as opposed to that gathered by technical means) because the enemy's methods of communicating and operating will not be easily susceptible to non-human penetration.

In the later nineteenth century British army officers began developing subtle humint and low-tech skills, mastering tribal languages and acquiring considerable ethnographic techniques – in India especially, sometimes on secondment to the Indian Political Service as district administrators or as advisers to the princely states (which were independent of the British Raj). They became players in 'the Great Game', the rivalry with Russia for control of Central Asia. Through this route, especially of secondments in administrative capacities, with the politically nuanced officer able to punch above his weight in dealing with feudal rulers, the notion of understanding and respecting the ways of native peoples entered the collective mind of the army. It remains there today, although in recent years professional jealousy in some of the agencies primarily concerned with civil affairs has curtailed this sort of activity, so that it has become increasingly a function of Special Forces.

It was on the North-West Frontier that the army learned, often in the brutal fashion that Kipling described, how to fight the 'low-tech' enemy. Soldiers could not rely on superior firepower – especially not on artillery – since the tribesmen would not oblige the gunners by presenting a suitable target, and the proximity of the civil population had to be taken into account, even if they were hostile. And so the idea of fire discipline developed further, with emphasis on the use of ground and marksmanship to outmanœuvre and outshoot the enemy. Stricter fire discipline has characterized the army's involvement in 'operations other than war' (peacekeeping and counter-insurgency), and with it has come the principle of 'minimum force'. This approach has often stood in marked contrast to, for example, that of the Americans, for the US army's formative experience was almost entirely that of 'warfighting', in which the dominating principle is 'overwhelming force'.

But when there was real war, instead of punitive expeditions or 'police actions', there was no lack of will to apply overwhelming force; the problem usually lay in bringing to bear men and materiel in overwhelming quantity in what was often the most difficult terrain. Before the US army was drawn into the Balkans in the late 1990s its generals would sometimes say, 'We don't do mountains.' When the Italian army was asked if it might help with the intervention in Sierra Leone in 1999, its generals replied, 'We do mountains and deserts, not jungles.' But in the last two decades of Victoria's reign the British army was doing mountains (Afghanistan), deserts (Egypt and the Sudan) and jungles (Burma).* This operational 'range' extended throughout the next century, with fighting in both world wars in all types of terrain, and has continued in conflicts of lesser intensity since.

Although British troops had been getting sand between their toes since the days of the Tangier garrison, the first major test of operating in real sand seas came with the invasion of the Sudan in 1896. The story began with the opening in 1869 of the Suez Canal, an Anglo-French investment venture. Eight years later, Egypt's finances were in such a state that the Khedive ('Lord') Ismail, the monarch-governor, was forced to accept an Anglo-French condominium, and was deposed soon afterwards in favour of his son Tewfik. This stirred resentment in the Egyptian army, not least because of Tewfik's economies (as well as his corruption), and in 1882 there was a coup. Gladstone resolved to intervene, though the French demurred. The 'Ashanti Ring' (Lieutenant-General Sir Garnet Wolseley and the favoured band of staff officers who had performed so well in the Ashanti campaign in West Africa the decade before) was assembled once more, and 'our only general' was sent with 40,000 troops to retrieve the situation. It was the largest expeditionary force the nation had assembled since Waterloo: 16,000 men were sent from Britain, testimony to the effectiveness of the Cardwell–Childers reforms, and the rest gathered up from the Mediterranean garrisons and India.

* Lieutenant-General Sir Robert Napier's troops in the 1868 campaign in Abyssinia had traversed arid, mountainous country, and Lieutenant-General Sir Garnet Wolseley's 1873–4 campaign against the Ashanti of the Gold Coast had also been in deep jungle.

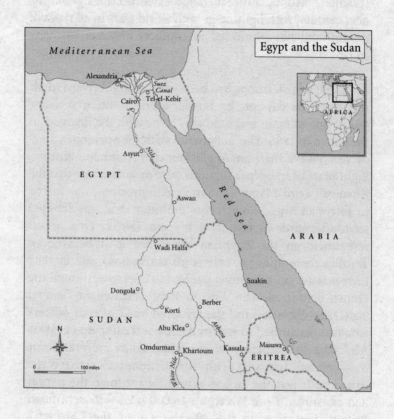

Egypt and the Sudan

Mediterranean Sea

Alexandria

Suez
Canal

Cairo   Tel-el-Kebir

AFRICA

EGYPT

Asyut

Nile

Aswan

Red Sea

ARABIA

Wadi Halfa

Suakin

Dongola

Korti   Berber

SUDAN

Abu Klea

Atbara

N

Omdurman   Khartoum   Kassala   Massawa

White Nile

ERITREA

0   100 miles

Egyptian army rebels held the approaches to Cairo from the port of Alexandria, and when these defences proved too strong Wolseley gave up attempts to break through and instead steamed down the canal and landed at Ismailia, 90 miles from the capital. Here he moved quickly and decisively, seizing control of the Sweet Water Canal (to give him a ready supply of drinking water) and railway rolling stock (to move his guns and supplies) before pressing 25 miles inland towards the rebel camp at Tel-el-Kebir. After a long and ambitious night march, Wolseley's two divisions attacked at dawn: the battle lasted barely half an hour, and the campaign was all but over before breakfast. The army had been moving by night since Marlborough's day, but at Tel-el-Kebir it learned the real psychological advantage of night operations, and ever since has favoured night approach marches and attacks.

Two days later Wolseley was in Cairo, and a month after that he was back in London, with a peerage. The army was to stay in Egypt for more than half a century, however, and learn to love the desert. Wolseley himself, as the saying went, having once drunk the water of the Nile, was bound to return. In the Sudan (with Egypt, a political entity since 1821) an Islamic fundamentalist 'warlord' styling himself 'Mahdi' or guide, Muhammad ibn Abdalla, led a revolt against Egyptian rule. In November 1883 he defeated a Turkish–Egyptian army under a British officer, William Hicks ('Hicks Pasha' – 'Lord Hicks' – whose head was presented to the Mahdi), and a relief column led by another former British officer, Valentine Baker.* The Egyptian

* Valentine Baker ('Baker Pasha') had commanded the 10th Hussars, was a favourite of their colonel, the Prince of Wales, and had been marked out for advancement, but in 1875 he was convicted of indecent assault on a young woman in a railway carriage and sentenced to a year's imprisonment. On his release he sought his rehabilitation with the Turkish army. The Prince of Wales mounted a long campaign to restore him to the Army List, and eventually the Queen relented, though Baker died – on the Nile – before the news could reach him.

government appealed to Gladstone for the help of General Charles Gordon.

'Chinese' Gordon had made his name during the Taiping rebellion of 1860 following the Second Opium War between Britain and China, and subsequently in the Sudan suppressing the slave trade. Gordon was an unconventional officer, to say the least. It used to be said of sappers – Gordon had been commissioned into the Royal Engineers – that they were either mad, married or Methodist. Gordon never married, and his mystical–evangelical Christianity would have been too ex-treme for many a Methodist. But he was entirely without fear – some thought, indeed, to the point of madness. Nor did he have much time for political direction, although he was said to be Queen Victoria's favourite general.

Gordon left London in January 1884 to take up his new appointment as governor-general of the Sudan, calling briefly in Cairo before going on by train and Nile steamer to the capital, Khartoum. But in the inevitable fate of a politically inspired venture not backed by adequate military resources (Gladstone had ordered the evacuation of the remaining British garrisons in the Sudan), by May Khartoum was cut off, with Gordon inside, and most of the country to the north, through which the Nile flowed, was in the hands of the Mahdi's followers – or the dervishes, as they were sometimes known.*

Sir Evelyn Baring, the government's representative in Cairo, had opposed the venture from the outset, arguing that if a British governor-general found himself in trouble in the Sudan there would be a British public clamour to send

---

* The term 'Dervish' – which refers to mendicant Sufi Muslim ascetics of any nationality – was used at the time and later (wrongly, students of Islamic sects would claim) as well as 'Mahdist'. 'Fuzzy-Wuzzy' was the usual soldier's term – a not unreasonable allusion to the distinctive way the warriors wore their hair, and a not unaffectionate one either.

Sir Garnet Wolseley – 'our only general' – the safe pair of hands
in many a campaign of Victoria's wars.

troops to extricate him – and adding that Gordon was just the man to get himself into trouble. And so it proved, though Gladstone at first refused to send help. It was not until August that, in part at the Queen's urging, he sent for Wolseley. A month later Wolseley was in Cairo with as many of the Ashanti Ring as he could muster, preparing to take an Anglo-Egyptian force of 6,000 men and 8,000 horses and camels 1,700 miles up the Nile – the most demanding logistical operation yet to be undertaken by the army.

But the Nile above Wadi Halfa proved agonizingly slow to navigate, despite the help of Canadian *voyageurs*, some of them from the native tribes, with whose help Wolseley had worked miracles in the similar if altogether less ambitious Red River expedition of 1870 against the Metis rebels in Manitoba. The official history puts the problem archly: 'Had British soldiers and Egyptian camels been able to subsist on sand and occasional water, or had the desert produced beef and biscuit, the army might, in spite of its late start, have reached Khartoum in November. But as things were, the rate of progress of the army was dependent on the rate of progress of its supplies.'*

An advance guard of Wolseley's men reached Khartoum by river on 28 January 1885, having taken the gruelling short cut across the desert from the great bend of the Nile at Korti to rendezvous with steamers sent by Gordon, fighting two sharp battles on the way at Abu Klea and Metemmah (and in red jackets still – the last time British troops would wear red on campaign)†. But it was too late: Khartoum had fallen to

* Written by Colonel (later Major-General Sir) Edward Colville. Authoritative accounts of campaigns began to be written at this time by the intelligence branch of the War Office. *The* 'Official History', now the responsibility of the Cabinet Office, only began with the First World War. The MoD still continues to write campaign narratives, however.
† Though in the uSuthu Rebellion of 1888 (sometimes misleadingly called the Second Anglo–Zulu War) some troops took to the field in red; but it was not a true campaign.

the Mahdists two days earlier, and 'Gordon Pasha' had been killed.

Wolseley returned to London to a consolatory viscountcy (and would later be promoted field marshal), for the opprobrium was turned entirely on Gladstone, who promptly resigned. But the aura of success that had accompanied 'our only general' for so long was somehow dimmed by the failure, not least because it emerged that he had dismissed local advice from soldiers like Sir Evelyn Wood (a VC and later field marshal) who had urged him to take not the Nile route but the far shorter desert one from Suakin on the Red Sea to Berber, and thence to Khartoum via the easily navigable 200 miles of the Upper Nile. And though it was Gladstone, the 'Grand Old Man', who carried the can (the popular press reversed the initials G.O.M. to spell 'Murderer of Gordon'), Wolseley lost some of the moral authority he needed in his long-running battle with the commander-in-chief, the duke of Cambridge, for army reform. It would be another ten years before the duke would retire (or, more correctly, be forced to accept retirement) and Wolseley would take over, by which time his intellectual power was fading. That would be too late for the army, which was about to have its biggest test since the Crimea, against the Boers (for a second time), in which it would falter as badly. But before that test would come the reconquest of the Sudan by the man who perhaps comes closest to ranking with Marlborough and Wellington – Kitchener.

The first lesson of Wolseley's Nile campaign had been clear enough: as ever, logistic preparation was everything. At the tactical level there had been an important lesson in firepower, too. Like the defenders of Rorke's Drift, Wolseley's infantry carried the Martini–Henry rifle whose blackpowder cartridges made concealment difficult – though this would be more of a problem in South Africa against the

Boers than in the Sudan (and in any case, a red coat was hardly an aid to concealment). The norm was still volley fire from a standing or kneeling position, followed by rapid (individual) fire. In the open, when the enemy obliged by charging en masse and without fire support, this was usually enough to defeat any attack, although at Abu Klea the Mahdists had broken a square – if only temporarily. But the presence there of a few Gatling guns (primitive machine guns)* suggested how the infantry might increase their defensive fire in the future, and how they might arrange their own supporting fire in the attack, which up to then had been the job of the artillery. For the machine gun, integral to the infantry company, could release the artillery to concentrate its fire for greater effect, not least against the enemy's artillery. An eye witness at the battle of Abu Klea describes how the partially broken square of the Royal Sussex Regiment managed to re-form, and that 'At last the Gatling guns were got into action, and that practically ended the battle. The Soudanese were simply mown down. Their bodies flew up into the air like grass from a lawn mower.'

When it came to the second expedition to Khartoum, twelve years later, time was not so pressing: there was no Gordon to relieve. But the neighbouring friendly colonial power, Italy, had suffered a major defeat in Abyssinia and there was every expectation that the Mahdists, under the late Mahdi's successor, the Khalifa or 'steward' Abdallahi ibn Muhammad, would move against the Horn of Africa, with all the consequential problems for Cairo and London of a heightened threat to Egypt itself and a menacing presence bordering the route to India.

Major-General (Horatio) Herbert Kitchener, who had

---

* For details of the Gatling gun and other developments in small arms, see 'Notes and Further Reading'.

served as a staff officer on Wolseley's campaign, and who was now sirdar (commander-in-chief) of the Egyptian army, was given a clear enough military task: defeat the Mahdists and re-annex the Sudan. Perhaps because he was a sapper, and certainly because of Wolseley's experience of inadequate logistics, he determined on a very deliberate advance up the Nile, consolidating his gains in Mahdist territory and building up his lines of communication before going further. In this there was nothing new – the approach was both Marlburian and Wellingtonian – but it had not been practised much in the lifetime of Kitchener's contemporaries. The sharp checks at the beginning of almost every colonial campaign to date had, however, reminded those with a will to see (and Kitchener's cold ambition made him a very careful observer) that there was a balance to be struck between the bold manœuvre on exterior lines, which can knock the enemy off balance and leave him open to defeat by smaller numbers, and the application of mounting pressure from the methodical build-up of strength along interior lines. For although the latter risks allowing the enemy freedom of action, the pressure can be steadily increased until it becomes irresistible. The modern colloquialism is 'the more you use, the less you lose'.

And Kitchener intended using everything he could lay his hands on. He began his advance from Wadi Halfa on the Egyptian–Sudanese border in March 1896 with the initial aim of clearing Dongola province. His force of 15,000 was predominantly Egyptian, with British officers in key appointments, plus six Sudanese regiments loyal to Cairo and a stiffening of British battalions (there would be 8,000 British troops by the end of the campaign). By September, after two minor actions at Akasha and Firket, he had secured Dongola. And here he made his base for a whole year while behind him a railway was built from the border to the junction of the Nile and the Atbara River just south of

Berber. Some 400 miles of track was to run through the desert to the bend of the Nile at Abu Hamed (which he took in August 1897), and then along the east bank to within 200 miles of Khartoum, where the river once more becomes navigable to gunboats during the flood season – for which they would now have to wait. Telegraph lines from Atbara to Cairo, and to several points east and west, made for timely and efficient movement of men and materiel and allowed Kitchener to rest his men in the rear areas rather than having them all at the furthest point of the advance.

In April 1898 he sent a reconnaissance force to determine the strength of the Mahdist defences south of Berber. These consisted essentially of a *zariba*, or thorn enclosure, with mounds and entrenchments. The force would have come to grief but for the initiative in extricating them of a young cavalry officer, Captain Douglas Haig. As it was, with the information gained Kitchener was able to launch an attack three days later which quickly overwhelmed the defenders, though he came in for criticism for unimaginative tactics and relatively heavy casualties (568 killed and wounded). They might have been even heavier but for the Royal Artillery's preparatory fire, which had evolved some way towards the model which has been used ever since, known as the 'fire plan'.* In Kitchener's fire plan the guns, rather than deploying side by side with the infantry, would fire on the objective from a different position, preferably at such an angle that the bombardment could continue until the infantry had advanced almost on to the enemy position.

The Royal Artillery's field guns had improved out of all recognition since the Crimea. They were Maxim–Nordenfeldts – breech-loading, rifled, highly mobile, and

---

* The NATO standard definition of 'fire plan' is: 'A presentation of planned targets giving data for engagement. Scheduled targets are fired in a definite time sequence. The starting time may be on call, [or] at a pre-arranged time, or at the occurrence of a specific event.'

firing fixed-case mechanically fused ammunition. They could hurl 9 pounds of Lyddite 5,000 yards with great accuracy, aided by optical sights.* The gunners fired direct – observing themselves where their shot fell – rather than indirect, for there was no need of concealment when they vastly outranged the enemy's small arms, and the Khalifa had no artillery (worth the name) of his own. Thus used, these guns wrought a great deal of destruction, although with better coordination the infantry might have been able to get into the *zariba* before the Mahdists had time to recover.

Even so, the British infantry at least were armed with a superior rifle, the new magazine-fed Lee–Metford, with smokeless cartridges and a much greater rate of fire than earlier models (though the Egyptian–Sudanese regiments carried the older single-shot Martini–Henry). The battalions still advanced in tight formations, halting to volley at long range and then firing independently when the Mahdists' Remingtons, captured from the Egyptians in earlier campaigns, took effect at 300 yards, before going the last 50 with the bayonet. Major Ivor Maxse, one of the Egyptian-brigade-majors who later interrogated the Mahdist prisoners, and who would command a corps during the First World War, concluded that 'What beat them was the steady, disciplined advance of our drilled battalions, things which they had . . . not seen nor imagined.' It was, indeed, a rather old-fashioned battle, if fought with modern weapons; and it would not be the last such.†

* Lyddite was a much improved explosive named after the Lydd ranges in Kent (near Sir John Moore's old training camp) where it was first tested.
† In fact the two Egyptian brigades, under command of British lieutenant-colonels, advanced a good deal more imaginatively than the British brigade under Brigadier-General Gatacre, a good trainer of men at the individual level (he was known as General Backacher) but with limited tactical sense despite his long years in India. Two years later, in South Africa, he was removed from command.

But still Kitchener was in no hurry to advance. After the battle at the *zariba* he thinned out his troops, sending the regiments back down the railway to summer quarters where they could rest and train while work continued on his lines of communication. By the middle of July 1898 the railway had at last reached Atbara itself, and Kitchener could begin bringing up his second British brigade, for the river was now high enough for the final push on Khartoum. With these reinforcements came a British cavalry regiment, the 21st Lancers, who were particularly eager for action. Since their re-raising thirty-four years earlier they had scarcely heard a shot fired in anger (their motto, said the wags, was 'Thou shalt not kill'). With them on secondment from the 4th Hussars was the 23-year-old Lieutenant Winston Churchill, who the following year would write a gripping and astonishingly mature account of the campaign, *The River War*. The railway also brought up several gunboats in sections, and more artillery including two massive 40-pounders and mule-drawn Maxim machine guns, which had by now replaced the Gatlings. Kitchener's firepower was now formidable, a match for anything the Khalifa could bring to bear, and heavy enough to batter down the fortifications of Khartoum itself.

Several of the Khalifa's emirs had in fact urged him to quit Khartoum and move west to the country of the Baggara, the tribe of which he was also leader, but he refused: he was first the leader of the whole country, he protested; and besides, he could not abandon the capital and the tomb of the Mahdi. He therefore decided to strengthen the Khartoum garrison and concentrate his field forces at the village of Omdurman on the opposite, western bank of the Nile. This, however, left the river free for Kitchener's gunboats to sail right up to the walls of the city.

By the middle of August, Kitchener's advance guards were within sight of the capital, their methodical advance marked by numerous supply depots, and his Anglo-Egyptian army had grown to 26,000. Captain Douglas Haig described the sight that greeted the cavalry as they crested the Kerreri ridge on the morning of 1 September:

> On reaching the hill a most wonderful sight presented itself to us. A huge force of men with flags, drums and bugles was being assembled to the west of the city; the troops formed on a front some three miles long, and as each body or 'roob' ['quarter'] was complete, it commenced to move northwards. With my glass I saw that they were moving very fast indeed. To my mind we were wasting time where we were.

Not wasting any more time, they withdrew south into Kitchener's own *zariba* a few miles north of Omdurman, which was supported by the gunboats.

Kitchener wanted to lure the Khalifa into a fight in the open, drawing him on to his superior firepower. As Haig and the cavalry observed from the Kerreri ridge, he played a clever psychological game by embarking a battery of 5-inch howitzers (high-angle guns which could drop shells over the walls) and despatching them with 3,000 irregulars under Major Edward Montagu-Stuart-Wortley (who would command a division at the Somme) to begin the bombardment of Khartoum – with specific aim at the Mahdi's tomb, striking symbolically at the heart of Mahdism. As the Khalifa's men watched helplessly, shells began tearing holes in the great white dome towering 90 feet above the tomb. One of the Coptic clerks wrote afterwards that 'There was a natural and embarrassed silence in the ranks of the army.'

There was also dissent among the emirs: should they attack, or await attack? And among those who urged attack there were some who wanted to do so at once and others who wanted to wait for darkness. The problem was the gunboats (and soon the other artillery, for the east bank quickly fell into Kitchener's hands): they made Omdurman untenable as a defensive position by day, and at night their searchlights swept the approaches to the *zariba*. In the end the Khalifa, who had never seen a living white man, let alone the effects of the Lee–Metford and the Maxim, was stung into action by his eldest son's jibe that to attack other than in daylight would 'be like mice and foxes slinking into their holes by day and peeping out at night'. They would attack at dawn on 2 September – almost incredibly, into the rising sun.

Even so, the Khalifa's plan did not lack subtlety. He would mount direct attacks against the *zariba* from due west and in the southern flank, supported by fire from the *mulazimin* (guard) who would take the Kerreri hills on which sat the Egyptian cavalry and the Camel Corps. With some 14,000 troops in the assault, he believed he had every chance of breaking into the *zariba* defences. But if he did not, he would fall back, luring the infidel out of the *zariba* and on to the plain, whereupon his reserve of 12,000 mounted men and foot soldiers concealed behind the Jebel Surgham would take them in the flank, while the *mulazimin* streamed down from the Kerreri hills to complete the destruction.

Kitchener's cavalry patrols (which included Churchill) reported the Mahdists' movement at first light – a host of mounted men and foot soldiers. In the 1972 film *Young Winston*, Churchill himself brings the information to Kitchener who is on the march with his army towards Omdurman. In reality, Kitchener's infantry were already stood to in the *zariba* with five brigades deployed forward on the mile-long perimeter, one in reserve and the artillery interspersed the length of the defences. But then *Young Winston* was one of Richard Attenborough's films. Zoltan Korda's 1939 epic *The Four Feathers*, an adaptation of A. E. W. Mason's novel, is a far more accurate portrayal – unsurprisingly, perhaps, for the screenplay was by R. C.

Sherriff, a decorated former infantry officer and author of *Journey's End*. In Korda's film a young soldier watching the great 'dervish' host coming on relentlessly asks the older soldier next to him, anxiously, 'When do we open fire?', to which the old sweat replies, 'When we're told!' Masterly dialogue, and the very essence of fire control.

At 6.45 the Maxim–Nordenfeldts of 32 Battery RA opened up, sights and fuses set for 2,800 yards. The Grenadier Guards volleyed shortly afterwards at 2,000 yards, the extreme of the Lee–Metford's range, and the Maxims burst into chattering life at about 1,700 yards (a fraction short of a mile). The bulk of the infantry joined in at 1,500 yards, and at about 800 most of the battalions ordered rapid fire.

The gunnery and small arms produced the most deafening noise that any soldier had ever heard – even those at Waterloo – for the high-velocity weapons packed huge explosive power in the breeches, and the fire was continuous. And suddenly, as if crazed by it, the Baggara horsemen, thousands of them tight-packed, broke from the great press of advancing Mahdists and charged flat out. Every rifle in the *zariba* turned on them, and every man and horse was brought down.* Not a single 'dervish', mounted or on foot, reached the perimeter, while the Khalifa's artillery, out of range, could not inflict a single casualty on the defenders. On the Kerreri hills the attack by the *mulazimin* was cleverly drawn away by the admirable Colonel Robert Broadwood (who would command a division on the Western Front and die of wounds in 1917) and his Egyptian cavalry, taking them completely out of the battle. The Khalifa's plan was fast unravelling.

With the initial attack repulsed, Kitchener now ordered a sortie by the 21st Lancers – four squadrons, some 320

---

* Interestingly, Adam Smith in *Wealth of Nations* gives the noise of battle as one of the reasons why warriors had to be regimented.

troopers – to harass the retreating Mahdists and head them off from Omdurman. In *Young Winston* the Lancers trot to their task with parade-ground smartness. Churchill, in *The River War*, paints a more realistic picture of troops on active service: the regiment was 'a great square block of ungainly brown figures and little horses, hung all over with water-bottles, saddle-bags, picketing-gear, tins of bully-beef, all jolting and jangling together; the polish of peace gone; soldiers without glitter; horsemen without grace; but still a regiment of light cavalry in active operation against the enemy'. *Brown* figures indeed – khaki: gone were the red of the infantry and the blue of the light cavalry. This was modern war. But would it be modern war for the Lancers?

As they rounded the eastern end of the Jebel Surgham they saw scattered parties of horsemen and foot soldiers. Churchill describes what happened next:

We advanced at a walk in mass for about 300 yards. The scattered parties of Dervishes fell back and melted away, and only one straggling line of men in dark blue waited motionless a quarter of a mile to the left front. They were scarcely a hundred strong. The regiment formed into line of squadron columns, and continued at a walk until within 300 yards of this small body of Dervishes. The firing behind the ridges had stopped. There was complete silence, intensified by the recent tumult. Far beyond the thin blue row of Dervishes the fugitives were visible streaming into Omdurman. And should these few devoted men impede a regiment?

Yet it were wiser to examine their position from the other flank before slipping a squadron at them. The heads of the squadrons wheeled slowly to the left, and the Lancers, breaking into a trot, began to cross the Dervish front in column of troops. Thereupon and with one accord the blue-clad men dropped on their knees, and there burst out a loud, crackling fire of musketry. It was hardly possible to miss such a target at such a range. Horses and men fell at once. The only course was plain and welcome to all. The Colonel, nearer than his regiment, already saw what lay behind the skirmishers. He ordered, 'Right wheel into line' to be sounded. The trumpet jerked out a shrill note, heard faintly above the trampling of the horses and the noise of the rifles. On the instant all the sixteen troops swung round and locked up into a long galloping line, and the 21st Lancers were committed to their first charge in war.

Two hundred and fifty yards away the dark-blue men were firing madly in a thin film of light-blue smoke. Their bullets struck the hard gravel into the air, and the troopers, to shield their faces from the stinging dust, bowed their helmets forward, like the Cuirassiers at Waterloo. The pace was fast and the distance short. Yet, before it was half covered, the whole aspect of the affair changed. A deep crease in the ground – a dry watercourse, a khor – appeared where all had seemed smooth, level plain; and from it there sprang, with the suddenness of a pantomime effect and a high-pitched yell, a dense white mass of men nearly as long as our front and about twelve deep. A score of horsemen and a dozen bright flags rose as if by magic from the earth. Eager warriors sprang forward to anticipate the shock. The rest stood firm to meet it. The Lancers acknowledged the apparition only by an increase of pace. Each man wanted sufficient momentum to drive through such a solid line. The flank troops, seeing that they overlapped, curved inwards like the horns of a moon. But the whole event was a matter of seconds. The riflemen, firing bravely to the last, were swept head over heels into the khor, and jumping down with them, at full gallop and in the closest order, the British squadrons struck the fierce brigade with one loud furious shout. The collision was prodigious. Nearly thirty Lancers, men and horses, and at least two hundred Arabs were overthrown. The shock was stunning to both sides, and for perhaps ten wonderful seconds no man heeded his enemy. Terrified horses wedged in the crowd, bruised and shaken men, sprawling in heaps, struggled, dazed and stupid, to their feet, panted, and looked about them. Several fallen Lancers had even time to re-mount. Meanwhile the impetus of the cavalry carried them on. As a rider tears through a bullfinch,* the officers forced their way through the press; and as an iron rake might be drawn through a heap of shingle, so the regiment followed. They shattered the Dervish array, and, their pace reduced to a walk, scrambled out of the khor on the further side, leaving a score of troopers behind them, and dragging on with the charge more than a thousand Arabs. Then, and not till then, the killing began; and thereafter each man saw the world along his lance, under his guard, or through the back-sight of his pistol; and each had his own strange tale to tell.

Stubborn and unshaken infantry hardly ever meet stubborn and unshaken cavalry. Either the infantry run away and are cut down in flight, or they keep their heads and destroy nearly all the horsemen by their musketry.† On this occasion two living walls had actually crashed together. The Dervishes fought manfully. They tried to hamstring the horses. They fired their rifles, pressing the muzzles into the very bodies of their opponents. They cut reins and stirrup-leathers. They flung their throwing-spears with great dexterity. They tried every device of cool, determined men practised in war and familiar with cavalry; and, besides, they swung sharp, heavy swords which bit deep. The hand-to-hand fighting on the further side of the khor lasted for perhaps one minute. Then the horses got into their stride again, the pace increased, and the Lancers drew out from among their antagonists. Within two minutes of the collision

* A cross-country fence like a hedge, which the horse jumps through rather than over.
† Churchill had been commissioned for barely four years (though he had seen a fair bit of service); nevertheless, the decidedness and authority with which he makes this statement is an indication of his confidence in his own military judgement.

every living man was clear of the Dervish mass. All who had fallen were cut at with swords till they stopped quivering, but no artistic mutilations were attempted.

Two hundred yards away the regiment halted, rallied, faced about, and in less than five minutes were re-formed and ready for a second charge. The men were anxious to cut their way back through their enemies. We were alone together – the cavalry regiment and the Dervish brigade. The ridge hung like a curtain between us and the army. The general battle was forgotten, as it was unseen. This was a private quarrel. The other might have been a massacre; but here the fight was fair, for we too fought with sword and spear. Indeed the advantage of ground and numbers lay with them. All prepared to settle the debate at once and for ever. But some realisation of the cost of our wild ride began to come to those who were responsible. Riderless horses galloped across the plain. Men, clinging to their saddles, lurched helplessly about, covered with blood from perhaps a dozen wounds. Horses, streaming from tremendous gashes, limped and staggered with their riders. In 120 seconds five officers, 65 men, and 119 horses out of fewer than 400 had been killed or wounded.

Death *and* glory. It was, indeed, a precipitate charge, and to no significant tactical effect. The one service the regiment might have performed – to discover the Khalifa's reserve behind the *jebel* – was now out of the question as the Lancers licked their wounds. And although medals were given out afterwards, the commanding officer was censured. The duke of Wellington would no doubt have cursed them for 'galloping at everything'.

But the battle was almost over: Kitchener's force was simply too strong. Even when the 12,000 horsemen of the Black Standard hurled themselves at the army's flank as the regiments surged from the *zariba*, they were stopped by the Martini–Henry fire of the 1st Egyptian–Sudanese

Brigade under the remarkable Colonel Hector Macdonald. A crofter's son who had risen through the ranks (shades of Sir Colin Campbell), as Major-General Sir Hector Macdonald he would command the Highland Brigade in South Africa.* By evening the army was in Omdurman, and two days later Kitchener entered Khartoum in ceremonial procession.

It would take another fifteen months of follow-up operations to finish Mahdism for good, however. In the end the Khalifa and his emirs died on their prayer mats facing Mecca at Omdibakarat, a battle won largely by the Maxim guns and their 600 rounds a minute. But in the eyes of the world the British army had regained its ascendancy. Kitchener too had gained an awesome reputation (and a peerage, as Baron Kitchener of Khartoum – 'K of K'); and Gordon had been avenged. Or, as Churchill puts it with rhetorical flourish, and in cadences that would become familiar to millions of listeners nearly half a century later:

> The long story now approaches its conclusion. The River War is over. In its varied course, which extended over fourteen years and involved the untimely destruction of perhaps 300,000 lives, many extremes and contrasts have been displayed. There have been battles which were massacres, and others that were mere parades. There have been occasions of shocking cowardice and surprising heroism, of plans conceived in haste and emergency, of schemes laid with slow deliberation, of wild extravagance

* Macdonald – 'Fighting Mac' – blew his brains out in a Paris hotel in 1903 after an article appeared in the *International Herald Tribune* saying that he was to be court-martialled for buggery with boys in Ceylon, where he was commander-in-chief (with much speculation on the fact that at fifty he was still unmarried). To the astonishment of everyone, his wife appeared in Paris to claim the body: he had married secretly in 1884, Kitchener disapproving of marriage by his 'Egyptian' officers on the grounds that it would be distracting (a common view in all good regiments even today – though not beyond the age of 27). See 'Notes and Further Reading'.

and cruel waste, of economies scarcely less barbarous, of wisdom and incompetence. But the result is at length achieved, and the flags of England and Egypt wave unchallenged over the valley of the Nile.

And the lessons? The 'River War' showed the army almost everything it needed to know about war in the future, except the use of aircraft (it was curious that Kitchener did not use observation balloons: they had been used in the first Sudan campaign and in Bechuanaland in 1885, and the Royal Engineers' Balloon School had opened at Chatham in 1888). But although many individual lessons were absorbed, in logistics especially, and small-arms and artillery improvements continued apace, the totality of these lessons escaped most theorists and practitioners alike. But just as the action at Rorke's Drift had set the gold standard for tenacity, fortitude and courage, so Kitchener's planning and management had set the standard for the *organization* of war. No one as yet seemed to imagine that war in future could be waged on an industrial scale – despite the lessons of the American Civil War to hand.* Even so, some germ of an idea had evidently been planted deep in Kitchener's mind, for when he was appointed secretary of state for war in 1914 he threw over the War Office's plans for the modest expansion of the army based on the Territorial Force and called instead for volunteers for 'New Armies' – an astonishingly prescient and bold course, for Britain had never before improvised a mass army.

But in the decade and a half between the end of the Sudan campaign and the outbreak of the First World War – when, in the words of the official history, the British Expeditionary

* That war was dismissed as irrelevant, a contest between militias, conducted by bumbling amateurs. It was the Franco–Prussian War, five years afterwards, in which they believed the lessons lay – and those principally in the area of improved small arms.

Force sailed for France as 'the best-trained, best-organized and best-equipped British army that ever went forth to war' – the army would suffer perhaps its greatest shock, from which it would emerge with an unprecedented passion for modernization. And again it would be a lesson taught by 'the cheaper man' – a mere 50,000 armed Dutch settlers or *boers*.

19

No End of a Lesson

South Africa, 1899–1902

AS QUEEN VICTORIA APPROACHED HER DIAMOND JUBILEE SHE HAD ruful cause for satisfaction in her earlier warning to Gladstone that the nation had to be prepared for attacks and wars, somewhere or other, 'CONTINUALLY'. The southern part of the African continent was as troublesome as the northern, and just as the nineteenth century had opened with fighting at the Cape, with British troops seizing the Dutch colony, so it would close with fighting on the Cape Colony's vastly extended borders.

Those Dutch settlers ('Boers') who could not accept British rule had migrated during the course of the century, first east along the coast towards Natal, and then north towards the interior where they established two independent but linked republics – the Orange Free State and the Transvaal.* The Cape Colony was expanding, too, and in

---

* Transvaal: literally 'across/beyond the Vaal', a tributary of the Orange River.

the footsteps of the Boer 'fore-trekkers'. Britain annexed Natal in 1845 but recognized the independence of the two new Boer republics in the 1850s, though ambiguously in the case of certain rights, notably over foreign policy.* But the discovery of diamonds in 1867 near the Vaal River changed the situation, triggering a 'diamond rush' that quickly turned Kimberley into a town of 50,000. In 1871 Britain annexed West Griqualand, the site of the bonanza, although the Boers disputed the claim since it lay inside what they considered were the natural, even if not recognized, boundaries of the Orange Free State. Six years later she annexed the Transvaal too.

The Transvaal Boers could do little to oppose the annexation since they were hard pressed fighting the Zulu on their north-east border; but in a magnificent example of the law of unintended consequences, they were able to impose terms in the aftermath of the British defeat of King Cetewayo in 1879. After a botched military operation by one of Wolseley's 'Ashanti Ring', Major-General Sir George Colley, resulting in a famously sharp rebuff at Majuba Hill, the British authorities shrewdly cut their losses by abandoning full annexation in exchange for the Boers' formal surrender of foreign policy. However, the discovery of massive gold deposits in 1886 soon put Transvaal back into the strategic limelight.

These annexations were not driven by simple cupidity; they were the logical response to what London and Cape Town perceived as the threat to British power in the region. This was the time of the 'scramble for Africa', with both French and German colonization giving growing cause for concern: gold- and diamond-rich anti-British (even pro-German) Boer republics on the borders of British South

---

* Such as it was, for the two land-locked farming-based nations hardly seemed forces to be reckoned with. Indeed, they were scarcely able to defend themselves against the Zulu.

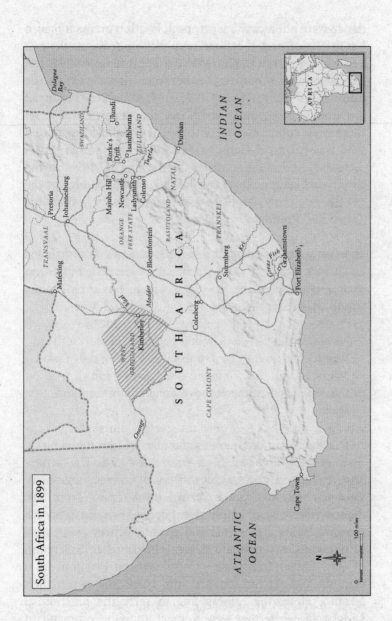

South Africa in 1899

Africa were not a pleasing prospect. Friction between British South Africa and the Boer republics increased, not least because huge immigration into the Transvaal goldfields put the Boers in the minority, and they were reluctant to concede immigrants' rights. From 1896 both republics began an armaments programme which was almost calculated to set them on a course for hostilities, given that the new-found gold was spent on Mauser rifles and Krupp artillery from Germany. And in the southern hemisphere's October spring three years later, no longer willing to concede foreign policy to Britain, and confident of his military capability, Transvaal's President Paul Kruger declared war.

To many a distant observer it looked like a war as ill matched as David's fight with Goliath had first seemed. But David started well: when the Boer War was at last over, Kipling would write:

> Let us admit it fairly, as a business people should,
> We have had no end of a lesson: it will do us no end of good.
> Not on a single issue, or in one direction or twain,
> But conclusively, comprehensively, and several times and again.

In fact the lesson began on the first day of fighting in Natal. Boer mobilization had been straightforward. The fiercely independent burghers, as the citizens of both republics were known, had no regular army units apart from the *Staatsartillerie*. They simply assembled in their districts, each group forming a *komando* and electing officers, each man bringing his own rifle and horses. The day after declaration of war, 12,000 Transvaalers under General Piet Joubert, a cautious old veteran, crossed into Natal and began their advance along the railway line to seize the junction at Ladysmith, where they were to meet up with Commandant

Marthinus Prinsloo advancing from the west with 6,000 Orange Free Staters. From Ladysmith they intended marching on Durban, and with Natal's capital and only port in Boer hands, they were confident that Britain would sue for peace.

At first Joubert met no opposition – there were not even demolitions along the railway – and occupied the town of Newcastle on the fifteenth after a leisurely 30 miles' march. From here he continued south towards Ladysmith in no great hurry (Prinsloo would not even cross the frontier for another three days), taking four days to cover only a further 20 miles. But the British scouts somehow failed to detect the strength of the advance, so that when shots were exchanged north of the mining centre of Dundee on the nineteenth, the commander there, Major-General William Penn Symons, believed the Boers to be a raiding party.

Penn Symons had a strong brigade at his disposal – four infantry battalions (two of them Irishmen, the Royal Irish Fusiliers and the Royal Dublin Fusiliers), three Royal Field Artillery batteries* and the 18th Hussars, plus 400 mounted infantry (MI) and the locally raised Natal Carbineers.† But Dundee was not an easy place to defend, surrounded as it was on all sides by hills from which fire could be poured into the town. One observer wrote that it felt like being in a chamber pot.

---

* The Royal Artillery had recently been divided into three entities: the Royal Garrison Artillery, manning coastal and fortress guns, and howitzers ('siege guns'); the Royal Field Artillery, the largest of the three, with the field, mountain and medium batteries; and the Royal Horse Artillery – drawn from the cream of the RFA – to accompany the cavalry.
† The idea of infantry battalions able to move from one fighting position to another by horse – in the manner of dragoons of old – had developed among regiments stationed in South Africa and Burma, and in 1888 an MI school was set up at Aldershot to train a small cadre of instructors. The Natal Carbineers were also mounted infantry, as were the Boers themselves: they almost invariably dismounted to fight.

Believing he faced only raiders, Penn Symons had withdrawn all his outlying pickets and patrols, so that before dawn on the twentieth 3,500 Boers under General Lucas Meyer – 'the lion of Vreiheid' – were able to climb the dominating Talana Hill undetected, though a mere mile or so north-east of the town. At first light they began shelling the British camp below with their Krupp field guns.

Penn Symons had not abandoned all field discipline, however: the brigade had stood to arms before dawn, with the 18th Hussars saddled and standing by their horses. He determined to attack the hill at once, with a short preparatory bombardment by his three artillery batteries, each of six Armstrong 15-pounder guns. And to cut off the Boer retreat, he ordered the 18th and the MI to make a wide flanking movement to the north of the hill. The plan was optimistic – the Boers outnumbered them and had all the advantage of the ground, as well as outstanding proficiency with the Mauser rifle – but it did at least have the merit of simplicity. Or, as one officer bitterly remarked later, all the originality of an Aldershot field day.

The 18th did, however, manage to get round the hill to a good position to enfilade the Boer line of escape, where they discovered 500 of the Boers' led horses and ammunition details waiting patiently. One of the squadron leaders, Major Percival Scrope Marling, who had won a VC with the 60th Rifles in the Sudan in 1884, urged the commanding officer to open fire at once. But the commanding officer could not bring himself to give the order: to Lieutenant-Colonel Bernhard Möller, who had been commissioned into the regiment as a cornet twenty-seven years before, the notion of firing on riderless horses was abhorrent. Instead, he took his hussars further round the hill, east, across the Boer line of escape. At this point the second-in-command, Major Edward Knox, whose father had re-raised the regiment after the Indian Mutiny, remonstrated with Möller: their orders were to cut off a retreat, not to engage the enemy from the rear, if that was the commanding officer's intention. They were heading into more trouble than they could deal with, Knox argued.

Meanwhile the infantry attack uphill against the Boers had faltered. With great but reckless courage, Penn Symons now galloped forward to urge on his men, and was himself killed. This, or perhaps the old infantry spirit returning, suddenly seemed to galvanize the brigade, which at last began to make progress. But although each man carried the Lee–Metford rifle, they did not advance in movement covered by fire. Instead, led from the front by the regimental officers, they went up the hillside exactly as their forebears had at the Alma heights in the Crimea – almost as if on parade.* And just as the Russians took a terrible toll of the lines of red that day at the Alma, so too did the Boers of the lines of khaki. They might even have thrown back the assault for good had it not been for the Royal Artillery, who managed eventually to keep enough Boer heads down for the infantry to make progress. And as the sweating riflemen at last neared the top, the *komando*, whose tactical doctrine did not include standing their ground and fighting hand to hand, gave up the position and began streaming back down the hill, east towards the waiting ponies.

Möller, now realizing the peril he was in, decided to split his force, sending Knox with Major Scrope Marling's squadron and another due south while he himself took the remainder (with the Maxims) north. Knox managed to manœuvre his half out of danger and rejoin the infantry, but Möller was caught 3 miles on by the Boer second echelon – some 2,500 men under General Hans Erasmus – who opened a very

---

* In fact, at the Alma the battalions advanced in close order and maintained their dressing quite remarkably, whereas at Talana Hill they were in more open order. The difference was quite academic to a Boer sharpshooter, however.

effective rifle fire from a commanding position on Impati Hill 3 miles north of the town. With casualties mounting by the minute, Möller decided to surrender. In a cruel irony, the regiment had wagered with another cavalry regiment on the passage south that they would be the first into Pretoria. They were, but they arrived by train as prisoners of war.

It had not been the best of starts – for either side. British losses were 200 wounded and over 50 killed on the slopes of Talana Hill, a high proportion of them officers, and 250 or so captured or missing, mainly from the 18th Hussars. On the other hand, although the Boers had suffered far fewer casualties they had quit the field. But if the British had shown the Boers that marching through Natal was not going to be easy, the Boers had exposed serious shortcomings in the British ability to coordinate fire with manœuvre. Courageous leadership from the front and the bravery of private soldiers were not going to be enough to defeat Boer marksmanship.

In the days after Talana Hill there was a deal of galloping and skirmishing, and though overall it was inconclusive, three 'names' of the First World War made their reputations. A column of Free Staters had managed to seize the station at Elandslaagte between Ladysmith and Dundee, cutting the Dundee garrison's lines of communication. Brigadier-General Ian Hamilton – who would command the ill-fated Mediterranean Expeditionary Force at Gallipoli – was sent to eject them with a brigade of three infantry battalions and artillery. In support was a cavalry brigade under command of Brigadier-General John French, who would command the British Expeditionary Force in 1914; his brigade major (chief of staff) was Douglas Haig. The Boers at Elandslaagte held their ground rather longer than Meyer's had at Talana Hill, but were eventually driven off with the bayonet, and then harried by sabre and lance – the only occasion during the war on which a cavalry charge made real contact. French was feted for the *élan* and shrewdness with which he had handled the cavalry (in which glory Haig shared), as well as for a deal of personal courage, and Hamilton was lauded for his skill and bravery in manœuvring his infantry. Hamilton, indeed, was recommended for the VC for a second time in his career, but the citation was apparently rejected on the grounds that it might encourage other senior officers to take too many personal risks; Penn Symons had, after all, been killed only the day before.

Fearing that Ladysmith would soon be attacked, the general officer commanding in Natal, Sir George White, who had earned his own VC in the Second Afghan War,* now recalled all outlying troops including the garrison at Dundee for the defence of the crucial railway-junction town. The Boers soon obliged him by famously laying siege to it – and to Mafeking, an altogether less important place strategically, and also Kimberley, a far bigger undertaking just across the Orange Free State border in Cape Colony. It was a terrible mistake. In tying down the better part of their field armies in sieges, the Boers surrendered their most potent weapon – mobility. For as long as the besieged garrisons could hold out, reinforcements could be mustered from both Britain and India.

The sieges had another, unexpected effect too. The British public could identify with them, not least because a siege was easier for war correspondents to describe. It also played to the perceived strengths of the national character – fortitude and dogged courage, the island spirit. The sieges of Gibraltar

* Major-General, later Field Marshal, White was educated at Bromsgrove, a school which can count five VCs won by Old Boys. The Victorian public schools were indeed strong on the ethos of imperial leadership and their medal tables showed it. From 1860 many had a cadet training corps, fore-runner of the Combined Cadet Force (CCF), which to this day is a significant factor in officer recruiting. 2010 was also the 150th anniversary of the Army Cadet Force, from which comes a very high proportion of adult army recruits.

in the eighteenth century had been similarly lauded as heroic episodes at home, as the sieges of Tobruk and Malta would be in the Second World War.

And so at Ladysmith, Mafeking and Kimberley the army was once again feted by a nation fed on rousing despatches by the correspondents of every newspaper in London – soon to include Winston Churchill, by now a civilian, for the *Morning Post* – aided and abetted by shrewd propagandists like Colonel Robert Baden-Powell at Mafeking. The British public had become thoroughly bloodthirsty after Omdurman and now, imbued with all the ill-informed fervour of jingoism, cheered the army with genuine enthusiasm. Volunteers in the militia and yeomanry could not wait to join them.

Another VC was sent from England to take overall command, General Sir Redvers Buller, one of Wolseley's protégés; and with him came 47,000 reinforcements. Buller's original intention was to concentrate his force in an advance on Bloemfontein, the Free State capital; but the precarious situation in Natal now seemed to have priority. He therefore sent Lieutenant-General the Lord Methuen with the 1st Division, 10,000 men, to relieve Kimberley, and ordered Major-General Gatacre, of Sudan fame, to remain on the defensive with a much smaller force just south of the Orange River, covering any Boer advance from Bloemfontein into the Cape. Buller himself would take most of the remainder, about 20,000, to Natal to relieve Ladysmith and Mafeking.

It was a classic example of division instead of concentration of force, and it failed dramatically. First, Gatacre exceeded his orders by trying to capture the important railway junction at Stormberg in the north-east of Cape Colony, where he was repulsed with heavy losses. The following day, 11 December, having forced a crossing of the Modder River, Methuen was defeated at Magersfontein, losing more than a quarter of his assault force. Four days later Buller, with four infantry brigades and one of cavalry, plus six batteries of artillery, attempted to cross the Tugela River at Colenso 20 miles south of Ladysmith. With no room to manoeuvre, and with Boers led by the able General Louis Botha on high ground commanding the drifts, the attack was defeated. Buller's losses were high: several hundred men taken prisoner and 1,000 casualties, 150 of them fatal. Among the dead was Field Marshal Lord Roberts's only son.*

It rapidly became known as 'Black Week', though some honour was salvaged at Ladysmith by Sir George White who, on receiving Buller's message that he could not break through and should surrender, chose to ignore it. In London, however, the government was already taking steps to relieve Buller, sending for the commander-in-chief in Ireland, 68-year-old Lord Roberts of Kandahar. 'Bobs', as he was known to every soldier in the army, accepted the appointment on the day he learned of his son's death. Six days later he sailed from Southampton, putting in at Gibraltar to pick up Kitchener, who was to be his chief of staff, and reaching Cape Town on 10 January.

Here he learned that Buller intended going on to the offensive again in Natal, so he sent instructions that all troops should remain on the defensive until he could assess the situation for himself. But Buller signalled back that the Boers had just made an unsuccessful attempt to take Ladysmith and would therefore be ripe for counter-attack. Reluctantly, Roberts allowed him to go ahead. The result was the murderous battle of Spion Kop, with a loss of almost 2,000 men, and a further defeat at Vaal Kranz a fortnight later.

Spion Kop was perhaps the nadir of the regular British

---

* Lieutenant Frederick Roberts of the King's Royal Rifle Corps, aged 27, was posthumously awarded the VC – one of only three sons of VCs ever to receive the medal.

army, for it had neither the technical means nor the tactics to cope with long-range artillery and magazine-rifle fire, though these had been in service for over a decade. Smoke rounds had not yet been developed, and at Spion Kop – just as at Talana Hill – use of ground was unimaginative and there was little coordination of fire and movement. Reconnaissance was also perfunctory, and communications not up to the demands of a fluid battle in the open (the heliograph and semaphore flags proving hopelessly in-adequate), so that commanders had only the most tenuous grip on the action. What was more, at Spion Kop junior leadership – the boast of the army – failed when the situation became confused.

The battle came about as Buller attempted to cross the Tugela to relieve Ladysmith. He had a much stronger force than at Colenso, Lieutenant-General Sir Charles Warren's 5th Division having arrived from Britain, and this time he had chosen to attack further west opposite the Rangeworthy Hills, of which Spion Kop was one. There is still dispute as to which of the two main forces engaged was the diversion and which was the force intended to break through to Ladysmith 20 miles or so to the north. Major-General Neville Lyttelton, who had commanded a brigade at Omdurman and would be appointed first Chief of the General Staff in the post-war reorganization of the army, was to cross the Tugela with his brigade of rifle regiments* at Potgeiter's (sometimes Potgieter) Drift to the east of the main attack where a loop in the river protected the crossing from enfilade fire. To divert Boer attention, Warren's division (13,000 men and 36 guns) was to cross further west at Trikhardt's Drift and advance on to the

---

* All infantry regiments were now equipped with the rifle, but the old distinction remained. It is one of the glorious anomalies of the regimental system that the obvious title of this brigade – the Rifles Brigade – could not be used because it would have been almost identical to the regiment of that name, one of whose battalions was in the brigade.

Rangeworthy Hills, while Buller, with the reserve of 8,000 men and 22 guns, would then cross at Potgeiter's Drift and follow up Lyttelton, who would push north for Ladysmith. However, the alternative suggestion is that Lyttelton's force was to be the diversion, and that Buller expected Warren's division to break through the Boer defence line and drive on for Ladysmith.

Warren's appointment to command was controversial, almost worthy of the Crimea. He was a sapper primarily interested in survey, a veteran of many a mapping and archaeological expedition. From 1886 to 1888 (during the 'Jack the Ripper' murders) he was Commissioner of the Metropolitan Police, and he had not heard a shot fired in anger since a native war in the Transkei in 1878. In 1898, after command in Singapore and the Thames District, he retired, but was recalled to the colours at the age of 59 when war broke out the following year, possibly on account of his topographical knowledge of South Africa. It could not have been for his experience of command in the field, nor for the want of other capable commanders, and it has led to suggestions of masonic influence.

He began crossing the Tugela on 17 January, and two days later was still bringing his column across the river while his artillery bombarded the ridge opposite Trikhardt's Drift. But General Botha, who had thrown back Buller's first attempt to cross the Tugela at Colenso the previous month, saw the threat to his extreme right flank and brought up reinforcements and guns to the hills commanding the drift. On 23 January, losing patience with Warren's lack of urgency, Buller rode forward and ordered him to begin the attack.

Warren saw Spion Kop as key to the Rangeworthy position and ordered the 44-year-old Major-General Edward Woodgate, who had fought in the Abyssinia campaign and the Ashanti and Zulu wars, to take it with his Lancastrian

Brigade,* reinforced by an unhorsed company of MI under Lieutenant-Colonel Alexander Thorneycroft and a company of sappers to dig trenches at the peak. The brigade climbed the steep sides of the Kop that night, arriving at dawn at what in the mist they mistook for the top. Accordingly the sappers began to entrench – with great difficulty, for the ground was rocky – while the rest of the brigade lay down (in an area not much more than an acre). Why they did not all put their backs into building *sangars* has never been adequately explained,† though the infantrymen had left their picks and shovels behind since these would have made for a noisy climb, and they believed that at the peak they were beyond the reach of Boer marksmen (though not beyond the reach of the Boer guns).

In fact, as they had been approaching the false crest the Boer picket had abandoned the hillside and rushed to warn Botha. The Lancastrians had not been long at their ease when every Boer gun in the area began to fire on them. And as the mist lifted, they saw to their horror that the true summit was several hundred yards off and in the possession of Boer marksmen – who now poured deadly accurate fire into what would become known as 'the murderous acre'. What few entrenchments the sappers had been able to dig were too shallow to give shelter or allow return of fire. Woodgate was soon dead, and Lieutenant-Colonel Malby Crofton, commanding officer of the general's former regiment, the King's Own, took temporary command and signalled that they were in trouble

---

* Comprising the 2nd King's Own (Royal Lancaster) Regiment, 2nd Lancashire Fusiliers, 1st South Lancashire Regiment and the 1st York and Lancaster Regiment.
† Afghan tribesmen would snipe from rocky crevices camouflaged and fortified with slabs of rock. The Afghan word for these 'tiny forts' is *sangar*.

and needed reinforcements.*

Warren now ordered Major-General John Talbot Coke's 10th Brigade (2nd Dorsets and 2nd Middlesex) to reinforce the Lancastrians, though Coke himself had a severe leg injury and could not reach the top. Convinced, too, that Crofton was being defeatist, Warren sent a message promoting Lieutenant-Colonel Thorneycroft of the MI to local brigadier-general. Thorneycroft's first job would be to stop the Lancashire Fusiliers from surrendering: they had already taken many casualties, their ammunition and water were running low, and no one could see a way out of their predicament (yet fifteen years later this same regiment would win six VCs in a morning during the Gallipoli landings).

Lyttelton's brigade of rifle regiments, which had crossed at Potgeiter's Drift without difficulty, now launched a supporting attack: the 2nd Scottish Rifles (the Cameronians) were ordered to climb Spion Kop to join Thorneycroft's troops while 1st Battalion the Rifle Brigade made a demonstration

---

* Crofton had heliographed that Woodgate was dead, and requesting reinforcements. The message got changed slightly in transmission (the heliograph was shattered by a piece of shrapnel, and the rest of the message was sent by semaphore flags), and to Warren it sounded less than resolute. After the battle, Crofton was placed on half-pay and sent home. The senior major (Yeatherd) would take command, but he too was to be killed a month later. The King's Own would lose 8 officers (4 dead, 4 wounded) at Spion Kop, with 56 other ranks killed and 90 wounded. In the fighting which followed Spion Kop, between 13 and 27 February, they would lose 2 officers and 28 men killed, and 8 officers and 145 men wounded – two-fifths of the battalion's fighting strength, and a particularly heavy toll of officers. Yeatherd would be succeeded by Lieutenant-Colonel John Moore Gawne, brought out from England; in December he too would be killed. Such was the price of good regimental leadership – and in the finest tradition of the regiment, seemingly, for at Sebastopol the commanding officer of the King's Own had been killed in the thick of the action. The adjutant that day at Spion Kop, Captain Dykes, who would win the DSO in the 'murderous acre', would be killed while commanding the 1st Battalion in the opening moves of the First World War (see ch. 21).

before the 'Twin Peaks' to the east, which helped the 60th Rifles to fight their way up the precipitous hillside of the Twin Peaks and take the summit of the ridge. But the fiercely hot day came to a close with Spion Kop still under artillery fire and Thorneycroft and his men at the end of their tether: Warren had sent him no further orders during the day, and failed to tell him that substantial reinforcements were on their way.

As darkness fell, the Boers at the summit of Spion Kop, dismayed by the abandonment of Twin Peaks, began slipping away too. But Thorneycroft did not realize this; so, instead of following up the retreating enemy, he resolved to withdraw with the demoralized remnants of the Lancastrian Brigade, the Middlesex Regiment and the Scottish Rifles. As he did so the reinforcements began arriving in the crowded acre and a vigorous dispute followed, their commanding officer insisting that the hill be held, Thorneycroft – acting brigadier-general – adamant that it could not. Down the hillside scrambled the unhappy force, and at dawn next day the astonished Boers reoccupied the peak.

Warren now recrossed the Tugela with his tail between his legs; Buller's second attempt to force his way through to Ladysmith had failed abysmally. Warren was sent back to England, and Buller was sidelined (a VC could not be humiliated entirely). Thorneycroft rejoined his MI, which he commanded during the later operations in the Eastern Transvaal, and was promoted colonel and appointed a Companion of the Bath. The scapegoating had for once found the right level.

One of Buller's 'gallopers' that day had been Winston Churchill, by now back in uniform with the hastily raised South Africa Light Horse. He made several ascents of Spion Kop carrying messages, writing afterwards in his report: 'Corpses lay here and there. Many of the wounds were of a horrible nature. The splinters and fragments of the shells had torn and mutilated them. The shallow trenches were choked with dead and wounded.'

It had indeed been a terrible drubbing, a stark and brutal reminder that, in Kipling's words, to the bullet 'All flesh is grass'. But after Spion Kop the steel began to re-enter the army's soul – as if the hill were a sort of watershed. The army carried the magazine rifle and wore khaki, but still at heart too many of them carried the musket and wore red. The sheer killing power of the rifle and artillery and the futility of close-order movement had been demonstrated once and for all. The British soldier and his officers knew now that what the bullet and the shell dealt to the dervish it would deal also to them. When the reckoning came after the war, Spion Kop – and the battles of 'Black Week' – would stand as an irrefutable pointer to the way ahead.

The army had seen it, the nation had seen it, the whole world had seen it. 'The vast majority of German military experts believe that the South African war will end with a complete defeat of the English,' wrote Count von Bülow, the German foreign minister, gleefully after Black Week. 'Nobody here believes that the English will ever reach Pretoria.' But news of Buller's further reverses did not dull Lord Roberts's determination to recover the initiative. And Boer jitters would soon work in his favour, for despite the British set-backs in Natal, the Free Staters were expecting a heavy offensive towards Bloemfontein from Colesberg, which the newly promoted Major-General John French and the cavalry division had all but taken back from a Boer commando the month before. Roberts now concentrated some 37,000 men plus 12,000 horses and twice as many more transport animals south of Kimberley while keeping up a pretence of advancing along the Colesberg axis. Within a month he had lifted the siege of Kimberley, and by a broad flanking move-ment and some spirited action by the cavalry division he outmanœuvred General Cronje (commanding in the west), bringing him to battle at Paardeberg on 18 February and forcing his complete surrender a week later.

On 1 March Buller was at last able to lift the siege of Ladysmith, and after a few brisk skirmishes Roberts entered the Free State capital on the thirteenth. Here he rested his forces to gather their strength (enteric fever was sweeping the army) and take in more reinforcements from Britain, before continuing the advance on the capital of the Transvaal, Pretoria, which he entered in procession in early June. On the way he had taken Johannesburg and drawn off men from the siege at Mafeking (which was finally relieved on 17 May after holding out for 217 days). On 28 May Britain formally annexed the Orange Free State; in August the last remaining Boer field force under Louis Botha was defeated, and on 25 October the Transvaal too was under the British flag.

The war had begun in a way that had become all too familiar: faulty strategy, faulty campaign planning, faulty tactics. It had then proceeded with the equally familiar combination of heroism (Queen Victoria ordered that a regiment of Irish Guards be formed to mark the bravery of the Irish regiments) and medals (seventy-eight VCs were awarded during the course of the war), with the self-healing regimental system averting total catastrophe before a capable pair of hands got a grip, took the fight back to the enemy and beat him. Roberts had done just that, and having done so he now handed over to his chief of staff, Kitchener, and returned to England where he was to replace Wolseley as commander-in-chief. But in the same way that the brilliant US-led invasion of Iraq in 2003 brought down Saddam Hussein and his army in short order, only to be followed by a wholly unplanned-for insurgency, so now the war in South Africa – which everyone assumed was over – took an unexpected turn.

A few months after Roberts's return to England, the most able of the Boer field commanders, Christian de Wet, began to wage guerrilla war. Over the following eighteen months he and others who had vowed to fight 'to the bitter end' would draw in the resources of a great part of the British Empire before their final defeat – nearly half a million men, indeed, and a great deal of treasure. Kitchener's response was as ruthlessly efficient as his Nile campaign had been. First he isolated the commandos from their sources of intelligence and supply – their farms, scattered the length and breadth of the veldt. Then he began a scorched earth policy – farm-burning – and interned the families of those 'on commando' in what were called concentration camps. Kitchener's true purpose in these camps remains ambiguous. They were meant as a means of isolation – and, indeed, some of the Boers were glad to be relieved of the responsibility of feeding and protecting their families while on commando* – but the conditions inside the camps rapidly deteriorated, and many of the inmates died from disease. There was outrage in the more liberal press in London, and indeed in Parliament, but Kitchener was undismayed, just as he was undaunted by the vastness of the veldt and the elusiveness of the *komandos*. Erecting thousands of miles of barbed wire and chains of blockhouses each garrisoned by a dozen or so men on the same principle as Hadrian's Wall, he conducted massive sweeps of the country in between, driving the *komandos* on to these stop lines in a constant war of attrition – techniques of containment, denial and harassment that would become familiar to later generations of British soldiers in Malaya, Cyprus, Kenya and southern Arabia.

When the war was at last truly over there was an inquiry into

---

* The word was consciously adopted by the army, and then the Royal Marines, for the raiding forces formed in 1940.

what had gone wrong. How had a rag-tag collection of settlers been able to humiliate the British army in a whole series of field battles? And how had it taken nearly half a million men to bring the war to a close (the 'bitter enders' fighting on until May 1902)? For there were gloating eyes in Europe and further afield: was this all it took to bring the British Empire practically to its knees? There were covetous as well as gloating French eyes on colonial Africa, too, and opportunistic Russian eyes on the North-West Frontier of India. Most menacing of all, there were calculating German eyes scrutinizing every detail of the fighting so that when it came to the contest in Europe – as the German general staff believed it must – they would be able to deal decisively with the British army if it should choose to set foot on the Continent.

When the contest finally came it would take more than Bismarck's policeman to deal with the British Expeditionary Force; but, observing the débâcle of the war in South Africa, Berlin was convinced it would not take a very great deal.

20

The Dynamic Decade

England, 1902–14

NOT ALL THE FOREIGN EYES WATCHING THE ARMY IN THOSE LAST years of the nineteenth century had been hostile. On 27 February 1890 the *New York Times* carried a piece about army reform:

> London. The report of the Marquis of Hartington's commission on the army and navy will appear next week. It is severe on the War Office system, which it finds to be extravagant, cumbrous, and inefficient. It dwells with emphasis upon the fact that the responsibility of all heads of bureaus is only nominal, and it proposes to abolish the position of Commander in Chief, now held by the Duke of Cambridge, and to substitute a military chief of staff, to be assisted by an advisory, to whom all heads of departments shall be directly responsible . . .

Nothing had come of the recommendations, however, and so the army had stumbled into the Boer War in much the way that it had into the Crimea. The Crimea had been a loud wake-up call; but the country had merely pressed the snooze button. And it had ignored the mild repeating calls in the years that followed. The war in South Africa had finally woken everyone with a start – with alarm, indeed. But in 'The Lesson', Kipling was as consolatory as he was excoriating:

> It was our fault, and our very great fault—and now we must
> turn it to use.
> We have forty million reasons for failure, but not a single
> excuse.
> So the more we work and the less we talk the better results we
> shall get—
> We have had an Imperial lesson; it may make us an Empire
> yet!

And empire was part of the problem, as well as being part of the answer. How was the Empire to be protected, the home-land secured and an army sent to the Continent all at the same time? The proponents of an ever stronger navy, the 'blue-water school', harked back to Admiral St Vincent in the invasion scares before Trafalgar, who, addressing the House of Lords, had declared with matchless irony: 'I do not say that the French cannot come; I only say that they cannot come by sea!' Except that after the various points of colonial dispute with France had been settled in 1904 (the 'Entente Cordiale'), it was the Germans who would or would not come by sea, the French now allies instead. Meanwhile there was a new crop of invasion novels warning of the country's unpreparedness for war with the Hun, of which Erskine Childers's *The Riddle of the Sands* (1903) was probably the most influential, and possibly the most readable.

Those who argued against the blue-water school that the Royal Navy alone would be insufficient safeguard in a war with Germany, and that the army must be strengthened, were derided as the 'blue funk school'. Inevitably a compromise was reached, which was itself another hark back to the invasion scares of a century before: when the regular army was deployed to the Continent, home defence would become the responsibility of the auxiliary forces. Meanwhile, imperial defence would continue to be a primary function of the regular army, boosted by better coordination of colonial resources through a Committee for Imperial Defence and the build-up of the colonies' own forces, many of which had shown exceptionally well in South Africa (the Australians and Canadians had fielded superb mounted troops). But Britain's auxiliary forces would also have to be modernized. Although they, too, had performed well at times (indeed, many of those who gave evidence to the subsequent inquiries spoke of the superior physical and intellectual condition of the volunteers over that of the regular soldiers), their organization and training were haphazard.

But all this reorganization could scarcely come about without the sort of radical reform at the top, at the War Office and the Horse Guards (as the commander-in-chief's headquarters was still known), that Lord Hartington's 1890 report had recommended, and which a further inquiry chaired by Lord Esher now urged.* In 1904 the post of commander-in-chief was duly abolished and replaced by a chief of the general staff (soon redesignated chief of the imperial general staff (CIGS) to reflect his primacy in planning with the staffs of the colonial and dominion forces); an Army Council was formed, modelled on the Board of Admiralty – a single collective body to determine policy, thus ending the confusion of responsibilities between the civil and military; and a professional army staff was established to support the military members of the Army Council.

The duties of the new general staff were to be shared

---

* Report of the War Office (Reconstitution) Committee, 1904.

among a director of military operations, a director of staff duties (responsible for all matters touching on the organization of the army, including the 'order of battle' – in particular where units were to be stationed and under what command arrangements) and a director of military training. The adjutant-general's staff (as opposed to the general staff) was given overall responsibility for the soldier's welfare and discipline, medical services, casualties, and what is now loosely called 'conditions of service'. Under him was a director of recruiting, a director of personal services, a director-general of medical services and a director of auxiliary services. Apart from manufacture, or what today is called procurement, all supply, accommodation, transport and movement matters became the responsibility of the quartermaster-general, whose staff consisted of a director of transport and remounts, a director of movements and quartering, a director of supplies and clothing, and a director of equipment and ordnance stores. The Master General of the Ordnance's staff was principally concerned with procurement of materiel – weapons, ammunition and warlike stores – with a director of artillery, a naval adviser and a director of fortifications and works.*

This division of responsibility was replicated at each subordinate level where appropriate, so that at the lowest level of headquarters, the brigade, the functions of the general staff were coordinated by a brigade major (today called a chief of staff), and those of the adjutant-general and the quartermaster-general were combined in an appointment with the longest title in the army: the deputy assistant adjutant and quartermaster-general, or DAA&QMG (usually abbreviated

---

* These arrangements endured until the centralization of the three separate service staffs, beginning in the mid-1980s, although their functions remain somewhere or other in the residual 'one-army' staff, as it is now called, or in the 'purple' (tri-service) staff – even the director of remounts.

to 'DQ' – and also a major) – today called the deputy chief of staff. Indeed, the system as a whole remains much as it was first devised.

The idea of drawing these lines of responsibility was to establish an administrative framework of military districts, each commanded by a lieutenant- or major-general, which would leave commanders of field units free to train for war. And it rapidly began to bear fruit, thanks to the intellect and energy of the war minister in the new Liberal government, Richard Burdon Haldane (1905–12), and to Major-General Douglas Haig, who as director of military training and then of staff duties was a prime mover in the changes.

But Haldane's biggest challenge was how to 'echelon' the regulars and auxiliaries at home – that is, how to organize and employ them in relation to each other. He decided to reduce the three existing categories – regulars, militia and volunteers – to two, and organize them into the 'Field Force' (of three army corps and a cavalry division, with the implied intention of continental service), and a 'Territorial Force' for service at home and, if the situation warranted it, and after further training, abroad. He set out his plan in a memorandum of February 1907:

> The Field Force is to be so completely organized as to be ready in all respects for mobilization immediately on the outbreak of a great war. In that event the Territorial or Home Force would be mobilized also, but mobilized with a view to its undertaking, in the first instance, systematic training for war ... The Territorial Force will therefore be one of support and expansion, to be at once embodied when danger threatens, but not likely to be called for till after the expiration of the preliminary period of six months.

The Territorial Force (TF) – renamed Territorial Army after the First World War – came into being in 1908, though with two critical concessions to the widespread opposition of the 'old and bold' – the retired officers who ran the county militia 'establishment'. First, the militia would not be entirely done away with: men in the county cadres would have the option of joining a 'special reserve' to supplement or support the Field Force. Second, the TF would be for home service only, administered by county associations, with its funding 'ring-fenced'; Haldane thought that the question of foreign service was a bridge better crossed when they came to it. The TF was, however, organized on a more complete all-arms basis than hitherto, with artillery, engineers and supporting services. On paper, it amounted to fourteen infantry divisions and fourteen mounted yeomanry brigades, with an overall strength of approximately 269,000. And to help over-come a shortfall in officer recruiting in both the regular and reserve armies, an officer training corps (OTC) was estab-lished with a senior division in eight universities and a junior division in the public schools.* If the TF (and for that matter the Special Reserve) never quite met its target strength or training standards, there was at least a coherent system; and as the years passed, everyone grew accustomed to the idea of 'home service only', seeing no prospect of any bridge to cross.

While this reform of the system as a whole was going on, in the field army there was unparalleled activity. For the first time in its history the army was provided with official manuals on how war was to be made: *Field Service Regulations Part I – Operations*, and *Part II – Organization and Administration* (1909). These were to be the basis of campaign planning for both British and imperial forces, and without them the mobilization of the Empire in the First World War could scarcely have been conceived. The Staff

* There had been various OTC schemes before the Boer War, none very satisfactory.

College was at once reinvigorated: with such clarity at the top, at last it now had a real sense of purpose. In turn, each arm and service received a new training manual too (the word 'training' was itself a significant change from the former 'drill'), and reworked its tactics and procedures.* The Esher Report had sought to decentralize administration from the War Office not simply to improve efficiency but because 'if the Army is to be trained to exercise the initiative and independence of judgement which are essential in the field, its peace administration must be effectively decentralized. The object should be to encourage the assumption of responsibility as far as possible.' Training thus became the direct responsibility of the officers who would command in the field. Hitherto in the cavalry, for example, training had been the business of the adjutant and the riding master; it now became unequivocally that of the squadron and troop leaders. And since the majority of officers and NCOs throughout the army wore the South Africa medal, there was no lack of understanding why the change was necessary. Any who doubted it had only to glance at the *Report of His Majesty's Commissioners on the War in South Africa* (1903), which found serious fault with every branch of the army save for the Army Service Corps.

The scale and zeal of change was at least comparable with the Cromwellian reforms and the formation of the New Model Army. Ironically, in fact, the New Model's red coat was now abandoned entirely except for ceremonial parades: soldiers now both trained and walked out in khaki service dress, which was in itself a psychological advance – in the

---

* The terms 'arms' and 'services' were always a little loosely employed: 'arms' were the infantry, cavalry and artillery, while 'services' embraced the engineers, medical, supply and administrative corps. Today the terms used are combat arms (infantry, armour, army aviation), combat support (artillery, engineers, signals, intelligence) and combat service support (supply, maintenance, medical and administrative).

infantry, especially. Marksmanship rather than drill became their obsession, for as old President Kruger had said – and as the infantry had learned to their cost – 'The Boers can shoot, and that is everything.' Indeed, the infantryman's pay was now linked to his skill with the improved Lee–Enfield rifle, a shorter version of the one that had entered service in South Africa and one that the cavalry could carry too, instead of the less powerful carbine. The 'rifle, short, magazine, Lee–Enfield' – the SMLE – had a fast-operating bolt action and a ten-round magazine which could be quickly recharged by clips of five. Fifteen aimed rounds per minute became the standard rate on the order 'rapid fire', though well-trained riflemen could manage twenty to thirty rounds a minute, making the Lee–Enfield the fastest bolt-action rifle of its day. In 1914 Sergeant-Instructor Snoxall of the School of Musketry at Hythe set a record which still stands for a bolt-action – thirty-eight rounds in a 12-inch target at 300 yards in one minute. Little wonder that in the opening weeks of the First World War German field intelligence reported that the establishment for machine guns in a British infantry battalion was twenty-eight. It was, in fact, two.*

Gone with the red coats was close-order drill for fire and movement alike (in the fighting at the crossing of the Tugela, the Irish Brigade had advanced shoulder to shoulder). As early as 1904 a foreign military observer noted the change in tactical movement: 'In their manoeuvres the British infantry showed great skill in the use of ground. Their thin lines of khaki-clad skirmishers were scarcely visible. No detachment was ever seen in close order within three thousand yards. Frontal attacks were entirely avoided.' And although a fierce

---

* The SMLE, along with the Vickers machine gun of both world wars, the Royal Ordnance 25-pounder field gun of the Second World War and the Centurion tank, is one of the four 'twentieth-century greats' of British armament. All but the Vickers remain in service in some part of the world still.

debate about the future of cavalry continued between those who saw its role as mounted infantry and those who advocated shock action in the charge with sword and lance (the *arme blanche*), and although the artillery still preferred direct to indirect fire (their experience of the veldt had convinced them of the advantage of seeing the target from the gun itself), the tactical atmosphere was of innovation, with technology an added driver. To a great extent the SMLE decided the *arme blanche* versus mounted infantry debate, for the cavalryman's pay was now linked to his shooting prowess, as the infantryman's was, and when it came to war in 1914 and he was *forced* to dismount, he was almost as good as an infantryman. But the artillery, although re-equipped with a superb field gun, would have a particularly rough time in the first six months of the war, losing many gun crews and horse teams in the direct-fire role before the end of the mobile phase determined the issue.

Nor was the petrol engine neglected: the motor car was fast becoming the means of liaison between headquarters, and increasingly too the motor lorry displaced the horse in the Army Service Corps, although horsepower would remain the primary motive means for the ASC and the Royal Artillery for another twenty years until the reliability and durability of motor vehicles, particularly on unmade roads, was proved.

The electric telegraph had linked London with the army in the Crimea, and the first telephone exchange had opened in the capital in 1879, yet twenty years later in South Africa there had been no field telegraphy, let alone telephony: a line paid out to headquarters during the climb up Spion Kop would have made all the difference to the course of the battle. Consequently, battlefield line-laying, and rudimentary radio, now joined the repertoire of the already versatile Royal Engineers. And if further proof of the Edwardian army's innovative spirit were needed, in 1914, just five years after Blériot's precarious pioneering flight across the English Channel, the four 'aeroplane squadrons' of the newly formed Royal Flying Corps (which had also taken over the Royal Engineers' balloon squadron) would fly to France with eighty aircraft. It had indeed been a dynamic decade.

The exam question, however, was 'How long would a continental war last?' In the opinion of the general staff and successive governments, based in part on observation of the Russo-Japanese War of 1904–5, it would be violent but relatively brief, with the advantage to the professional rather than to the conscript army. This was of course an analysis that underpinned the remodelling of the British army; or, put another way, it made something of a virtue of necessity. The TF, though under-recruited, had some fine men and fine units in 1914; the regular army was without question the best equipped, organized and prepared army that Britain has ever sent abroad at the beginning of a war. The trouble was that the war it went to fight was not the war the prevailing orthodoxy envisaged. Violent it was; brief it was not.

Before the fateful pistol shot in Sarajevo, however, a most extraordinary crisis would convulse the upper echelons of the army. It would lead to the resignation of the CIGS, Sir John French, as well as the war minister; it poisoned the wells of trust between senior officers and politicians; and it threw a match into the kindling wood beneath the great bonfire that was Ireland. Yet few today know much if anything about the 'Curragh Incident', or 'Curragh Mutiny' as it is sometimes called (and not without reason).

It began with that sacred cause of the old Liberal Party, Irish home rule – what would today be called devolution – which had twice defeated Gladstone and would now very nearly undo the Liberal prime minister Herbert Asquith. In

1912 he introduced the Third Home Rule Bill,* which was passed in the House of Commons by ten votes but rejected overwhelmingly by the Lords. In 1913 it was reintroduced with the same result. Asquith intended bringing it forward for a third reading in 1914, after which – expecting a further defeat in the Lords – he would use the provisions of the new Parliament Act to over-ride the Lords and send it for royal assent.

But how would it be implemented? In Ulster the Protestants, who were in the main strongly Unionist – opposed even to home rule, let alone independence – were in a slight numerical majority. Whatever their coreligionists in the south were prepared to live with, the staunch Ulster Unionists were not prepared to be ruled from Dublin by a Catholic majority. In January 1913 the paramilitary Ulster Volunteer Force (UVF) was formed to resist implementation of the Bill (estimated 80–100,000-strong). They were well armed (Germany actually supplied more Mauser rifles to the ultra-loyal UVF in 1913 than to the IRA during the First World War) and well organized, and enjoyed the active support of a good number of senior army officers both retired and serving – including Lord Roberts.

To reinforce the garrisons in Ulster and guard the military armouries and ammunition depots, and perhaps support the Royal Irish Constabulary there, the commander-in-chief in Ireland, Lieutenant-General Sir Arthur Paget (grandson of Lord 'One Leg' Uxbridge), had at his immediate disposal a cavalry and an infantry brigade at the Curragh camp outside Dublin. What actually happened in the run-up to the 'mutiny' – what was said and the exact sequence of events – has never been definitively established. It is at least known

---

* The first Irish Home Rule Bill was introduced in 1886; it was defeated in the Commons and never introduced in the Lords. A second bill was introduced in 1893, to be passed in the Commons but defeated in the Lords.

that the secretary of state for war, Jack Seely, a colonel in the yeomanry with a DSO from South Africa, asked Paget if he thought the two brigades would be prepared to go north, for it was well known that in the 3rd Cavalry Brigade there were a good many officers from the Anglo-Irish ascendancy, not least its commander, Hubert Gough, who might be opposed to home rule. Seely suggested a certain latitude be given to those Anglo-Irish officers who would find duty in the north objectionable – that they might be allowed to disappear on leave before any actual order was given.

Paget was perhaps the least suitable man to be commander-in-chief in these circumstances. Now 63, he had seen service, but contemporaries said that he spoke as if he were thinking aloud. His obituarist in *The Times* in 1928 wrote: 'Had he only devoted to Military Study a fraction of the time which he gave to the observation of trees and shrubs he might have ranked as a learned soldier.' On 20 March he called the brigade commanders to his office and spoke with sufficient vagueness to give the fiery Gough the impression that he would be asked to 'coerce Ulster' into accepting home rule, and that his officers' views were to be sought. When Gough put the question to the assembled officers of his brigade later that day – saying that the decision was for each of them, but that he would prefer dismissal to taking up arms against Unionists – all but two voted with him.

Paget informed the War Office by telegram that all was not well. The press learned of it at the same time and a full-blown crisis followed in which hardly anyone had a clear idea of what anyone else had said, or even in key respects what the actual issues were. The press thundered, the *Daily Chronicle* reporting: 'For the first time in modern English history a military cabal seeks to dictate to Government the Bills it should carry or not carry into law. We are confronted with a desperate rally of reactionaries to defeat the democratic movement and repeal the Parliament Act. This move by a few aristocratic officers is the last throw in the game.' The *Daily Express* announced in thick, black, funereal type that 'the Home Rule Bill is Dead'. The *Daily News* asked 'whether we govern ourselves or are governed by General Gough'. The chancellor of the exchequer, David Lloyd George, who as prime minister overseeing war strategy after 1916 would find himself at odds with most senior officers, practised his famed demagogy: 'We are confronted with the greatest issue raised in this country since the days of the Stuarts. Representative government in this land is at stake. In those days our forefathers had to face a claim of the Divine Right of Kings to do what they pleased. Today it is the Divine Right of the aristocracy to do what it pleases.'

In a week it was all over. Asquith said there had never been any intention of using the army to 'coerce Ulster'; Seely gave Gough and his fellow officers the assurances they sought, and those officers who had already resigned were reinstated – while at the same time all who had refused the hypothetical order said they would of course obey any specific order to help maintain law and order in Ulster. On 28 March a new army order concerning discipline was issued, beginning: 'No officer or soldier should, in future, be questioned by his superior officer as to the attitude he will adopt in the event of his being required to obey orders dependent on future or hypothetical contingencies.'

But however deftly the cracks were papered over, damage had undoubtedly been done – and very publicly. The following month both the CIGS and the secretary of state resigned, and Asquith himself took the War Office portfolio. French was replaced by the 64-year-old Sir Charles Douglas, inspector-general of the forces (whose place French took in a face-saving exchange). Remarkably, therefore, as the crisis over the assassination of Archduke Ferdinand mounted, the war department was in the hands of a man whose maxim as prime minister was 'wait and see', and the army was headed by an unlikely general who would die of the strain of office almost as soon as war broke out.

The legacy of suspicion between the 'brass and the frocks'* was an enduring one, and mutual distrust would dog the conduct of the war. The distrust was born of many things, but the intemperate language and precipitate actions of the Curragh Incident unquestionably fuelled it. Even Churchill, who as a Liberal MP and first sea lord had been actively involved in championing the home rule Bill, and had made sanguinary pronouncements against the Unionists, was to feel that distrust. In the end, the Curragh Incident made no difference to the passage of the Bill, which was quietly laid aside when war broke out with Germany in August. But it was grist to the mill of violent Irish nationalism: with all confidence in parliamentary procedure now lost, a resort to arms by the most extreme nationalists was virtually unavoidable. On Easter Monday 1916, under a banner proclaiming 'We serve neither King nor Kaiser, but Ireland', an armed rebellion broke out which ever since has shaped the army and its thinking.†

---

* 'Brass hats' referred to the gold wire embellishments on the peaks of general officers' hats, while 'frocks' referred to the frock-coats worn by politicians.
† The Bill was passed in the Commons at the end of May, and after the expected defeat in the Lords it was sent under the provisions of the new Parliament Act for royal assent. In July, however, the Ulster Unionists forced an amendment for the exclusion of Northern Ireland from the workings of the Act, with the precise details (the number of counties – four, six or nine – and whether exclusion was to be temporary or permanent), to be further negotiated. The haggling continued until the Kaiser, in effect, put an end to it.

21

The Lights Go Out

France, 1914

THE FOREIGN SECRETARY, SIR EDWARD GREY, NOT A MAN FAMOUS for his hard work, was nevertheless working late in August 1914 as the nation headed inescapably for war with the 'Central Powers' – Prussia (or 'Germany'), Austria-Hungary (or simply 'Austria'), the Ottoman Empire (or 'Turkey') and, eventually, Bulgaria. As he stood at the window of his office watching the lights being lit in the street below, he made his memorable remark: 'The lamps are going out all over Europe; we shall not see them lit again in our lifetime.' Politically, his words would prove all too prophetic. Militarily, they implied something that neither his cabinet colleagues nor the army and navy chiefs had believed, and therefore something they had not prepared for: that the war would be a long one.

'Over by Christmas' was the prevailing opinion – so much so that the Staff College practically shut up shop so that its staff and students could join the British Expeditionary Force (BEF) to see action. Two men besides Grey, who perhaps did not himself see the military implications of his gloomy prediction, believed it would be otherwise. One was Lieutenant-General Sir Douglas Haig, though as a corps commander in the BEF his opinion was as yet inconsequential. The other was the new secretary of state at the War Office, Field Marshal Lord Kitchener, whom Asquith appointed the day after declaring war on 4 August.

Kitchener predicted a costly war lasting three years at least. Crucially, he calculated that Britain's full military potential could not be reached until 1917, but that its weight would then be decisive. He set aside a dozen years' strategic thinking in a mere few hours, but his was nevertheless a clear strategic analysis on which to plan; and, of course, it was proved right. Although the cabinet was not yet ready to introduce conscription, within two days Kitchener got parliamentary approval for an initial increase in the army's strength of half a million men. And up went his famous *Your Country Needs You!* posters for the first 100,000 recruits, to be enlisted 'for the duration'.

The scheme was controversially different from anything the War Office had hitherto conceived. No one had thought in terms of such a colossal expansion of the army, only of getting the Territorial Force (TF) ready for overseas service if that proved necessary – though legally it was still liable for home service only. Kitchener did not even plan to expand the army through the county Territorial associations: his half a million, when they were all eventually recruited, would be regulars, not Territorials, and they would serve in formed units – fourth, fifth, sixth battalions and so on of existing regiments rather than as individual reinforcements for the BEF. Indeed, his intention was simply to raise a series of 'New Armies', each mirroring the BEF, complete in all its branches (though the yeomanry would provide the cavalry). He both mistrusted the quality of the Territorials as a basis for expansion and feared the complications of the legal position. If Territorial battalions volunteered en bloc to serve overseas their offers would be accepted, although he saw their role primarily as 'backfilling' – relieving regular units at home and in overseas garrisons for service in France. In fact the Territorials soon had battalions in action in France and began their own expansion programme alongside that of the 'New Armies'.

How large an army Britain would need eventually, Kitchener was unsure. By mid-September he was speaking of fifty divisions in the field; by July 1915 he had upped the figure to seventy. This phenomenal expansion raised two further questions on top of how so many men were to be recruited: how were they to be trained and commanded, and how were competent divisional and corps staff to be found? At the outbreak of war the regular army stood at some 247,000 (about a third of them in India), while the Special Reserve (the old militia) and the TF together could muster roughly twice this number. By November 1918 in France alone there were a million and a half British soldiers, not including colonial troops, a further half a million in other theatres overseas, and getting on for a million and a half at home (including Ireland) in training or garrison duties. (The population of Britain and Ireland in 1914 was estimated at 46 million.) A fourteenfold increase in its size, and in so short a period, inevitably redefined both the character of the army and therefore the nature of the operations it could undertake. But it might not have been so traumatic a process – so steep a learning curve for the New Armies in particular – had not the old regular army been effectively destroyed by the first year's fighting.

In August 1914 the Kaiser had allegedly mocked the BEF as a 'contemptible little army', a jibe which its members were soon wearing as a badge of pride, calling themselves 'the Old Contemptibles' (the Kaiser's chief of staff, Colonel-General

Helmuth von Moltke, is supposed to have been more circumspect in ascribing weakness to its small size, however, describing the BEF as 'that perfect thing apart'). As soon as war was declared, the cabinet met to decide where the BEF would be used. The German strategy was one of rapid offensive in the west to knock out France while the Russian army was still mobilizing, after which troops would be transferred from the western to the eastern front via the excellent Prussian railway system to deal with the second great power of the 'Triple Entente'. Britain, the third member of this informal alliance, hardly figured in the German war plans, as the Kaiser's assessment of the BEF suggested (although he knew that relative naval strength would not be to Germany's advantage in a long war).

To overcome the formidable defences on the Franco-German border, in the early years of the century the Grosser Generalstab under Count Alfred von Schlieffen had devised a plan to make a broad outflanking march through neutral Belgium (it was the violation of Belgian neutrality that tipped Britain into the conflict), enveloping Paris from the west and thence threatening the French armies from the rear. It was compared with a revolving door swinging through Belgium with its pivot on Metz in Lorraine. The French counter-plan – 'Plan XVII' – was for an immediate offensive in Lorraine to throw the German army off balance and to threaten the rear and lines of communication of the German right flank swinging towards Paris. The revolving-door effect would have been completed by the German armies in turn taking the French in the rear.

Both plans were ludicrous. French belief in the power of *l'offensive à l'outrance** took little account of the effects

---

* The all-out offensive. The new French service regulations of 1913 had tried to do away with the defensive-mindedness that had become the prevailing doctrine after the Franco-Prussian War: 'The French army, returning to its traditions, henceforth knows no law but the offensive.'

of modern weapons; and the Schlieffen Plan required more troops than were available, as well as prodigious rates of advance, even against light opposition, and feats of logistics, command and control that were only realistic on a map. The plan had been devised in the early 1900s and worked up in subsequent years by Schlieffen's successor, Moltke, who as commander-in-chief in the west put his own final version into action at the beginning of August. Schlieffen's 'dying words' to Moltke had reportedly been 'only make the right wing strong!', for he knew that whatever else was to be changed, the plan could work only if there were enough troops on the extreme right wing to 'brush the coast with its sleeve' and continue the swing west of Paris. But Moltke could not bring himself to take the risk of weakening the 'revolving door' elsewhere, particularly that inner part against which the French counter-offensive was expected. Nor dare he leave East Prussia too weak, for the Russian railway system was improving with every year, and with it therefore the speed of Russian mobilization.

A young subaltern in Britain's newly formed Intelligence Corps, Rollestone West, writing in his diary as the BEF moved to its pre-arranged positions in France, marvelled at 'the astonishing completeness of it all'. That was, indeed, the reaction of many who observed the British army's arrangements for mobilization. Two complete army corps had been assembled (though initially the BEF would consist of four divisions, not five as originally planned, plus the cavalry division), reservists had been called up, equipped and despatched to their units, and many thousands of horses had been requisitioned from private owners, their whereabouts recorded in the Army Horse Register compiled annually by the district remount officers. And the BEF's move by rail and sea to the Continent was an undertaking so unprecedentedly large and executed so smoothly that many could scarcely believe the transformation from the old days of hasty improvisation and muddle. On 5 August, the first full day of mobilization, commanding officers received files of documents marked 'Top Secret' detailing their regiments' movements to their unnamed ports of embarkation. They were not over-wordy: 'Train No. 287Y will arrive at siding D at 12.35 am August 15th. You will complete loading by 3.40 am. This train will leave siding E at 9.45 am, August 15th. You will march onto the platform at 9.30 am and complete your entraining by 9.40 am.' Many officers thought it decidedly un-British.

But the BEF's deployment plan had never been politically endorsed by formal treaty with the French. Driven in large part by the personal determination of the director of military operations, Major-General Henry Wilson, the army staff had worked on the assumption that the BEF would operate on the left flank of the French army to counter the expected Schlieffen swing through Belgium. At the war cabinet's first meeting, other ideas had been put forward – such as reinforcing Antwerp (the Scheldt estuary still loomed large in British minds, as it had at the Congress of Vienna a hundred years earlier), and even landing on the German coast. But in the end, it was the complexity of the deployment plans – not least the slots for troop trains on French railways – as much as anything else that determined the BEF's role. And so, just as Austria, Germany, Russia and France had stumbled into war because of the inflexibility of their mobilization and deployment plans, each depending on a movement schedule that could be cancelled but not modified, the BEF entrained for its place in what that great populist historian A. J. P. Taylor called 'war by railway timetable'.

When the 18th Hussars' transport came into Calais harbour the troopers burst into song with a rendition of 'Here We Are Again' – much to the puzzlement of their officers, for the regiment had not set foot on the Continent since Waterloo. But a hundred years is nothing in the mind of a British soldier: the regiment had been here once before, and that was enough. More to the point, they had won their battle the last time. After unshipping their horses the regiment boarded their trains for Maubeuge, where later in the day they detrained and marched north, finally coming to a halt at Mons, 10 miles beyond the Franco-Belgian border. Indeed, had they continued another twenty-five miles they would have reached the field of Waterloo; and when the 18th's bugles blew 'ceasefire' on 11 November four years later, they would find themselves in almost exactly the same spot on which they first heard the sound of the German guns. That the French were on their right, rather than in front of them as at Waterloo, and the Germans in front of them instead of on their left was of no matter. The BEF were regulars, professionals: they did not yet hate 'the Hun'. They were here to do a job, as they had trained to do, and do it they would.

But allied relations had not got off to the best of starts. Unlike Wellington, Field Marshal Sir John French – brought out from under the Curragh cloud to command the BEF – spoke little of the language, and attempts to liaise with the commander of the flanking 5th Army, the brilliant but acerbic General Charles Lanrezac, did not bode well. Did *le Général* think the Germans (who were reported to be moving to their north) proposed to cross the Meuse at Huy? asked Sir John through his interpreter. 'Tell the Marshal,' snapped Lanrezac, 'that in my opinion the Germans have merely gone to the Meuse to fish.'

Given that French's orders from Kitchener stressed that the BEF must not be hazarded in any offensive action unless the French themselves were making the greatest commitment, the exchange was hardly conducive to heroic British efforts. In Lanrezac's disdain there was almost an echo of the Kaiser's contempt for the BEF's small size, both of them failing to make due allowance for quality, what today would be called the 'force multiplier'.

Sir John French's liaison officer at Lanrezac's headquarters was a subaltern of the 11th Hussars, Edward Spears (who in 1940 as a major-general would again play a major role in Anglo-French liaison). On 21 August, as French continued to advance in spite of Lanrezac's *hauteur*, Spears drove into the British sector and got his first glimpse of the BEF: 'The first I saw were a small detachment of Irish Guards, enormous, solid, in perfect step.' Next came a column of artillery:

I thought I should burst with inward gratification at the smartness of those gunners. They were really splendid, perfectly turned out, shining leather, flashing metal, beautiful horses, and the men absolutely unconcerned, disdaining to show the least surprise at or even interest in their strange surroundings . . . I said nothing, but stole a glance at the French officer who accompanied me and was satisfied, for he was rendered almost speechless by the sight of these fighting men. He had not believed such troops existed. He asked me if they were the Guard Artillery! [In fact there was no such thing.]

Soon after this we received a shock, and my French companion was further impressed, but in a way he did not much like, for we drove headlong into a most effective infantry trap. At a turn in the road we were suddenly faced by a barrier we had nearly run into, and found that without knowing it we had been covered for the last two hundred yards by cleverly concealed riflemen belonging to the picquet. Had we been Germans nothing in the world could have saved us.

The following evening the BEF reached the line of the Mons–Condé Canal, where they halted, Sir John French having received reports from the Royal Flying Corps and the cavalry that there were Germans directly to his front, and from Spears that Lanrezac's 5th Army was in severe trouble around Charleroi. He decided that to advance further was to risk being cut off, and so ordered his two corps commanders – Haig and General Sir Horace Smith-Dorrien – to take up defensive positions along the canal. Next morning, Sunday, the advance guards of General Alexander von Kluck's 1st Army clashed with Smith-Dorrien's II Corps (3rd and 5th Divisions) as Haig's I Corps (1st and 2nd Divisions) were still coming up on II Corps' right. It was 'encounter battle' – fluid, confused, and still very much a place for cavalry: the 15th Hussars, reconnoitring east of Mons for the 3rd Division at first light, were cut up badly by rifle fire and shelling, and 22-year-old Corporal Charles Garforth began his run of three heroic actions that led to the award of the VC.* It would be many weeks yet before the lines solidified and the cavalry were sent to the rear.

The pressure on Smith-Dorrien's II Corps increased throughout the day, and by evening they had pulled back to a second position just south of Mons. At about midnight French learned that Lanrezac's 5th Army was withdrawing, and so he sent out orders for the BEF to do likewise. Haig's corps, which had not been much engaged, was able to break clean without difficulty next morning, but II Corps had a much harder time of it: one of the rearguard battalions, the 2nd Cheshires, was surrounded, ran out of ammunition and had to surrender; and L Battery RHA, supporting 1st Cavalry Brigade, fired more rounds that day than it had during the whole of the Boer War (a few days later the battery was totally destroyed, its gunners winning three VCs in the process). The casualties mounted steadily, II Corps in all losing 2,000 men killed, wounded or missing.

---

* Charles Garforth was in fact recommended on three separate occasions for the award of the VC. Remarkably, he survived the war, dying in 1973.

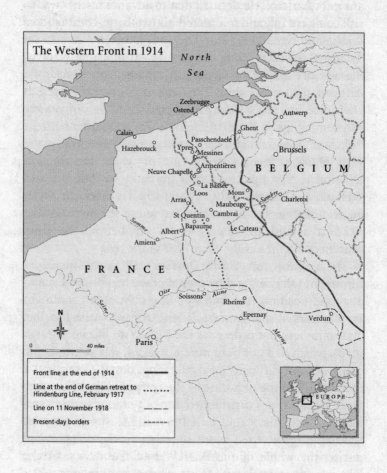

The Western Front in 1914

North Sea

Zeebrugge
Ostend
Antwerp
Ghent
Calais
Passchendaele
Ypres
Hazebrouck
Messines
Brussels
Armentières
Neuve Chapelle
BELGIUM
La Bassée
Loos
Mons
Arras
Maubeuge
Charleroi
St Quentin
Cambrai
Bapaume
Le Cateau
Albert
Amiens

Somme

FRANCE

Oise   *Aisne*
Soissons
Rheims

Seine
Epernay
Verdun

Paris

Marne

N

0 ——— 40 miles

Front line at the end of 1914
Line at the end of German retreat to Hindenburg Line, February 1917
Line on 11 November 1918
Present-day borders

EUROPE

But the fighting along the canal had inflicted far heavier casualties on the Germans, many of whose units, whatever their sophisticated tactical manuals said, attacked in column or shoulder to shoulder (it was said that the reservists gained confidence from *das Tuchfühl* – the 'touch of cloth'). The German novelist Walter Bloem, a 46-year-old reservist captain in the 12th Brandenburg Grenadiers, recorded how he spent the whole day trying to see the enemy, but felt only their rifle fire, his regiment losing almost all its officers and 500 men. It was not just the losses that hit hard but also the shock of finding themselves up against a more formidable enemy than the 'contemptibly little' one they had been expecting: 'my proud, beautiful battalion ... shot to pieces by the English, the English we laughed at'. It is always a severe blow to morale to discover that the enemy is better than had been rated, and makes for a double dose of wariness subsequently. Bloem's commanding officer issued orders for the night, showing just what a shock it had been: 'Watch the front very carefully, and send patrols at once up to the line of the canal,' he told his company commanders. 'If the English have the slightest suspicion of the condition we are in they will counter-attack tonight, and that would be the last straw.'*

Bloem was astonished when he learned next morning that 'the English' were withdrawing, and that his Brandenburgers were to advance. When they did they found more evidence of the BEF's shocking capability – a sand-bagged machine-gun emplacement in a wood from which a devastating enfilade fire had thinned their ranks the day before: 'Wonderful, as we marched on, how they had converted every house, every wall into a little fortress: the experience no doubt of old soldiers gained in a dozen colonial wars; possibly even some of the butchers of the Boers were among

---

* Bloem, *The Advance from Mons*, Eng. edn (1930).

them.' No doubt, but it lay, too, in a tradition reaching much further back – to the loop-holed walls of Hougoumont and the 'little fortress' of La Haye Sainte.

Indeed, II Corps' commander, Smith-Dorrien, would now reach back even further, however unconsciously – to Sir John Moore and the retreat to Corunna. His men were dog-tired after the march up to Mons, the battle the following day along the canal and the fighting withdrawal over the next two in the stifling August heat. They had not been able to break clean, the Germans harrying them every mile of the retreat, and Smith-Dorrien feared for the cohesion of his corps.

On the evening of the twenty-fifth, with the 4th Division (which the war cabinet had belatedly agreed to send to France) detraining at Le Cateau 30 miles south-west of Mons and placed by French under II Corps, Smith-Dorrien decided to stand and fight. He intended giving the Germans what he called a 'smashing blow' so that his corps could properly break clean. French neither approved nor forbade it, instead sending a half-hearted message that Smith-Dorrien must judge things for himself.

The twenty-sixth dawned even hotter after a rainy night, and the battle got off to a bad start on the extreme left flank where 4th Division were still taking up positions. They had had little rest in the cattle trucks bringing them up from Calais; they had then marched north to relieve 5th Division as rearguard and fought their way back again all through the night, so that by first light they wanted rest as badly as II Corps' other two divisions. With them was 1st Battalion the King's Own, commanded by Lieutenant-Colonel Alfred Dykes who had won the DSO as adjutant of the 2nd Battalion at Spion Kop. The searing experience of small-arms fire poured into the close-packed ranks of riflemen that day in South Africa appears to have failed him now, for he ordered the companies to pile arms, remove equipment and rest, despite being on a forward slope. The country was more open than the coal-mining landscape of Mons, but Dykes believed they were screened by French cavalry, although the battalion could clearly see the Lancashire Fusiliers digging in on the crest behind them on their left. At about six o'clock a party of horsemen appeared from the trees 1,000 yards ahead of the battalion, but Dykes took no notice, assuming them to be French. Soon after, a horse-drawn vehicle came out of the woods, and moments later a machine gun opened up in a long, raking burst which killed 83 men, including Dykes himself, and wounded 200.* Artillery accounted for another hundred in the scrambling minutes that followed, and those King's Own who emerged unscathed were extricated only by the spirited support of the 1st Royal Warwicks, one of whose platoon commanders was Lieutenant Bernard Law Montgomery.†

Nothing could have demonstrated better the new reality of war than this killing burst against one of the most experienced regiments in the army. If the Germans had had a rude shock at Mons, the BEF were now having theirs at Le Cateau. But the shock would not demoralize them: the 'experience of old soldiers gained in a dozen colonial wars', which Bloem so admired, was also the experience of coping with setbacks and accepting casualties. The BEF was not only a well-trained army, it was a toughened one. Although the King's Own had lost their commanding officer and half their fighting strength in a few minutes, they reorganized and fought throughout the day as doggedly as the rest of the brigade – just as the second battalion had done in South Africa after Spion Kop.

* Their bodies were buried some days later in a mass grave, the exact position of which is unknown but, according to the regimental history, can be accurately surmised.

† Six weeks later, near Ypres, Montgomery was shot through one of his lungs, and a grave was dug for him. Recovering, however, he was awarded a DSO for inspiring leadership and posted as brigade major to one of Kitchener's new divisions.

The fighting at Le Cateau intensified throughout the morning, but the Germans paid a heavy price for their gains, for Smith-Dorrien had chosen a position obliging them to advance down a long forward slope which gave his marksmen and gunners full opportunity to show their skill. Private F. W. Spicer of the 1st Bedfordshires (who would be commissioned and rise to lieutenant-colonel, such were the quality in the ranks of the BEF and the need for officers in Kitchener's New Armies) wrote in his diary:

> It must have been midday or later when the enemy infantry began to attack our immediate front. We had a real hour's hard work firing our rifles. Luckily we had brought plenty of ammunition with us, and we needed it. Line after line of German infantry advanced only to be mowed down by our rifle and machine gun fire. The battery of the 15th Artillery Brigade in the dip of ground [close behind] did great execution. The enemy suffered such enormous losses that they were unable to force us from our positions, and themselves had to withdraw from our front for a time until they were reinforced.

But although II Corps was able to repel the Germans' frontal attacks, soon after midday pressure on the right flank became too much, for a gap had opened up between Haig's and Smith-Dorrien's corps the day before while they were withdrawing to Le Cateau either side of the Forêt de Mormal. The 2nd Suffolks were cut off, their commanding officer was killed and the Germans called on them to surrender. They refused and the battle continued until they were surrounded and their ammunition exhausted, by which time all but 100 of the Suffolks' 1,000 officers and men who had detrained at Maubeuge five days earlier had been killed, wounded or taken prisoner.

By now, however, Smith-Dorrien was confident that he had given the Germans enough of a smashing blow to be able to order his divisions to start slipping away and putting some distance behind them. And indeed by nightfall he had been able to disengage, although the rearguard paid heavily, as did the artillery, for the guns had all fought forward with the infantry, exactly as they had at Waterloo, and when the horse teams came up to haul them out of action they were shot down in their hundreds. The Royal Artillery lost thirty-eight guns that day, more than in any action of the army's 250-year history. It would be the beginning of the end for artillery firing over open sights: from now on they would seek cover and fire indirect, with the fall of shot adjusted by observation officers with field telephones and, much later, radio.

But although they had broken clean, Smith-Dorrien's men were exhausted. Serjeant John McIlwain of the 2nd Connaught Rangers records how at

> about 4 am our officers, who were wearied as ourselves, and apparently without orders, and in doubt what to do, led us back to the position of the night before. After some consultation we took up various positions in a shallow valley. Told to keep a sharp lookout for Uhlans [German lancers]. We appeared to be in some covering movement. No rations this day; seemed to be cut off from supplies. Slept again for an hour or so on a sloping road until 11 am. We retired to a field, were told we could 'drum up' as we had tea. I went to a nearby village to get bread. None to be had, but got a fine drink of milk and some pears. Upon the sound of heavy gunfire near at hand we were hastily formed up on the road.

McIlwain might have been writing the diary a hundred years earlier, for it has that same trusting, make-do attitude of the British soldier before Waterloo. They knew a much-vaunted army was bearing down on them, now as then, and yet they were unconcerned enough to go off foraging. Le

Cateau was indeed the army's biggest set-piece battle since Waterloo. And when at the end of the day's fighting Sergeant McIlwain's platoon joined back up with the rest of the battalion, he found that the casualty toll had been just as grievous: 'By evening we mustered half-battalion strength. "C" had about half a company. Over the whole Major Sarsfield took command, the Adjutant, Captain Yeldham, was also present. Colonel Abercrombie and 400 or so others were captured by the Germans.'

The losses throughout the BEF had indeed been heavy: 7,812 men were accounted killed, wounded or taken prisoner, most of them from II Corps, and for this reason Le Cateau, although celebrated as a British victory (and a name much favoured for post-war barracks), is still a controversial battle. Sir John French's intense dislike of Smith-Dorrien had not helped.* Although he wrote in his despatches of his subordinate's 'rare and unusual coolness, intrepidity and determination', privately he blamed Smith-Dorrien for disobeying orders, and believed the battle to have been both unnecessary and at the same time more costly than in fact it was.† From then on Smith-Dorrien was a marked man, and in May 1915 he was sent home. Or rather, sent ''ome', for the news was given to him by the BEF's chief of staff, Lieutenant-General Sir William ('Wully') Robertson, who had risen from the rank of private in the 16th Lancers (and would rise even further, to field marshal) and delighted in dropping his

---

* The root cause of the dislike almost certainly included jealousy (although French was Smith-Dorrien's senior, the latter had greater experience of action), but ostensibly it was professional – a profound disagreement over the role of cavalry, Smith-Dorrien (an infantryman) being a proponent of the 'mounted infantry' school. Lieutenant-General Sir James Grierson had been commanding II Corps when the BEF sailed for France but had died of a heart attack in his train to Amiens. French had wanted Sir Herbert Plumer to replace him; Kitchener sent him Smith-Dorrien instead.

† The casualty figures for Le Cateau are notoriously difficult to pin down.

aitches to emphasize his origins.* And so the only man in the BEF other than Haig capable of taking over as commander-in-chief when French was himself relieved at the end of 1915 would be dismissed with a matter-of-fact ''Orace, you're for 'ome.'

For the moment the retreat continued. But the French commander-in-chief, the massive, imperturbable General Joseph Joffre, had not been idle. Despite the poor progress of the counter-offensive in Lorraine he was able to cobble together a new army (the 6th) to cover Paris, his efforts at personal liaison much aided by his being speeded along the roads of north-west France by Georges Boillot, the Peugeot racing team's lead driver.

And, indeed, it was the petrol engine that now came to the rescue, first on 2 September, when a lone pilot on a reconnaissance flight from the Paris garrison brought the news to Joffre that von Kluck's 1st Army was turning *east* rather than continuing south and west to envelop the city, thereby presenting a flank into which the allies could attack.† And then the military governor of Paris, the wiry veteran General Gallieni, rounded up every taxi cab he could find and sent the garrison forward in comfort and the high spirits that go with clever improvisation and the prospect of turning the tables. But Joffre had to persuade the BEF to join in, and

* Examples such as Colin Campbell, Hector Macdonald and 'Wully' Robertson, though perhaps few, give the lie to the notion of strict exclusivity in the Victorian and Edwardian army. Men such as these would not have gained advancement so easily in the Prussian, Austrian, Russian or even French armies of the period. Only the US army showed an equal disregard for social background, although in many ways it could be even more rigid in its reverence of West Pointers than ever the British Army was of Sandhurst alumni.

† The day before, papers had been found on a dead German staff officer indicating that von Kluck was so alarmed by the gap opening up between his army and von Bülow's that he saw no option but to close with him and give up the Schlieffen idea of a western envelopment of Paris.

Lanrezac's words before Mons still rankled with Sir John French, who the day before had been dissuaded from withdrawing the BEF from the fighting only through Kitchener's personal intervention. His meeting with French was at first frosty, but then 'Papa Joffre' played to the heart: 'Monsieur le maréchal, c'est la France qui vous supplie.' Sir John French, reduced to tears, tried to reply but language failed him. Turning to his interpreter he said, 'Dammit, I can't explain. Tell him all that men can do, our fellows will do.'

In any case the BEF had had enough of retreating. They had twice done what the British army used to do so well in Wellington's day – occupy a position and defy eviction – but it was now time to do what they had also done well for even longer, since Marlborough's time: go to it with the bayonet.

The main weight of the counter-attack was to be in the valley of the Marne (indeed, the turning back of the seemingly unstoppable German advance on Paris would be dubbed 'the Miracle on the Marne'). However, the BEF's advance was at first hesitant, for Sir John French's head had not quite caught up with his heart, although as the Germans began to give way before the wholly unexpected onslaught of three French armies Smith-Dorrien's and Haig's corps were able to increase the pressure too. By 13 September the Germans had fallen back to the River Aisne, with the allies snapping at their heels, though they were never quite able to turn harassing pursuit into rout. Then the swelteringly hot weather broke at last and the torrential rain swelled the rivers and streams which lay in the path of the BEF, the bridges blown by the retreating Germans who now dug in on the high ground north of the Aisne.

Many a gallantry medal was won by the BEF's sappers and infantry as they threw assault bridges across the Aisne and attacked the heights – but to no avail. The Germans had seized the best ground, and the allies were running out of steam. Indeed, the nature of the battle in north-west France was changing, as Sir John French noted in a letter to the King: 'The spade will be as great a necessity as the rifle, and the heaviest types and calibres of artillery will be brought up on either side.'

But neither side was ready quite yet to concede that the war of manœuvre was finished: there remained the open flank – the gap between General Manoury's improvised 6th Army, which Joffre had placed on the BEF's left, and the sea. Both the allies and the Germans would now begin two months of increasingly desperate scrabbling to turn each other's flank. But with mobility still determined by the marching pace of the infantry, the race for the flank succeeded only in drawing out the trench lines north across the uplands of the Somme, and on in front of the ancient city of Arras, and then down into the polder and sandy lowlands of Flanders – and eventually to the dunes at Nieuport (thereby preserving a symbolically important piece of Belgian sovereign territory after Antwerp eventually surrendered in October). On the way, the BEF would have the worst of its fighting so far, and a battle that would all but extinguish the old pre-war regular army, the men who had sailed for France only two months before.

Reinforcement had been continuous since those heady days in August. Regulars redeployed from both home defence and overseas garrisons, and volunteer units of the TF as well as British and Indian troops from the subcontinent, swelled the BEF to five corps – four of infantry (predominantly), including the Indian Corps, and one of cavalry. By the middle of October Sir John French had some 250,000 men at his disposal, a match at last for von Kluck's army opposite him. And it looked as if he had at last found the German flank – at Ypres. Urged on by General Ferdinand Foch, whom Joffre had sent north to re-energize the fighting, French put the

BEF on to the offensive along the Menin Road on 21 October. They soon ran into trouble, however, for under the forceful leadership of Erich von Falkenhayn, who had replaced the broken Moltke, the Germans had reinforced their right flank by bringing Prince Rupprecht of Bavaria's 6th Army from Alsace-Lorraine, and had themselves gone on to the offensive.

A month's hard fighting followed. In defence the BEF's regiments held their ground with all the insolent resolution of Wellington's men at Waterloo. In the attack they went forward with the steady resolve of Raglan's troops at the Alma – and assaulted with all the ferocity of the Peninsular infantry storming the walls of Badajoz. In innumerable local counter-attacks they threw themselves at the Germans with that same zeal to get to close quarters that King William's bayonets had shown at Steenkirk. And it was 'all hands to the pump', the cavalry corps sending the horses to the rear and taking up their rifles to hold Messines Ridge south of Ypres. The first TF regiments were blooded too – the London Scottish, the first Territorials to go into action, losing half their strength in the process. And the Germans had their first sight of the turbans, pugarees and Gurkha pill-boxes of the Indian Corps; Sepoy Khudadad Khan of the Duke of Connaught's Own Baluchi Regiment won the first ever Indian VC at Hollebeke on 31 October, when it looked as if the Germans might break the line. These were indeed desperate days, when individual soldiers in individual regiments made a difference. The whole BEF line was probably saved in fact by the 2nd Worcesters, the only troops left in front of Ypres as the Germans captured Gheluvelt on the Menin Road on that last morning in October; the battalion had already been reduced to 400 (less than 50 per cent of its embarkation strength in August), and lost another 200 retaking the vital ground.

On 11 November there was another crisis, when the 1st

and 4th Brigades of the Prussian Guard attacked astride the Menin road, breaking through the defence line and getting to within 2 miles of the city. Only the bayonets of the 2nd Oxford and Buckinghamshire Light Infantry and a scratch force of Grenadiers, Irish Guards and the Royal Munster Fusiliers were able to restore the situation – but at shattering cost, including the loss of the brigadier, Charles FitzClarence, who had won the VC at Mafeking.

The attack of the Prussian Guard, however, like that of the Garde Impériale at Waterloo, was the high-water mark of the German offensive at Ypres, and in the days that followed the fighting slackened. From now on, indeed, the armies went to ground – what was left of them, for while the Germans had fallen in droves, often shoulder to shoulder (and even singing), so too had the British. The BEF lost 58,000 officers and men in what would become known as the first battle of Ypres, and most of them were the irreplaceable regulars who could march all day, use cover and fire fifteen aimed rounds a minute. By Christmas in many battalions there were no more than a few dozen soldiers and a single officer left who had been at Mons. The battles of late summer and autumn 1914 were the breaking of the 'Old Contemptibles', the old regular British army. For them the lights had indeed gone out. The army would have to be remade by Kitchener's men. Meanwhile – until 1917, when by Kitchener's calculations Britain's full military potential could be reached – the fort would have to be held by Territorials and by troops cobbled together from every corner of the Empire.

And the 'fort' would be the trenches. From the North Sea to the Swiss border the armies on both sides were digging, wiring and sandbagging, sometimes within 100 yards of each other. A continuous line of trenches – continuous *lines*, indeed: fire trench, support trench, reserve trench, with communication trenches linking each – soon lay across the French countryside like giant railway tracks. For the most

DIAGRAMMATIC SKETCH OF PORTION OF A FRONT LINE, WITH SUPERVISION TRENCH, LIVING DUG-OUTS AND SHELL TRENCHES.

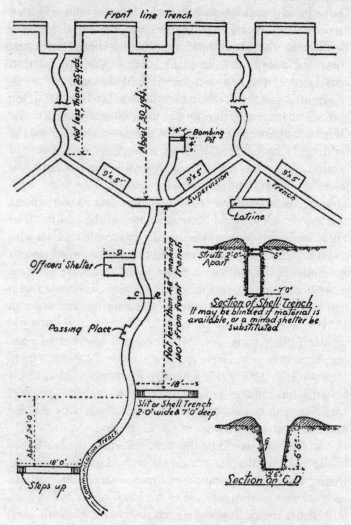

Front line Trench

Not less than 25 yds.

About 30 yds.

4'×4' Bombing Pit

9'×5'   9'×5'   9'×5'

Supervision   Trench

Latrine

Officers' Shelter 9'

Struts 2'0" Apart   6"

-7'0"

Section of Shell Trench.
It may be blinded if material is available, or a mined shelter be substituted

Passing Place

Not less than 48' making 140' from front trench

18'

Slit or Shell Trench 2'0" wide & 7'0" deep

6'6"

3'6"

Section on C.D

About 24'0"

18'0"

Steps up.

Communication trench.

This sketch is from 'Notes for Infantry Officers on Trench Warfare' (1917).

part, the Germans had the best of the ground – higher, for good observation, and drier, so that they could dig their shelters deep. When Kitchener's New Armies were ready to go to France, they would have to face the problem of how to cross the ground between the lines – 'no-man's-land' – before the Germans could come out of their holes with machine guns, and how then to break through the deep defensive belt into the open country beyond before the Germans could seal the breach with troops brought from reserve. And many hundreds of thousands of Kitchener's men would die trying to find the answer.

22

Your Country Needs You!

1915–16

KITCHENER'S NEW ARMIES HAVE NO TRUE PARALLEL IN HISTORY. The response to the call to arms was remarkable, and the mass improvisation no less astonishing. While some of the recruits were doubtless thankful for an excuse to exchange a life of hardship or tedium for one which promised excitement and a little local fame, the motive of the vast majority appears to have been simple patriotism – for 'King and Country' – or else a sense of local solidarity, of 'mateship'. For at the same time that individual recruits were flocking to the army's recruiting centres, Kitchener gave local grandees leave to raise battalions en bloc (often initially at their own expense) with the simple promise that those who enlisted together would be allowed to serve together.

The promise of serving together proved powerful. In Liverpool the earl of Derby, the great mover and shaker of the commercial north-west (and subsequently director-general of army recruiting), called on the office workers to form a battalion: 'This should be a battalion of pals, a battalion in which friends from the same office will fight shoulder to shoulder for the honour of Britain and the credit of Liverpool.' And so were born the 'Pals' battalions. But instead of one battalion, between 28 August (a Friday) and 3 September Liverpool raised *four*, the clerks of the White Star shipping company forming up as one platoon, those of Cunard as another, and so on through the warehouses of the Mersey waterside, the cotton exchange, the banks and the insurance companies – men who knew each other better than their mothers knew them, and now doing what their mothers would never ordinarily have countenanced: going for a soldier – at least 'for the duration'.

Across the Pennines in Barnsley two battalions of Pals were raised, officially designated the 13th and 14th (Service) Battalions, York and Lancaster Regiment (the prefix 'Service' – abbreviated from 'General Service', as opposed to 'Home Service' – was appended to all New Army battalions). Most of the recruits here were miners, many of whom had worked underground since they were 13 and so were not averse to the prospect of three square meals a day and work in the open air for a year or so. In the cities north of the border the response was the same. Glasgow raised three Pals battalions for the Highland Light Infantry, which were named after the institutions from which they had spontaneously sprung: Glasgow Tramways (15th Battalion, Highland Light Infantry), Glasgow Boys' Brigade (16HLI) and Glasgow Commercials (17HLI). Newcastle boosted the Northumberland Fusiliers by two such institutional battalions – the Newcastle Commercials (16NF) and the Newcastle Railway Pals (17NF) – and by no fewer than twelve 'tribal' battalions: six of Tyneside Scottish and six of Tyneside Irish, forming two whole brigades. Perhaps the most exclusive-sounding of the Pals was the Stockbrokers' Battalion (10th Battalion, Royal Fusiliers), and the most endearingly unmilitary (to modern ears, at least) the 21st

Battalion, West Yorkshire Regiment – the 'Wool Textile Pioneers'. John Keegan in *The Face of Battle* (1976) writes:

> The Pals' story is of a spontaneous and genuinely popular mass movement which has no counterpart in the modern, English-speaking world and perhaps could have none outside its own time and place: a time of intense patriotism, and of the inarticulate elitism of an imperial power's working class; a place of vigorous and buoyant urban life, rich in differences and in a sense of belonging . . . to any one of those hundreds of bodies from which the Edwardian Briton drew his security and sense of identity.

But although the Pals battalions were of their own time and place, and although their experience was frequently tragic, the sense of unit identity which came from recruiting from a single locality gave them instantly both high morale and a self-regulation which the army has sought to emulate ever since. The alacrity with which recruits came forward was confirmation that the Cardwell–Childers regimental system, based on geographical identification, had been the right model for the infantry, even if in a way they had never imagined. But how were these Commercials, Railwaymen, Miners, Accrington Pals, Grimsby Chums and Wool Textile Pioneers to be made into soldiers capable of fighting the German or the Turk?

The answer, in truth, was improvisation. It was one thing to form a command structure – there were retired officers and NCOs ready to step forward,* and the Pals battalions

---

* One of the best and most successful (perhaps against the odds) examples was the earl of Derby's brother, the Honourable Charles Stanley, at 43 a decorated captain in the Reserve of Officers with service in the Sudan and South Africa. With Derby's intervention he was made temporary brigadier-general and took command of the four Liverpool Pals battalions comprising the 89th Brigade.

found it easy enough to 'elect' their company officers and NCOs – but quite another to find the cadre of regular or even Territorial officers and NCOs to instruct them. For while Kitchener might reasonably have expected the regular army to be a ready source of training staff when he issued his call for volunteers, the mounting losses in the BEF soon put paid to this idea. And so reservist NCOs with the most recent service were diverted to the new battalions, as were officers on home leave from India, usually with the 'sweetener' of promotion – anything, indeed, that the authorities could think of to attract them.

So successful was the call to arms that there was not only a shortage of trainers but also a chronic shortage of uniforms, rifles, ammunition and accommodation. Men drilled with broom handles in the clothes in which they had enlisted; they slept under canvas, or in seaside lodging houses. Only boundless good spirits and forbearance seemed in plentiful supply. There were New Army battalions that formed with only one regular officer – in the case of the 8th (Service) Battalion East Surrey Regiment a captain, who drew up his new command and asked those who felt they could control six to eight other men to step forward (a bold step which clearly worked well, since several of these self-selected NCOs went on to win commissions, and the battalion's war record proved second to none).

The battalions were conceived of not as reinforcements for existing brigades and divisions but as parts of entirely new Kitchener formations (brigades, divisions, corps), each successive 100,000 men forming a replica BEF, termed unofficially K1, K2 etc. Some Kitchener formations were already thoroughly militarized. In Northern Ireland the Ulster Volunteer Force, which earlier in the year had threatened armed resistance to the implementation of the home rule Bill, in effect turned itself into the 36th (Ulster) Division, with an unsurpassed fighting record on the

Western Front.* In the south of Ireland, too, the traditional source of manpower for virtually every regiment of the army at one time or another, the spirit of volunteering was no less strong, if somewhat more complicated. The 'Irish Volunteers' had been formed as a counter to the UVF, and their leaders – not so extreme as those who would take up arms in 1916 – now decided that the cause of home rule would be best advanced by a show of loyalty to the King, exactly as the UVF's leaders believed that volunteering safe-guarded Unionism. Two divisions were formed – the 10th and 16th – predominantly but by no means wholly Catholic (as the 36th Division was by no means exclusively Protestant). The 5th (Service) Battalion of the Connaught Rangers was raised in Dublin in August – unusually, for Dublin is in Leinster, not Connaught – and a second service battalion the following month in Cork (Munster)†. The 5th were fortunate enough to have the excellent Serjeant McIlwain posted to them as a serjeant-major after Ypres, when his much depleted 2nd battalion of the Connaughts was amalgamated with the 1st.

Before the war it was reckoned that it took about eleven months to make a soldier – and this with the full resources of the regular army. The first of Kitchener's divisions, the 18th, was judged ready to deploy to France in May 1915, even though it was a K2 formation (i.e. raised as part of the second 100,000). Its early readiness was in part due to the imagination and energy of its commander, Major-General Ivor Maxse, who had made his name in the Sudan campaign and South Africa and had commanded the 1st (Guards) Brigade from Mons to the Aisne. Maxse's ideas about training were innovative – 'drilling for initiative', as he

---

* Four out of the nine VCs awarded on the first day of the Somme were won by the 36th (Ulster) Division.
† with many volunteers from West Belfast.

called it, insisting that all tactical movements must be planned and rehearsed with nothing foreseeable left to chance (what would later be known as 'battle drill'), so that the commander's mind was then free to apply the drill to the particular situation. He was impressed by the quality of the recruits, and his training concept had obvious resonance with men used to making decisions for themselves. Not all Kitchener divisions were so fortunate, and some, though in theory fully trained, would be plunged into action before they were able to get used to conditions in the field.

On 2 January 1915 Sir John French received a letter from Kitchener that could have been penned by Pitt a century or so before: 'The feeling here is gaining ground that, although it is essential to defend the line we now hold, troops over and above what is necessary for that service could better be employed elsewhere. The question *where* anything effective could be accomplished opens a large field and requires a good deal of study. What are the views of your staff?'

The letter revealed the usual symptoms of the chronic British strategic disease: seeking easier victory somewhere other than where victory might achieve the strategic ends – through what would later be derided as 'stratagems of evasion'. Kitchener's question was reasonable enough in that it recognized that battering the army's head against the brick wall of the German defences on the Western Front was going to be both costly and unproductive; but the idea that the Germans could be defeated other than by defeating the German army did not recognize the reality of the German state. 'Finding the flank' was one thing at the tactical level, but quite another at the strategic.

But the opportunity seemed to present itself after Turkey entered the war on the side of Germany and Austria in November 1914. The war cabinet's thoughts – in particular those of the first lord of the admiralty, Winston Churchill –

turned to the eastern Mediterranean and the twin prospect of opening up warm-water lines of communication to Russia and defeating Germany by 'knocking away the props'. If Turkey were defeated, he argued, Italy, Greece and Bulgaria, who were still sitting on the fence, would come in on the allied side. This could be achieved at little cost, he further argued, believing that the Dardanelles could be forced by a handful of obsolescent warships. Kitchener had misgivings but the war cabinet gave way, and in February a naval bombardment of Turkish forts commanding the Dardanelles Straits began. By the middle of March the efforts to force the straits had come to grief, with both French and British warships destroyed by Turkish shore batteries and mines.

At this point the stratagem's 'economy of effort' was abandoned and Kitchener agreed to commit ground troops. The Mediterranean Expeditionary Force (MEF) was formed under command of General Sir Ian Hamilton, consisting of the 29th Division – the country's only strategic reserve – and the Australian and New Zealand Corps (Anzac), which was assembling in the Suez Canal zone.

Hamilton's record is as controversial as French's. The campaign would be a costly failure, but the margin between failure and brilliant success at Gallipoli was narrower than is sometimes supposed. The peninsula was by no means impregnable, although the Turkish army was experienced and its discipline formidable. The problem lay in the quality of some of the MEF's subordinate commanders and the lack of training in many of its battalions, those of the New Armies especially. Indeed, Hamilton himself had only thirty-five days to prepare for what was to be the first machine-gun-opposed amphibious assault in the history of warfare.

His plan for multiple landings on 25 April, given that the months of naval bombardment had forfeited strategic surprise, was feasible enough. He intended the 29th Division to land at Helles on the tip of the peninsula and then advance on the forts at Kilitbahir; the Anzacs were to land some 15 miles up the coast on the Aegean side and advance across the peninsula to cut off any Turkish retreat and prevent reinforcement of Kilitbahir; the French (for this was an allied operation) would make a diversionary landing on the Asian shore, and the Royal Naval Division would make a demonstration in the Gulf of Xeros to confuse the Turkish high command as to where the main effort lay.

The execution of the plan was not up to its conception, however. The Anzacs, under the exceptional 49-year-old British officer Major-General William Birdwood, a Bengal Lancer, gained tactical surprise by landing before dawn without a preliminary bombardment. Thereafter, in crude terms, he simply ran out of luck. The 29th Division, under the far less exceptional Aylmer Hunter-Weston, simply impaled themselves on the cliffs and machine-guns at the southern tip of the peninsula in broad daylight after a preliminary naval bombardment that would have left the dullest Turk in no doubt as to what was about to happen.

The Gallipoli peninsula thereby became a salient every bit as lethal as Ypres, and with the added complications of supply across open beaches, water shortage, intense heat and insanitary conditions. Despite suicidal gallantry on the part of British, Anzac and Indian (including Gurkha) troops, none could get any further than 3 miles from the beaches. As on the Western Front, barbed wire, machine guns and artillery put paid to any thoughts of tactical manœuvre: the only alternatives were head-on attack or evacuation. The excellent Serjeant-Major McIlwain whose battalion of Connaught Rangers was part of the 10th (Irish) Division in 'K1' recounts one such assault:

Attack begins at 4 p.m. Small parties of 'A' 'B' and 'C' Coys attack after heavy bombardment. We [D Company] are in reserve. Our people have heavy casualties. Major Money commanding attack from sap [trench extending towards enemy lines] where I am with him. After dark when fighting has slackened, and the trenches chock full of Irish and Turkish dead and dying, seemed owned by no acting force. Capt Webber takes up us the reserve (about 50 strong) to occupy the position. At bifurcation of trenches the Captain goes north and sends me to command right and occupy where practicable. I did not see him again. The dead being piled up quite to the parapet I take my party over the top in rear and with about 20 men occupy portion of trench nominally Australian as many wounded Anzacs are there. Not long there when Turks bomb us [throw grenades] from front and left flank, also snipe us along the trench from left. My men with few exceptions panic stricken. By rapid musketry we keep down the bombing. My rifle red almost with firing. By using greatcoats we save ourselves from bombs. Turks but ten yards away drive us back foot by foot. Have extraordinary escapes. Two men killed beside me following me in the narrow trench and I am covered head to foot in blood. Casualties alarming and we should have fought to the very end but for the 18th Australian Battalion a party of whom jumped in amongst us and held the position until reinforced. When able to look about me I find but two Rangers left with me. The rest killed, wounded, or ran away before or after the Anzacs had come.

Hamilton was relieved of command in the middle of October and replaced by Sir Charles Monro. In this, at last, the troops were lucky: Monro was fresh from command of the 3rd Army on the Western Front, and as commandant of the Hythe School of Musketry after the Boer War he had been the architect of the infantry's education in marksmanship and the new tactics of fire and movement. He was independent-minded and his reputation stood high, and he lost no time in assessing the situation and reporting to the war council that there was no future in the campaign –

which advice they accepted, if grudgingly, for it had also been Hamilton's opinion.

By 8 January 1916 the peninsula had been evacuated. And, just as at Corunna and later at Dunkirk, the evacuation was a sort of brilliant parting shot in an otherwise dismal campaign: remarkably, 134,000 men were taken off without a single loss, and with the Turks oblivious of what was happening almost literally under their noses – a masterpiece of deception, organization and command that would be studied at the Staff College for years afterwards. If only the same degree of ingenuity had been there from the outset. If only the same guileful tactical methods could have been used six months later on the Somme, indeed: for if surprise could be achieved in the withdrawal, there was no reason in principle why it could not be achieved in the offensive.

The cost of the Gallipoli campaign had been massive. British casualties were at least 75,000, including 21,255 dead, a great many from dysentery and enteric fever.* The New Army battalions suffered particularly high losses, especially of officers – none more than the 6th (Service) Battalion, the Yorkshire Regiment (Green Howards), where only the quartermaster survived unscathed. And these losses were all the more calamitous for their cost to operations on the Western Front: the lack of reserves at the battle of Loos in late September and early October 1915, for example, was a critical factor in the failure of the British autumn offensive in Picardy. Nor did the casualty lists do anything for recruiting, which took a dip towards the end of 1915 – rather as yellow fever in the West Indies had hampered recruiting during the Napoleonic Wars. Confidence in British generalship among the Anzac troops (and, just as important, their

---

* Australian casualties were 28,000 (7,600 dead); New Zealand's 7,250 (2,700 dead); French (estimated) 27,000 (10,000 dead); and Indian Army (including Gurkhas) 5,000 (1,400 dead).

governments) fell dangerously, although regard for Birdwood remained high; and a general sense of failure hung upon the nation. On the positive side, the failure at Gallipoli put the Western Front back at centre stage as the decisive theatre of war, although it would take another year at least for the politicians to accept the full consequences of this.

Perhaps the most dispiriting aspect of the campaign was its predictability (for all that taking the peninsula was by no means an impossible concept). Its planning took place after a full six months of fighting on the Western Front, by which time there could have been few illusions about the nature of modern warfare. Aircraft would have been particularly useful in reconnaissance and artillery spotting, and Hamilton's chief of staff had asked for them early on during the planning; but the request was turned down flat by Kitchener. The navy had aircraft, some of them ship-based, but they were not well integrated with the action ashore. The greatest lesson, however, that amphibious operations were fraught with special dangers and required planning and resources of an exceptional kind, would eventually see a spectacular pay-off, though with another high-priced lesson in between, thirty years later in the Normandy landings.

If several military heads rolled after Gallipoli (though not enough: Hunter-Weston showed no more acumen when in command of a corps at the Somme later in the year), none rolled so famously as that of one politician – Winston Churchill. In November 1915, in characteristic form, he left the government, put on soldier's uniform and went to France – to command one of Kitchener's battalions of the Royal Scots Fusiliers.* But if Churchill's time at the Admiralty had brought about one of the most calamitous and futile campaigns in the

---

* Churchill commanded 6th Battalion Royal Scots Fusiliers until May the following year, when the battalion was amalgamated with the 7th. He saw service in the trenches and was recommended for brigade command.

army's history (an experience which did not, however, entirely cure him of strategic gambling) it also did a very great deal to place in its hands a weapon that would do much to help it overcome the stalemate of the trenches.

The tank was neither Churchill's invention nor his conception, as is sometimes over-enthusiastically claimed, but without him its potential might not have been recognized so early or its development pressed so effectively. And it came about from that same thrusting impatience which had placed him so often in the middle of the action as a young man. As war was declared the first lord of the admiralty had sent Royal Marines and the ad hoc Royal Naval Division* to help defend Antwerp, and in the weeks before the trenches reached the Belgian coast the 'jollies and blue-jackets' were able to range along the flat roads of west Flanders in a veritable circus of improvised armoured cars and London buses. When Antwerp fell and the division was re-embarked, its buses went to the army as troop transports and many of the surviving armoured cars were sent to the Middle East. The instinct for mobility remained, though, and Churchill set up the Admiralty Landships Committee out of which emerged the machine whose image the new Tank Corps would wear, and still wears, as its cap badge.

The English-speaking world calls them 'tanks' because that is what they were called in the deception measures during their despatch to France: they were shipped as 'water tanks'. The French, when they were told of the invention and developed machines of their own, called them with Gallic panache *chars* – chariots. The Germans, when they learned

---

* There was a long tradition of 'spare' seamen serving ashore as infantry-men or gunners. Occasionally they took to it rather well and decided to make a career of it – like Midshipman Evelyn Wood, who served with the Naval Brigade in the Crimea, transferred to the 13th Light Dragoons and rose to the rank of field marshal.

of them the hard way and made some of their own, called them with Teutonic pedantry *Panzerkampfwagen* – armoured fighting vehicles (later simply *Panzer*). Much is revealed in military language. And largely for the same reasons of deception, the Tank Corps was originally known as the Heavy Section, Machine Gun Corps, and the first men in tanks wore crossed machine guns as their badge. But before the tank was ready for service the British army would receive its greatest ever shock, whose effects would change the course of the war and are still felt, consciously or otherwise, even today – the Somme.*

Through 1915, then, the Kitchener battalions had begun to enter the field. By the spring of 1916, K1–5 (Kitchener's first 500,000) had either seen action or were deployed and ready. But the BEF's commanders had understandable reservations about their capability, for the Western Front had become largely a business of siege warfare. To some extent its routine of constant labour – digging, repairing, wiring, carrying forward ammunition and stores – was easy enough for the New Army battalions to cope with. In its turn in the line, a battalion would typically spend four days in the fire and support trenches, four in close reserve half a mile or so to the rear, and four resting out of range of field artillery. Out of the line they would be at training and recreation. But in 1916 the training of the New Army battalions was directed more towards individual trench skills and fitness than to

---

* This is a contentious assertion, but ever since the Somme the army has been casualty conscious in a way that it never quite was before 1916. The losses in 1914 and 1915 were generally seen as the price to be paid for fighting in the big league, and largely unavoidable. The Gallipoli débâcle sounded a warning bell, but did not adversely affect operations on the Western Front. The casualties on the Somme changed everything, however. And they haunted politicians and soldiers alike in the inter-war years and during the Second World War.

battalion and company battle drills and tactics (and 'musketry' occupied nothing like its place in the pre-war army). In short, Kitchener's men could look after themselves, but they had scant practice in how to advance in the face of the enemy. The techniques of fire and movement seemed to be yesterday's story – the experience of the few months of 1914 before the trenches reached the coast. Now it was down to the bomb (grenade) and the machine gun, and occasionally the bayonet. The routine of trench warfare seemed to operate as a suppressant even to thinking about mobility. To an extent senior officers had willed this, or at least accepted it, but many of them were all too aware of the limitations it now imposed. The only part of the army still movement-minded (other than, of course, the Royal Flying Corps) was the cavalry, and they could only manœuvre once the infantry had punched a hole in the defences. To the very end, Haig saw the winning of the war in terms of the collapse of a part of the German front line so that mobile troops – horse-borne – could exploit tactical success and restore the war of manœuvre by which the decision would be gained. It was loosely, and often irreverently, referred to as 'galloping through the "G" in Gap'. Knowing the want of training in the New Army battalions, however, senior officers sought a simpler way of crossing no-man's-land to break open the German defences; and as the last of the K5 battalions were arriving in France in the spring of 1916, some thought they had found one in the experience of a battle in March the previous year – at Neuve Chapelle.

By 1915 the BEF had grown to such a strength that it had been reorganized into two armies – 1st Army, commanded by the newly promoted General Sir Douglas Haig, and 2nd Army, commanded by Horace Smith-Dorrien (until in May 'Wully' Robertson told him he was 'for 'ome'). Haig's stock stood particularly high after he had kept his head and held the line at First Ypres, and so Sir John French delegated to him the spring offensive in Artois (Pas de Calais). Although the offensive came to nothing other than a long casualty list, Haig had a measure of success at Neuve Chapelle. Here, after a huge but brief artillery bombardment, the attacking divisions (two regular British and two from the Indian Corps with a number of integrated British units) made progress, though not without cost: 11,200 casualties were taken during the three days' fighting, including six battalion commanding officers. But if the attack was in the end a disappointment – the Germans were able to recover more quickly than the British could exploit their own local success – it seemed to show what artillery could do when there was enough of it. In fact, Haig had been able to concentrate 340 guns – as many as the BEF had taken to France the previous August – against the German salient, a ratio of one gun to every six yards of front attacked. It followed, by his reasoning, that as great a concentration of guns firing for longer against the German line elsewhere would achieve even better results.

There were two flaws in this analysis, however, the first cruelly hidden. The bombardment at Neuve Chapelle had been brief – thirty minutes on the fire trenches and then a further thirty on the support lines – because of a shortage of shells (which would soon become the great 'Shell Scandal', forcing Asquith to form a coalition government and a separate ministry of munitions under Lloyd George). The brevity of the bombardment was regarded not just as the consequence of weakness but as *the* weakness, though in fact a degree of surprise was achieved by that very brevity, and it was this that made for success more than the actual damage inflicted by the guns. A prolonged bombardment naturally forfeited surprise; it served notice on the defenders that they would have to stand to and repel attack, and to senior commanders that they must move reserves ready to reinforce and counter-attack. The longer, too, that a battery fired, the more likely it was to be detected and put out of action by counter-battery fire. The success at Neuve Chapelle was more the result of surprise and the shock of the hurricane bombardment – what the Germans afterwards described as 'the first true drum-fire [*Trommelfeuer*] yet heard' – than of obliteration. If the artillery did not keep German heads down long enough, the assaulting troops would never make it across no-man's-land. Indeed, this much was obvious from one sector of the Neuve Chapelle attack, where a howitzer battery had been delayed getting into position and failed to fire on its designated target: the Germans in the untouched trench had been so quick to their machine guns when the noise stopped that 1,100 men fell to their fire. The lesson was clear either way, therefore: if the bombardment did not destroy the defenders, when it lifted the battle would be a straight race for the parapet.

The other flaw in the analysis lay in the assumption that the Germans – and their barbed wire – would be as vulnerable to artillery fire in the future as they had been at Neuve Chapelle. But in that part of Artois the water table is high (whence the 'artesian well'): the German trenches had had to be built up as much as dug down, and in consequence were more vulnerable to artillery. In the valley of the Somme, however, where Haig, who by 1916 had replaced Sir John French as commander-in-chief, had decided to mount that year's allied offensive, the Germans had burrowed deep (30 feet in places) into the chalk. And not only were the Germans dug in deep, in the second full year of static warfare the barbed wire defences in no-man's-land across the whole of the Front were thicker and deeper than ever. In other words, Neuve Chapelle was to the Somme as the English Channel is to the Atlantic.

Haig delegated the planning and execution of the Somme offensive to the man who had commanded IV Corps at Neuve Chapelle, Lieutenant-General Sir Henry Rawlinson (who had briefly been sent to Gallipoli to help plan the evacuation), and who now commanded 4th Army. Rawlinson's plan of attack would rely on a long obliterating bombardment which would destroy both the Germans in their dug-outs and the barbed wire which protected them, enabling Kitchener's under-trained infantry simply to march the several hundred yards across no-man's-land and occupy what remained of the enemy trenches. And because they would not advance across no-man's-land at more than walking pace (the rate at which the artillery barrage would 'creep' – a new technique – from the fire trenches to the support trenches) they would be able to carry the extra ammunition, defence stores and rations needed before resupplies could be brought forward – a load per man of some 60 pounds (27 kilograms).

Haig had also hoped that his attack would be able to use the new secret weapon – the tank. He was impressed by what had emerged from the Admiralty Landships Committee – an armoured rhomboid box on caterpillar tracks which did indeed look like the water tank of its cover name. It was armed with either machine guns or quick-firing 6-pounders mounted on sponsons either side, had a crew of eight and moved at walking pace. Haig had at first been sceptical, but began to think the tank might have potential after seeing a demonstration of its barbed-wire crushing and trench-crossing capability, and therefore its potential to help his infantry cross no-man's-land and punch the hole through which the cavalry could gallop. Unfortunately, technical problems meant that none would be available before September 1916. He had been willing to wait, but the Germans had opened a surprise offensive against the border fortress of Verdun in Lorraine, and Joffre urged him to make the attack no later than the end of June, else 'the French army would cease to exist'.

*   *   *

The seven-day preparatory bombardment on the Somme looked and sounded impressive – 1,500 guns firing 200,000 rounds a day* – but because of the length of front to be attacked (12 miles), the inadequate number of guns for the task, their varying accuracy and the failure rate of the ammunition (perhaps as many as one in three shells proved either duds or misfires), the weight of fire was proportionately only *half* that at Neuve Chapelle.

John Masefield, poet laureate from 1930 to 1967, had served on the Western Front as a medical orderly before going to the Dardanelles to write propaganda for the Foreign Office. The result, *Gallipoli*, a work of great lyrical beauty, apparently so raised the public spirits in lauding what there was to be proud of in an otherwise inglorious episode, that he was sent to France in October 1916 with a brief to do the same for the Somme.† He looked at what he could of the old battlefield, spoke to whom he could, including Haig, and read what he could. And of that first morning of the British army's most debilitating battle ever, 1 July 1916, Masefield would write:

> It was fine, cloudless, summer weather, not very clear, for there was a good deal of heat haze and of mist in the nights and early mornings. It was hot yet brisk during the days. The roads were thick in dust. Clouds and streamers of chalk dust floated and rolled over all the roads leading to the front, till men and beasts were grey with it.
>
> At half past six in the morning of 1st July all the guns on our front quickened their fire to a pitch of intensity never before

---

* In the end, the shell count was nearer 1.7 million, since the French added their weight on the right.

† In *Gallipoli* Masefield wrote, for example: 'On the body of a dead Turk officer was a letter written the night before to his wife, a tender letter, filled mostly with personal matters. In it was the phrase, "These British are the finest fighters in the world. We have chosen the wrong friends."'

attained. Intermittent darkness and flashing so played on the enemy line from Gommecourt to Maricourt that it looked like a reef on a loppy day. For one instant it could be seen as a white rim above the wire, then some comber of a big shell struck it fair and spouted it black aloft . . .

In our trenches after seven o'clock on that morning, our men waited under a heavy fire for the signal to attack. Just before half-past seven, the mines at half a dozen points went up with a roar that shook the earth and brought down the parapets in our lines. Before the blackness of their burst had thinned or fallen, the hand of Time rested on the half-hour mark, and along all that old front line of the English there came a whistling and a crying. The men of the first wave climbed up the parapets, in tumult, darkness, and the presence of death, and having done with all pleasant things, advanced across the No Man's Land to begin the battle of the Somme.

The 'race for the parapet' had begun. But a race it scarcely was, for on the one side lines of smart new khaki advanced at the march, while on the other, when the bombardment shifted on to the reserve trenches, files of fossorial field grey sprinted up the steps of the deep dugouts to man the machine guns in the contest for the green fields beyond. It was – remains – the greatest set-piece attack, the greatest battle, in British history: eleven divisions in line (and two more in an adjacent diversionary attack) – 13 × 18,000 men.

The German Maxims opened up in long raking bursts, the arcs of fire interlocking so that every inch of ground was swept by bullets. The silent, waiting artillery batteries sprang into life, the guns laid on pre-adjusted lines just in front of the barbed wire, so that there was no need of corrections. And into this hail of lead, high explosive and shrapnel marched that glorious if part-trained infantry: pre-war regulars who had somehow made it this far, Territorials toughened by a year at the front – and, above all, the New

Army battalions, Kitchener's men. One way or another, every man a volunteer.

Some to begin with seemed to be entering into the spirit of the expected walk-over. In the 8th East Surreys, the battalion formed by the single regular captain who had sensibly asked for those who felt they could control six to eight other men to step forward, one company went into the attack dribbling footballs across no-man's-land. The game did not last long: 147 officers and men were killed and 279 wounded before they reached the German fire trench. But as one of the few battalions to gain its objective that day, it reaped an impressive array of gallantry awards: two DSOs, two Military Crosses, two Distinguished Conduct Medals and nine Military Medals. The 'Football Attack' caught the imagination of the country, with graphic accounts in the press; and the regimental museum in Guildford has one of the tragic leather balls.

The 1st and 2nd Barnsley Pals (13th and 14th Battalions, York and Lancaster Regiment), attacking side by side in 31st Division further north, got nowhere. The 1st Pals went over the top 720 strong; by the middle of the afternoon there were but 250 of them left. The 2nd Pals fared slightly better, losing 300 before the brigade commander called them off. Both battalions were lucky, however, for besides being in the support wave their brigadier was the 34-year-old Hubert Conway Rees, a regular infantryman robust enough to halt their attack. Elsewhere, the plan continued as if nothing untoward were happening. In front of the Barnsley Pals had been another Pals battalion – the 12th East Lancashires, the 'Accrington Pals', 730 men from the close-knit cotton mill towns of 'Blackburnshire'. After the first half an hour, 600 of them were killed, wounded or 'missing' (in other words, there were no remains of them after shelling). The brigadier wrote of them and the rest of his brigade:

The result of the H.E. shells, shrapnel, machine-gun and rifle fire was such that hardly any of our men reached the German front trench. The lines which advanced in such admirable order, melted away under fire; yet not a man wavered, broke the ranks or attempted to go back. I have never seen, indeed could never have imagined such a magnificent display of gallantry, discipline and determination.

The poet Robert Graves was lucky to have missed the first day (he had been sent back to England for an operation on his nose to allow him to breathe properly in the new gas mask): three out of five of his fellow officers in the 1st Battalion Royal Welch Fusiliers (who were very particular about the spelling) were killed. His fellow subaltern-poet Siegfried Sassoon, who had just won the MC, would have been one of them too, except that his company was held in reserve. Sassoon had spent a good deal of the day and night before the battle crawling about in no-man's-land trying to make wider gaps in the barbed wire with his new wire-cutters, bought on leave at the Army and Navy Stores, so that the New Army battalion of the Manchester Regiment which was to attack from their trench might have a better chance: 'it seemed to me that our prestige as a regular battalion had been entrusted to my care on a front of several hundred yards,' he wrote later. Wilfred Owen, who had enlisted in the ranks in 1915 and would soon be commissioned into the Manchesters, was lucky, too, to be away from the front in an officer training battalion. (It was here that he wrote much of his poetry: 'My subject is war, and the pity of war. The poetry is in the pity.') He would be killed a week before the armistice leading an attack with the Manchesters, for which like Sassoon he would win the MC.

Isaac Rosenberg, one of the few acknowledged greats of the war poets not to have been commissioned, was serving with the 11th Battalion, the King's Own, a 'Bantam battalion' (consisting of men under the 1914 minimum height of 5 feet 3 inches) and therefore not assigned an assault role on 1 July, being employed instead on fatigue duties and burials. In the days that followed he would write what the American literary critic (and Second World War infantry officer) Paul Fussell has called the greatest poem of the war: 'Break of Day in the Trenches'.* It is a short, stark meditation on mud, rats and poppies, with none of the spirit that had motivated men to flock to the recruiting offices in their hundreds of thousands – including Rosenberg's own 'Bantams', pint-sized men determined to do their bit. Yet it was these very men who went over the top on 1 July, their ardour perhaps just a little diminished by recent experience, but not much:

> The darkness crumbles away.
> It is the same old druid Time as ever,
> Only a live thing leaps my hand,
> A queer sardonic rat . . . †

And Sassoon was no doubt thinking of the men of the Kitchener battalions of Rosenberg's regiment, the King's Own – the Royal Lancasters – when he wrote:

> But to the end, unjudging, he'll endure
> Horror and pain, not uncontent to die
> That Lancaster on Lune may stand secure.‡

At the end of the first day of the battle of the Somme the

---

* Fussell, *The Great War and Modern Memory* (1975).
† Isaac Rosenberg was killed while serving with the 11th King's Own on 1 April 1918.
‡ 'The Redeemer' (from *The Old Huntsman*).

casualties numbered 57,470 men, including that fearful category 'missing'. Of these, 19,240 were soon listed as 'killed in action'.

23

Never Again

1916 and beyond

SOLDIERS AND POLITICIANS HAVE WALKED EVER DEEPER IN THE shadow of the valley of the Somme. The casualty lists which for months filled the newspapers imprinted themselves equally indelibly on the mind of the nation. And the lists included only the names of the officers.

The casualties on the first day of the battle were almost as great as the duke of Wellington's in the entire Peninsular campaign. Yet at the end of that first day's fighting no single person had a complete grasp of the scale of the disaster. The day after the battle Haig recorded in his diary that the estimates were 'over 40,000 to date', adding that 'this cannot be considered severe in view of the numbers engaged, and the length of front attacked'. Rawlinson had expected 10,000 a day. But quite apart from the under-estimating of its cost, the attack had not been successful, and as yet Haig did not grasp this either. On 8 July he wrote that 'In another fortnight, with Divine Help, I hope that some decisive results may be obtained.' He did not in fact finally abandon ('close down' is the rather euphemistic term) the Somme offensive until the middle of November.

Yet even as the true scale of the losses became widely known, morale did not crack, as it did in the French army the following year in the wake of the failures of their renewed offensives. There were, however, some 'indications of a loss of fighting spirit': Haig records his dismay, for example, at the 49th (West Riding) Division, made up primarily of Territorial units, who appeared to have made no gains for proportionately fewer casualties than other formations. On 9 July the 11th Battalion of the Border Regiment, a 'Pals' unit formed in 1914 by the earl of Lonsdale and known throughout the army as 'the Lonsdales', simply failed to carry out a night attack when ordered (or rather, what was left of the battalion failed, for out of 800 men they had lost 500 killed, missing or wounded, including twenty-five of their twenty-eight officers on the first day). It was not mutiny: there was no outright refusal to obey orders, no desertions; the attack just didn't happen.

When the divisional commander learned of it he ordered exemplary punishment for the commanding officer (a captain who had been brought in from another battalion in the same brigade after the losses of the first day) and what few other officers were left, including even the regimental medical officer (RMO). The disciplinary papers wound their way up the chain of command, gathering more and more exclamations of outrage at the acting commanding officer's failure, and that of the RMO who had reported the men mentally unfit for duty, castigating also the Lonsdales' pathetically small remaining band of NCOs. Eventually the papers arrived at GHQ (Haig's headquarters) where the chief medical officer, the wonderfully named Surgeon-General Sir Arthur Sloggett, pointed out one or two facts that had been overlooked in the atmosphere of scapegoating after 1 July: the internal cohesion of the brigade had been severely shaken; the brigade commander himself appeared unfit for duty, as indeed did the Lonsdales' acting commanding officer; and the RMO had been asked his opinion of the state of the battalion's nerves, which was not within his capability to judge. Haig, sitting outside the 'bubble' of 5th (Reserve) Army, which was commanded by the ever-certain Gough of Curragh fame and which had taken over responsibility for that part of the front, was able to agree with Sloggett, and matters were quietly rearranged. The Lonsdales were rested and reinforced, and they fought well again later in the year; by the end of the war, the officers involved had all been decorated.*

It was a demonstration that though its limits are indistinct, endurance is finite: the Lonsdales' experience was perhaps extreme, but there were other cases of battalions having lost not so much their nerve as their confidence. Still, most persevered as before, continuing to lose men in steady numbers, particularly young officers. After the Somme, however, they were reinforced increasingly – and for the first time in the nation's history – by conscripts. A Military Service Act had been passed in January 1916 making virtually all men between the ages of 19 and 41 liable for call-up – except, controversially, men in Ireland.

The losses did not at first deter offensive action, certainly not the following year, although Lloyd George, who in December 1916 replaced Asquith (whose son Raymond had been killed on the Somme) as prime minister, tried to apply the brake: the huge attacks around Ypres, popularly known as Passchendaele, were both costly and just as futile as those of the year before. Only after the war did the impact of the Somme take full effect in the collective professional mind of the army – and, indeed, in the minds of individual officers.

* The story is told in full in an article by (Surgeon) Brigadier Timothy Finnegan in *The British Army Review*, Spring 2008.

It was not that Haig and his senior officers were in-different to the losses, that they had all somehow undergone compassion bypass surgery – the imputation of the *Oh, What A Lovely War* school of history. They knew their resources of manpower were finite – but also that they could not just stand on the defensive. Nor were they hidebound. When, for example, tanks at last made their appearance on the Somme, at Flers on 15 September – a mere handful of them, because of production problems and breakdowns – Haig was finally convinced of their potential in helping the infantry across no-man's-land, and began pressing for as many as possible to be sent to France. And when a diversion-ary amphibious landing in Belgium was discussed the following year, he suggested flat-bottomed landing craft with ramp fronts from which tanks could be disgorged to over-come the enemy's defences. The British army on the Western Front was, indeed, technically innovative – more so than the German, if not always as tactically agile.

Just occasionally there were glimpses of what could be done with smart tactics. An assault on the German salient south-east of Ypres, the Messines Ridge, by Sir Herbert Plumer's 2nd Army in June 1917 was the first successful large-scale offensive operation of the war, a meticulously planned and cunning all-arms battle making use of extensive mining, precision artillery fire, tanks, gas and an assault before dawn. And in November that year 450 tanks attacked at Cambrai on firm, unbroken ground north of the Somme, penetrating the German line by sheer surprise and firepower, although as in previous attacks the success could not be exploited before the Germans were able to recover and seal the breach.

Spring 1918 brought the massive German offensive known as the *Kaiserschlacht* ('Kaiser's battle'), for Wilhelm had personally ordered Ludendorff, effectively commander-in-chief, to add more men to the planned offensive to make them up to one million. It failed, and the allied counter-offensive began in August. This was a sustained three-month effort (the 'Hundred Days') which included American troops just arrived in France, and Italian pilots, and involved sophisticated all-arms cooperation and close air support. Mechanical, aeronautical and ballistic technology had all advanced apace: Blériot had spluttered across the English Channel not ten years before, yet by November 1918 over 1,000 aircraft were flying in close support of Haig's armies – on reconnaissance, artillery spotting, and in ground attack sweeps with bombs and machine guns. If only radio technology had advanced as quickly (Marconi had sent radio signals across the Channel in 1899) commanders would have had the means to exploit the full potential of tanks, aircraft and artillery. Nevertheless, the Hundred Days was undeniably Haig's victory, and one which until relatively recently had not been fully acknowledged. Some historians now maintain, in fact, that it was the British army's greatest victory.* Certainly in November 1918 the British army stood second to none. The French generalissimo Ferdinand Foch, who had been hastily appointed to overall allied command when the German offensive opened – the first time there had been a unified command in the whole course of the war – wrote to Haig after the armistice:

> Never at any time in history has the British Army achieved greater results in attack than in this unbroken offensive . . . The victory was indeed complete, thanks to the Commanders of the Armies, Corps and Divisions and above all to the unselfishness, to the wise, loyal and energetic policy of their Commander-in-Chief, who made easy a great combination and sanctioned a prolonged gigantic effort.

* The battle of Amiens which began the offensive in earnest on 8 August with a surprise tank attack by 4th Army (Rawlinson) broke through the German lines, destroyed six divisions and forced the line back 9 miles in one day. Ludendorff called 8 August the 'Black Day of the German Army'.

In fact, in those final Hundred Days the BEF engaged and defeated 99 of the 197 German divisions on the Western Front, taking nearly 200,000 prisoners – almost 50 per cent of the total taken by all the allied armies in France in this period.

But before the opportunity that the *Kaiserschlacht* gave Haig and Foch to get out of the trench system which had dominated strategy and tactics for almost four years and 'back' to mobile warfare, an ambitious plan was being worked up for the following year to attack deep into the German rear areas with faster, lighter tanks supported by aircraft – 'Plan 1919'. Victory in November 1918, which was in large measure attributable to the enormous increase in professional capability in all parts of the army (the artillery especially) and in inter-arm cooperation, consigned the plan to the archives, but its concepts were not wholly abandoned in the minds of the protagonists of armoured manœuvre. And they would be eagerly studied by what remained of the German army after the war – and in due course refined as *Blitzkrieg*.

For the time being, however, the British army would have no need of mechanical toys. The Great War had been the war to end all wars, certainly on the Continent, and the treaty conference which followed at Versailles aimed to bring about a settled European peace as enduring as that at Vienna had done a century before. The army could therefore return to its late Victorian priorities: home security – which meant aid to the civil power, since the Royal Navy could prevent any invasion – and imperial policing. And although the two brigades which comprised the British Army of the Rhine after the great bulk of the occupation forces left at the end of 1919 remained in Germany until 1929, the notion of a 'limited liability', as it was called, in continental warfare soon became in reality a policy of no liability at all. With deterrence provided by the navy and increasingly by the bomber force of the RAF, there was no need of an expeditionary force for Europe.

In 1919 the new coalition government adopted a 'ten-year rule', a rolling assumption that there would be no major conflict in Europe in the coming decade. The army, which had stood at three and a half million men in November 1918, contracted within two years to one-tenth that size, and shrank even more in the years that followed through further savage cuts in public spending (the so-called Geddes Axe of 1922) and because of the difficulty of finding recruits even during a time of rising unemployment, conscription having ended with the war. A decade later, the secretary of state for war, Lord Hailsham, was calling the army 'a Cinderella of the forces'.*

But what had the Great War actually taught the army? Had it in any way reshaped it, in body or in mind? Or did the army collectively deal with the memory of its heavy losses by putting the experience of 1914–18 to the back of its mind – as the CIGS himself (Sir George Milne) seemed to do when in 1926 he said that the Great War had been 'abnormal'?

To most outward appearances the army did indeed seem to be doing just that. Regimental soldiering returned with a renewed vigour – some might say with a vengeance. Robert Graves, convalescing with the 3rd (Special Reserve) Battalion Royal Welch Fusiliers in Ireland in January 1919, wrote of the regular officers' dismay on seeing some of the 'Manchester cotton clerks' who had been accepted for commissions in the desperation to find subalterns for the Kitchener battalions, and who were now returning for demobilization:

---

* For detail on how the composition of the army had changed, see 'Notes and Further Reading'.

The latest arrivals from the New Army battalions were a constant shame to the senior officers. Paternity-orders, stumer cheques, and drunkenness on parade grew frequent; not to mention table manners at which Sergeant Malley stood aghast. We now had two mess ante-rooms, the junior and the senior; yet if the junior officer happened regimentally to be a gentleman (belonged, that is, to the North Wales landed gentry, or came from Sandhurst) the colonel invited him to use the senior ante-room and mix with his own class.*

Terms like 'officer and temporary gentleman' and 'for the duration only' peppered the conversation of 'proper officers'.

Yet quite obviously not all the 'temporary gentlemen' had put shop girls in the family way, bounced cheques or been unable to hold their drink. And some had proved themselves capable officers – despite, perhaps, holding their knives like pencils at the mess table. The discovery, for the first time in the army's history, that a man from the lower middle class who possessed a good mind and leadership qualities might be trained directly as an officer rather than work his way through the ranks might be inconvenient, but it could not be forgotten. Certainly it was a most useful discovery for the growing technical and logistic arms.

The reduction of the army also meant the re-emergence of the cavalry as the coequally pre-eminent arm after their marginalization on the Western Front. In Palestine under General Sir Edmund Allenby mounted troops had had dramatic success, and with attention now turning increasingly to the Middle East – drawn thither by the Suez Canal and oil – and once again to India, the mounted arm could expect to be back in the thick of things. And although the newly ennobled Field Marshal the Earl Haig was to leave the stage soon after the war (and throw his remaining few years

---

* Robert Graves, *Goodbye To All That*.

into the welfare of ex-servicemen), in his final despatches he made clear his opinion that a strong cavalry arm would remain essential in *any* future warfare – and so authoritative an opinion as that of the recent victor of the war in the West could not be lightly disregarded. The petrol engine was still not universally reliable, and the cross-country mobility of the horse was undeniably superior to anything the proponents of mechanization could point to – when, of course, the horse was not being fired at. Nevertheless there were cavalry amalgamations when the Geddes Axe swung, if no immediate appetite to part the men from their mounts. The twenty-six battalions of the new Tank Corps were reduced to a cadre of four, and the tanks themselves were left to rust in a field near Bovington in Dorset until Kipling urged the authorities to improvise a museum.* (This fate was at least better than that of the US Army Tank Corps which was disbanded altogether, with further tank development limited virtually to the drawing board.)

The Royal Artillery, which had 'gone heavier' with every year of the war, now put aside most of its larger-calibre guns to return to field artillery, although direct fire except *in extremis* was now firmly a thing of the past (the RA, RHA and RGA were merged into one in 1924). The Royal Flying Corps and the Royal Naval Air Service had already morphed into the Royal Air Force – on 1 April 1918, and to some opposition (with taunts of the 'Royal April Foolers') – and henceforth 'army cooperation' – reconnaissance, artillery spotting and the like – would have to take its place in the queue as the RAF set out to demonstrate that there was in fact little need for an army at all, that the Empire could be policed through 'air control' – knocking dissident tribesmen into line by the simple expedient of dropping bombs on their villages – and that future war

* The museum today is a vast, newly built complex housing the largest collection of armoured vehicles in the world.

could be won by strategic bombing. 'The bomber will always get through' was soon the prevailing doctrine (and fear). So the army, starved of funds and with no pressing role in Ireland after the creation of the Irish Free State in 1921, became once again a far-flung patchwork of regiments.

However, the army now knew – unquestionably – how to organize for war; and both the unit and the staff organization that emerged from the First World War remain substantially the same today. The battalion of infantry, between 700 and 1,000 strong, had proved itself the enduring tactical unit for fighting, the irreducible minimum in the field. Commanded by a lieutenant-colonel, it had a major as second-in-command; a captain as the executive officer (adjutant); one or two specialist officers in charge of, for example, signalling or machine guns; a long-service quartermaster commissioned from the ranks responsible for every aspect of equipping, supplying, feeding, clothing and housing the battalion; and a second executive, the regimental serjeant-major (the patriarch of the NCOs), who knew how to get things done. Within its constituent 'rifle' companies – usually four, each commanded by a captain or, increasingly as time went on, a major – the platoon system had also proved its strength. Before the war the subalterns had usually been simply additional company officers, with no fixed responsibilities; the experience of the trenches had shown the value of the officer figurehead for the fifty or so men who comprised the platoon, which was in turn divided into sections of eight to ten under a corporal. The platoon could act as a unit of fire and movement within the company, and with proper training the section could act likewise within the platoon, especially after the Bren light machine gun was introduced in the 1930s, one to each section. The young platoon commander and his older, more experienced platoon serjeant became one of the most hallowed, if perilous, relationships in human existence.

The battalion's regimental identity, itself rooted in a county, a clan or a historical function (such as Rifles), had also proved itself the most effective means of assimilating replacements for battle casualties – indeed, at times, of wholesale regeneration. By 1916 it had become the practice for a 'battle margin' to be formed before any offensive: a reserve of two or three officers and the RSM, who were left out of battle to form the nucleus around which the battalion could rebuild in case of heavy casualties. The ability of battalions to recover after major setbacks was one of the most remarkable aspects of the First World War, and that resilience – an attitude of mind as well as a system – has remained a distinctive feature of the army. Rarely were battalions broken up during that war, except towards the end when manpower shortages forced a re-organization of the K5 battalions. It was better for morale to rebuild rather than redistribute.

All-arms cooperation, however, had only really been effective at brigade level. Although artillery observation officers, usually captains, deployed increasingly with the infantry battalions, it was usually to adjust the fall of shot and help the commanding officer conform to the brigade or divisional fire plan; it was not primarily to provide him with the means of calling for fire for his own purposes. An artillery regiment consisted of three or four batteries and might be placed in support of an infantry brigade, though it remained – as today – a divisional asset. Indeed, it was the division, commanded by a major-general, that emerged definitively as the 'permanent' all-arms grouping, with its constituent brigades (usually three) allocated divisional support units on an ad hoc basis. The number of divisions in a corps – more correctly 'army corps', commanded by a lieutenant-general – varied from two to four depending on fighting requirements. The division therefore came to be regarded as the highest-level 'tactical' formation, one that focused from day to day on fighting the enemy. The corps, Janus-like, looked in both directions

– to the headquarters concerned with overall campaign planning, as well as to the divisions during both the battle's planning and its execution.

When the BEF sailed to France in August 1914 it had consisted of two relatively small corps, with a force headquarters – G(eneral)HQ – commanded by Field Marshal Sir John French, plus the cavalry division and lines of communication troops working directly to him. By 1915 a subordinate level to GHQ had been created – the 'army', commanded by a full general – for the first time in British military history. By 1916 there were five armies, the composition of each depending on the fighting requirement but consisting of a minimum of two corps. It was at army level that operations of significance to the campaign as a whole – and thus to allied war strategy – were planned, with direction from GHQ, which is to say Haig (a field marshal), whose job it was, inter alia, to interpret the strategic direction from London. There was no allied headquarters superior to GHQ until the German spring offensive in 1918, when the imminent collapse of the front required a supreme commander (and Foch was appointed 'generalissimo') rather than merely consultation.

All this was, so to speak, learned on the job. The resulting structure would become the model for the army in the Second World War – although after twenty years of gathering dust, since with the disbandment of the whole higher organization of the BEF there was no opportunity periodically to put it through its paces. The experience of command and staffwork at divisional, corps and army level was, however, preserved in the magisterium of the staff college at Camberley, and also that of the Indian Army at Quetta. The young Lieutenant Bernard Montgomery, given up for dead at Ypres, had subsequently learned his staffwork as a brigade major in K1, had seen what meticulous planning could achieve in the attack by Plumer's 2nd Army at Messines, and as chief of staff of 47th (London) Division in 1918 had experienced the exhilaration of the

'Hundred Days'. He and others like him in turn became the teachers at the staff colleges, so that the knowledge of how to organize, supply, maintain and train these formations, and above all take them into battle, was not entirely lost when in 1939 the political penny at last dropped and the commitment to continental warfare was again embraced.

But tactical ideas, as opposed to battle procedure, did not advance much during the inter-war years. In fact for the most part, ironically, they regressed: the Hundred Days were soon largely displaced in the collective mind of the army by the previous four years of linear warfare. 'Lines', so essential to the daily life of soldiers, became once again a dominating operational concept (and not only with the British), with near-catastrophic results in France and Belgium in 1940, and again in the North African desert. Indeed, linear thinking had not been wholly eradicated by the time of the D-Day landings, nor yet in the advance from Normandy to the Baltic. In fact, for the best part of the four decades which followed the end of the Second World War it was the dominant feature of the plans for the defence of Western Europe.

There were, however, after the First World War some officers determined to develop the embryonic theories of mechanized warfare which Plan 1919 had suggested and whose possibilities the Hundred Days had tantalizingly shown. The War Office, if only perhaps at first out of weariness, allowed them to proceed experimentally. Two men in particular – 'Boney' Fuller and Basil Liddell Hart* – wrote extensively on the potential of the petrol engine on land and

*John Frederick Charles Fuller had been a regular officer before the war, transferring from the infantry to the new Tank Corps at its inception. Liddell Hart had joined the infantry in 1914, and although he saw little active service, being dogged by a poor constitution and gas wounds, he applied his mind to warfare in a theoretical way. He remains controversial, with a reputation for a certain 'military shamanism' or even charlatanism. But such accusations are not uncommon when it comes to revolutionary military doctrine.

in the air to generate operations of great intensity and free manœuvre, provoking heated debate over what was otherwise in danger of becoming once more a pedestrian army. The politicians' mantra was 'never again' – never again the mass slaughter of a Western Front, and therefore never again a continental commitment. In the Great War, of the eight million men called up almost a million had been killed and two million wounded: 'never again' must mean the avoidance of war, not its cleverer prosecution. On the other hand, Fuller, Liddell Hart and theorists like them, in their own revulsion at mass casualties and with no faith in treaties, sought a military alternative in their doctrines of the 'indirect approach'.

The dominant idea in the minds of many officers, however, had come to be the simple question 'Is it worth it?' – an instinctive cost–gain calculus: no longer would they press home attacks at huge loss, 'at all costs', for that phrase had frequently seemed to reflect an unwillingness or inability to calculate the costs. Montgomery, for one, was quite clear about this: 'The frightful casualties appalled me. The so-called "good fighting generals" of the war appeared to me to be those who had a complete disregard for human life. There were of course exceptions and I suppose one was Plumer.' Montgomery would evolve in his own mind a doctrine that encompassed a higher regard for human life – though not to the exclusion of gaining a truly worthwhile objective. And therein lay the difference between Montgomery and the Great War generals he criticized. Minimizing casualties, as opposed to simply despairing of a big 'butcher's bill', became a defining factor in the mental approach of British officers. It continues to characterize the approach today, together with what is called 'force protection', and occasionally appears to stand in contrast to the attitude of US army officers who, while using every technical means available, retain the sense that fighting is an inevitably bloody business.

Alongside the theoretical debates, in practice the Royal Tank Corps (RTC) – it was 'royalled' in 1923 – kept alive the flame of armoured warfare when official policy did not acknowledge any likelihood of fighting an enemy that had tanks. In 1927 Milne (CIGS), despite or perhaps because of his view that the Great War had been an aberration, authorized an Experimental Mobile Force consisting of two battalions of the RTC, a regiment of field artillery, another of light artillery, a company of Royal Engineers and a machine-gun battalion, the whole force to be controlled by radio. Its purpose was to determine the ideal organization of a mechanized division – even if the requirement was still hypothetical. And two regiments of cavalry were unhorsed and equipped with armoured cars – the 11th Hussars and the 12th Lancers, who as the most junior of the regiments remaining unamalgamated after 1922 drew the short straw in the eyes of their fellow cavalrymen. To the surprise of almost everyone, none of the officers resigned; no one seemed to have noticed that in these two particularly exclusive regiments almost every officer had his own motor car, and several had their own aeroplanes.*

When senior officers did agree on the need for tanks, however, they frequently disagreed on what the tanks were needed for – and therefore on their specifications. Most viewed them (as did their American counterparts) as supports for the infantry in an all-arms battle along the lines that had developed in France in 1918. A minority saw tanks as a decisive element in themselves, as in a fleet action at sea. Tank design was not, therefore, very innovative. New models tended to be slow-moving, infantry-paced. They were heavily armoured because at these slow speeds they were vulnerable, and the heavy armour in turn limited their

---

* At the same time, several yeomanry regiments were equipped with armoured cars under the auspices of the TA RTC.

speed. The tank carried only a small gun because its purpose was to overcome infantry opposition, not destroy other tanks. Faster, minimally armed and lightly armoured 'cruiser' (or 'cavalry') tanks were also developed for reconnaissance; these were in effect little more than armoured cars on tracks, though they could exploit a breakthrough as long as they were not expecting much opposition. Both types would prove ineffective when up against the panzer forces developed in the late 1930s by German armoured warfare theorists who had studied and developed the views of Fuller and Liddell Hart along lines of reasoning that the British army had largely rejected.

For most of the inter-war period armoured warfare remained a hypothetical requirement because of the government policy of 'limited liability' for continental warfare. As late as 1936 the army's annual expenditure on horse feed was £400,000 – nearly four times what it was spending on petrol. When, however, Hitler seized absolute power in 1934, and the following year formed three panzer divisions and unveiled an embryonic Luftwaffe, the prospect of armoured warfare suddenly looked rather more real. But even then, when the ten-year rule was genuinely abandoned and re-armament begun (the rule had officially been dropped three years earlier, but the defence budget had not increased), no significant extra funding came the army's way except for anti-aircraft guns for the air defence of the United Kingdom; the RAF and the Royal Navy took priority. Indeed, so starved of cash was the army equipment programme that the domestic arms industry had almost disappeared when at last in 1938 Neville Chamberlain's government turned to the army's own rearmament. Fortunately, thanks in no small measure to Milne's foresight in forming the EMF, there was at least an understanding of the requirements of mechanization, and the cavalry at last began exchanging their horses for armoured cars and light tanks; the petrol engine – on paper at least – became the army's motive power.*

The army of 1938 was still an army that saw itself in the mirror of the Great War, however. It wore practically the same uniform, the same canvas web equipment, the same steel helmet; and it carried essentially the same Lee–Enfield rifle. It knew what its predecessor in the trenches had endured (indeed, most officers above major, and many senior NCOs, wore Great War medal ribbons), and viewed improvisation in the midst of confusion as par for the course. The same is true, if perhaps less consciously, today. Recruits are taken to the battlefields of the Western Front to give them an idea of what their regimental forebears faced, with the implicit message 'remember before complaining what those who have worn the same cap badge endured' – and, of course, the unspoken reminder of duty 'even unto death'.

Although fascism might not have emerged, and the German army might not have been equipped and trained as it was, had the First World War not been fought, rendering the comparison futile, there is little doubt that the British army would have been in no condition to fight in 1940 without its experience of 1914–18. Those nations that had

* In 1939 the Royal Armoured Corps was formed, an umbrella organization for the RTC and the mechanized cavalry. Two regiments remained horsed, the 1st Dragoons (Royals) and the Scots Greys, as well as the two regiments of Household Cavalry (and much of the yeomanry). The Royals and the Greys were the senior of the line, which indicated the instinct that remaining horsed was a privilege. In fact the Greys' commanding officer had written to all Scottish MPs to lobby for the retention of horses, which act earned him 'an expression of the Army Board's severe displeasure'. Ironically, one of the MPs who raised the Greys' case in Parliament was William Anstruther-Gray, member for North Lanarkshire (at a time when there still were Tory MPs in Scotland), who would go on to win an MC with his (armoured) yeomanry regiment, the Lothians and Border Horse. In hindsight he cannot have regarded his October 1938 question to the secretary of state for war as his finest moment.

remained neutral in the First World War were rapidly defeated in 1939 and 1940, though hard fighting in that war did not of itself guarantee fighting power in the second, as the French and Belgians found to their cost. The First World War had forged a modern army whose foundations remained fundamentally sound even though much of the super-structure had become derelict in the two decades that followed. Not least it gave certain commanders a priceless sense of the reality of war, and the moral authority that comes of the experience of hard fighting. It is unthinkable, for example, that Montgomery could have fought the battle of El Alamein as he did, or the Normandy battles, without – in crude terms – all that came with his DSO for the fighting at Ypres.*

In February 1939 the war minister, Leslie Hore-Belisha (of the eponymous beacons from his time as secretary of state for transport), was at last authorized to instigate planning for a continental field force of four regular and four Territorial divisions. The following month he doubled the size of the TA (which was itself mechanizing apace), and in April – even more to the consternation of many of his cabinet colleagues – introduced conscription, or rather universal military training. But if the BEF of August 1914 had been, in the official historian's words, 'the best-trained, best-organized and best-equipped British army that ever went forth to war', that of September 1939 was in many

* There were three other developments in the First World War that would have a profound bearing on the way the army would fight in the Second World War – and have done ever since. First was the systematization of operational analysis, doctrine, training and rehearsal. Second was the recognition of the 'moral' (non-materiel) element of combat, and therefore the need for the spiritual sustenance of the soldier – from which the Chaplains' Department emerged as a significant element of the 'moral component' of fighting power. And third was the realization that there was a role for women in khaki. See 'Notes and Further Reading'.

respects no better found than the army that had sailed to the Crimea. The shortages of equipment were staggering: of the 1,646 heavy infantry tanks required, for example, there were just 60. Elsewhere the improvisation was pure 'Dad's Army'. The 13th/18th Royal Hussars had handed over their horses in India and returned to Britain in November to mechanize: all that greeted them was a £10 training grant with which to set up a driving and maintenance school. Fortunately the local garages of Kent were generous in lending them old engines to study. Their Vickers light tanks did not arrive until June; and when they did they were found to be worn-out 'demonstration models', most of which broke down between the railway sidings and the barracks. With Hitler tramping into the Sudetenland, the sight of a trail of broken-down tanks cannot have been inspiring for either the troopers or the onlookers. By September, when he lit the final fuse by marching into Poland, half the regiment's tanks still had no proper gun mountings, and the gunners had had only three days' range practice. Only on the day before embarkation for France did the Hussars receive their full complement of transport – bakers' vans and grocers' lorries in every conceivable colour. The last evening was spent painting them khaki.

24

The Shock of War

1939

ONCE AGAIN, BELGIAN NEUTRALITY WAS AT THE CENTRE OF events. In 1914 it had been a strategic priority; in 1939, however, it was an operational hurdle in the allied defence plan, the Achilles heel indeed of the entire French defensive strategy. Between 1930 and 1935 France had built a modern-day Hadrian's Wall along the 200 miles of its border with Germany from Luxembourg to Switzerland: the 'Maginot Line' (named after the defence minister), a complex of obstacles, linked forts and gun batteries forming a defensive belt in places 15 miles deep. But unlike Hadrian's Wall, it had an open land flank: the Belgian border. And it was a huge one – as long as the Maginot Line itself.

The reason was a toxic cocktail of parsimony, political cowardice and wishful military thinking, not of course unique to France in that decade (or since). In fairness to the Maginot strategists, the left flank was supposed to rest on the Ardennes, which were thought to be impassable to armoured vehicles (though there is no record of any trial in which tanks became stuck in those wooded hills; the Ardennes were not, after all, the Dolomites), and to be protected beyond this pseudo-obstacle by the strong line of Belgian border forts. But in 1936 the Belgians repudiated the treaty on which the strategy depended and declared neutrality (little wonder that the likes of Liddell Hart put no trust in treaties). The French began hastily to extend the line along the Franco-Belgian border, but in nothing like the strength or depth of the original line on the eastern border. That original line had absorbed so much of the military budget (a great deal more than predicted) that the post-1935 extension proved beyond the reach of the French public purse, not least because it would have to run through the heavily industrialized Lille–Valenciennes sector, and would encounter a significantly higher water table. Besides, some argued, to do much more might send a signal of no confidence in the Belgians to defend their own territory, and isolate them on the wrong side of the defensive line in the event of invasion. And with the eastern border secured by the Maginot Line, what was the threat through Belgium that could not be dealt with by the mobile elements of the French army?

The Maginot concept derived from a more fundamental logic, however – or, at least, had done at the time of its conception. After the losses of the First World War France was left holding a demographic time bomb: in the 1930s there would simply not be enough conscripts to maintain the size of army which the general staff calculated it would need. But the overwhelming conclusion that the staff had drawn from the great battles of 1916 – Verdun and the Somme – was that impregnable defences could be constructed, and with a German offensive thereby halted, artillery could do the rest. Such a line of defences would also buy time for general mobilization – recall of reservists – as well as protecting Alsace-Lorraine, which had been reacquired through the

Versailles Treaty. There was no longer talk in French military circles of *l'offensive à l'outrance*; this was a defensive-minded strategy, a vision of warfare along the lines of the middle years of the First World War. It was, indeed, not so different from the British perception, although until 1939 that perception had been academic since war on the Continent was – in the estimation of the cabinet – out of the question.

And so in May 1940, just before the Germans struck, what passed for the Anglo-French manœuvre force was waiting on the Franco-Belgian border. It did not expect to fight there, however. In the event of a German thrust into Belgium it intended sprinting across the border like an enormous Le Mans start to take up positions alongside the Belgian army on the River Dyle (Dijle), which linked the Ardennes in the south with Antwerp in the north. So sensitive were the Belgians about their neutrality, none the less, that reconnaissance of the Dyle by British and French officers had had to be done in plain clothes. It did not augur well.

The so-called 'phoney war' from 3 September 1939 (the actual declaration of hostilities) to mid-May the following year (when the Germans at last launched their offensive into the Low Countries) had been a vital breathing space for the BEF. On 23 August, Neville Chamberlain's government had finally been disabused of any thoughts of preserving peace by – somewhat ironically – the signing of a Nazi–Soviet non-aggression pact, even while its most alarming protocols remained secret. Two days later Britain signed the Anglo-Polish treaty of mutual assistance, though in military terms it was Britain, rather than Poland, that would prove the beneficiary. On 31 August, as wearily expected, Hitler gave orders for his troops to cross the Polish frontier the following day. After expiry of the obligatory ultimatum, Chamberlain sombrely announced on the BBC: 'I have to tell you that no such undertaking [to withdraw troops from Poland] has been received, and that consequently this country is at war with Germany.' The air-raid sirens sounded almost immediately when a lone plane approaching London was mistaken for a bomber. But except at sea there was little immediate military action that Chamberlain could take: he was not willing to provoke retaliation by sending Bomber Command to Berlin, and it would be months – by any reckoning – before the BEF was in a position to take the offensive.

Besides making up the shortfall in equipment, the army had to address the organizational problems created by its rapid expansion, much as in 1914. At the beginning of 1939 there were 56,000 British troops in India, 14,000 men in garrisons from the West Indies to Hong Kong and 107,000 regulars in the United Kingdom. In addition there were 21,000 in the Mediterranean and Middle East, though because of the uncertainty over Italy's intentions and Japanese expansionism in the Far East they were not available for operations in France, as they had been in 1914.* Yet by the end of September 160,000 troops had already crossed to France, and by the following April, just before the Germans struck, the BEF had grown to over 380,000. Even so, tanks were few and none of them were integrated with the divisions, which had only a 'divisional cavalry regiment' for reconnaissance equipped with the woefully inadequate Vickers light tanks armed with twin machine guns.

A good many of the officers and NCOs with Great War medals had been promoted to command and staff posts in the expanded army. In consequence, and in contrast to 1914, when only a dozen years had passed since the South African War, there was at regimental level a dearth of fighting experience, except of colonial skirmishing. The colonial

* Germany, Italy and Japan were to sign the Tripartite Pact in September 1940 – and become known as the Axis powers – although Japan did not begin active military operations against Britain until late the following year.

experience was by no means valueless, but it did not translate automatically to the needs of the *Blitzkrieg* which was about to befall the BEF. The period of 'sitzkrieg', as inevitably the military wags dubbed the winter of waiting, was therefore a blessing in providing time to get the BEF trained.

Unfortunately not all its commanders were up to the job, as one of them who was – GOC (general officer commanding) II Corps, Lieutenant-General Alan Brooke (later to be Churchill's CIGS) – explained in his memoirs:

> The First World War had taken the cream of our manhood. Those that had fallen were the born leaders of men, in command of companies or battalions. It was always the best that fell by taking the lead. Those that we had lost as subalterns, captains and majors in the First World War were the very ones we were [now] short of as colonels, brigadiers and generals.*

France had a demographic time bomb; Britain had damp powder.

The defect at brigade level and above – though nearly every one of the BEF's generals had a DSO, and there were several VCs too – would be brutally exposed in the withdrawal to Dunkirk, but it was evident even during the 'sitzkrieg'. It was revealed also at the regimental level, with some units appearing not to have appreciated the demands of a war of manœuvre as distinct from one fought on the solidified lines of the old Western Front. Major-General Bernard Montgomery, commanding 3rd Infantry Division, complained for example that his divisional reconnaissance regiment, the 15th/19th Hussars (equipped with the Vickers light tanks), 'attached more importance to billets and messes

---

* Sir Arthur Bryant, *The Turn of the Tide* (1957). This was, in effect, the first volume of Field Marshal Lord Alanbrooke's memoirs; the second, *Triumph in the West*, was published in 1959.

than to organized rehearsal for battle'. They were by no means alone in attracting the general's displeasure, though they would undoubtedly have had the best officers' mess.

In the regiments these defects were treatable by ruthless pruning and training. Higher up it was not so easy; and at the very top, next to impossible. The BEF's commander-in-chief was General the Viscount Gort VC, DSO and two bars, MC.* There can be no arguing with such an unsurpassed record of valour, and yet Gort soon appeared out of his depth, derided as 'the best platoon commander in France'. On 22 November Brooke wrote of him in his diary,

> Gort is a queer mixture, perfectly charming, very definite personality, full of vitality, energy and joie de vivre, and gifted with great powers of leadership. But, he just fails to be able to see the big picture and is continually returning to those trivial details which counted a lot when commanding a battalion, but should not be the main concern of a Commander-in-Chief.

Perhaps the medals had dazzled during the inter-war years when, as in the late Victorian period, 'regimental qualities' had come to be seen as everything: no promotion board could have denied the next step on the ladder to those post-nominal letters. Or perhaps as a Grenadier and very much a 'guardsman's guardsman' he had risen by that private Household elevator which had once been a feature of the army. Either way, it did not bode well.

There was a distant overture to the *Blitzkrieg*, however – in Norway. In early April, for a mix of political and military reasons including Russia's invasion of Finland, the Nazi–Soviet non-aggression treaty, Swedish natural

---

* A 'bar' to a medal – indicated by a rosette on the medal ribbon – means that the decoration has been awarded a second time; two bars means that it has been awarded three times.

resources, the attraction of Norwegian bases for U-boats and the uncertainty of Anglo-French intentions, the Germans invaded Norway by invading Denmark en route, and also via Sweden which they traversed by the simple expedient of buying railway tickets. At Churchill's urging (from the Admiralty, to which he had returned as first lord when war was declared) a hastily assembled landing force sailed for Norway in an attempt to forestall the seizing of the northerly ports. The force, also comprising French *Chasseurs d'Alpin* and Polish infantry, arrived too weak and too late, the Germans having already mounted a full-scale and highly effective invasion, although allied troops did help extricate the Norwegian royal family and much of the country's gold reserves (for which the Christmas tree in Trafalgar Square each year is a memorial gift of the Norwegian government). Perhaps the greatest consequence of the débâcle, for so the landings had been, was Chamberlain's resignation and his replacement by Churchill at the very moment the Germans struck in the Low Countries – in the early hours of 10 May.

At a stroke the strategic situation now changed: Belgium was an ally once more. The flag dropped and the BEF sprinted across the Franco-Belgian border for the Dyle, and by the early evening its leading units had reached the river – including the 15th/19th Hussars who had presumably left behind their encumbering mess silver. They were not there long. The Belgian forts fell more quickly than in 1914, for the Germans did not tend to repeat obvious mistakes: they took Fort Eben Emael, for example, in a brilliant *coup de main* by parachute and glider troops. To make matters worse, on 14 May the Dutch, who did not escape invasion this time, capitulated, leaving the Dyle's northern flank open. With the Belgian army rapidly disintegrating, and the French surprised – shocked – by armoured penetration of the Ardennes which threatened to outflank the Dyle in the south, the BEF began withdrawing on the evening of 16 May, re-establishing the defensive line west of Brussels along the River Escaut, with an intermediate line along the Dendre. It was pretty well a textbook operation, the divisional reconnaissance regiments covering the withdrawal much as they had done in 1809 at the beginning of Sir John Moore's run for Corunna, although the 15th/19th Hussars (one of whose antecedent regiments had distinguished themselves covering Moore's escape across the Esla) would pay dearly for the time they bought the infantry to get clear on the Belgian flank. Their Vickers light tanks were shot to pieces, and thereafter the regiment practically ceased to exist.

To the south the Germans, having slipped through the Ardennes, were making astonishing progress under General Heinz Guderian, the architect of the panzer forces. They had brushed the French aside at Sedan and crossed the Meuse with little difficulty, and by 19 May Guderian's tanks – over 800 of them, faster than the British tanks yet with a better balance of protection and mobility and better armed, because they were intended for independent action, not as close support for the infantry – were at Abbeville near the mouth of the Somme.

That day, Gort signalled London that he had been 'unable to verify that the French had enough reserves at their disposal south of the gap to enable them to stage counter-attacks sufficiently strong to warrant expectation that the gap would be closed'. London replied that the BEF should fall back south-west along their lines of communication, but Gort, believing (rightly) that the French had little fight left in them and that such a move consequently risked en-circlement, decided instead to withdraw northwards towards the Belgian coast – much to the anger of the new prime minister, though later, in his speech to the House of Commons after Dunkirk, Churchill would acknowledge the truth:

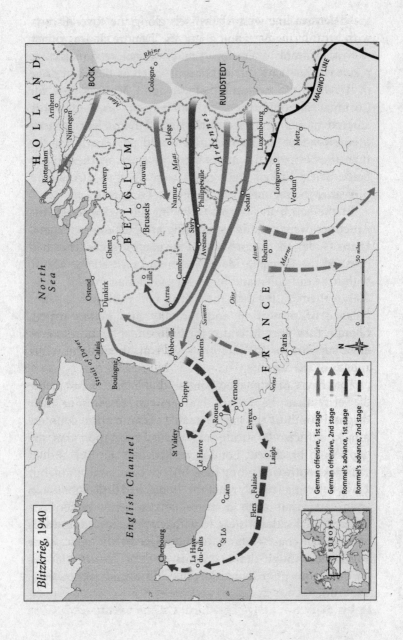

Blitzkrieg, 1940

the German eruption swept like a sharp scythe around the right and rear of the Armies of the north. Eight or nine armoured divisions, each of about four hundred armoured vehicles of different kinds, but carefully assorted to be complementary and divisible into small self-contained units, cut off all communications between us and the main French Armies. It severed our own communications for food and ammunition, which ran first to Amiens and afterwards through Abbeville, and it shore its way up the coast to Boulogne and Calais, and almost to Dunkirk. Behind this armoured and mechanized onslaught came a number of German divisions in lorries, and behind them again there plodded comparatively slowly the dull brute mass of the ordinary German Army and German people, always so ready to be led to the trampling down in other lands of liberties and comforts which they have never known in their own.

The foreign secretary, Lord Halifax, who had been one of Chamberlain's strongest supporters in the policy of appeasement but who had latterly embraced rearmament and taken a firm stand against Hitler and Mussolini, put it more prosaically in his diary on 25 May, but with the raw truth that had shaken Whitehall: 'The mystery of what looks like the French failure is as great as ever. The one firm rock on which everyone was willing to build these last two years was the French army and the Germans walked through it like they did through the Poles.'

On 21 May elements of 50th (Tyne and Tees) Division in Brooke's II Corps, with all the available tanks – seventy-four under-gunned death-traps – had mounted a counter-attack of sorts at Arras which checked the Germans' ardour for a while (through surprise rather than destruction), but the following day Guderian invested Boulogne and Calais. 'Nothing but a miracle can save the BEF now,' wrote Brooke in his diary on the twenty-third. On the twenty-sixth Gort was told to evacuate as much of the BEF as he could, and the Royal Navy began preparing to do what it had done 130 years earlier at Corunna – rescue from the Continent not *a* British army but *the* British army.

And so began Brooke's miracle, Operation Dynamo. But the 'winnowing' of senior officers – from above and below – had already begun. Gort, Brooke and the commander of III Corps, Sir Ronald Adam, were ordered home – Gort because it was unthinkable to allow a commander-in-chief to be captured, Brooke because he was wanted for the crucial command of the defences of southern England in the German invasion that would surely follow, and Adam to take on the Northern Command. The perimeter defences at Dunkirk were to be held by I Corps under Lieutenant-General Michael Barker, who was told by Gort that as a last resort he should surrender himself and what remained of his corps to the Germans. Montgomery, however, who had taken Brooke's place at the head of II Corps on 30 May, went to see Gort and told him frankly that Barker was not up to the job; that 'what was needed was a calm and clear brain, and that given reasonable luck such a man might well be able to get 1st Corps away, with no need for *anyone* to surrender; he had such a man in Alexander'. Gort took Montgomery's counsel and appointed Major-General Harold Alexander (GOC 1st Division) to command I Corps. And 'Alex', as he was known universally, would indeed get them all out.

Brooke recalls in his diary how, that evening of 30 May, he was carried to an open boat and paddled out to a destroyer which took him aboard and then had to wait until after midnight before she could risk running the gauntlet of the Channel, grounding on a sandbank in the early hours while dodging German torpedo boats, and finally reaching Dover just after seven in the morning. It was a story that, with variations, many more would soon be able to tell as the BEF was taken off the beaches by all manner of craft in the days that followed. The 13th/18th Hussars, Alex's divisional reconnaissance regiment, seem to have had a rather better time of it than most (although they were strafed by the Luftwaffe on the beaches):* having disabled their remaining light tanks, recovering what weapons they could, they were taken off that same night directly from the mole at Dunkirk aboard a cross-channel steamer. The commanding officer, who would go on to command a brigade, recalled: 'I remember little of the crossing except a very welcome whisky and soda before dropping to sleep in a luxurious double cabin with Captain Harrap (Adjutant) in the other bed. We were called by a steward in the morning with a cup of tea, who said "We are just approaching Dover Harbour, sir"'.

By 4 June, 210,000 British and allied troops, and 130,000 French, had been taken off by the Royal Navy, augmented by the 'little ships' immortalized in William Wyler's 1942 film *Mrs Miniver* and in fellow American Paul Gallico's short story 'The Snow Goose', published in the *Saturday Evening Post* within months of the evacuation. Churchill spoke ringingly of a 'miracle of deliverance', and the Dean of St Paul's perhaps more authoritatively of 'the miracle of Dunkirk'. It had certainly been a triumph of the national spirit, a masterly plan by the admiral commanding, Bertram Ramsay, and a bravura performance by the Royal Navy – and notable too for some equally masterly generalship, and for military discipline that held up remarkably well under the circumstances. But there had been a very great deal to be dismayed with in the weeks leading up to Dunkirk.

Perhaps, however, Montgomery's subsequent view of Gort, uncharacteristically charitable to a general whose ability he deprecated, is the proper epitaph for the man who

---

* After one of the Luftwaffe's strafings Major John Cordy-Simpson had his squadron form line and pick up litter: the 'Dunkirk area-cleaning fatigue' passed into regimental legend.

had had to play the part of a Sir John Moore. On 30 May Montgomery had written that 'Gort was incapable of grasping the military situation, and issuing clear orders . . . He had "had it". But later, on reflection, he added:

> He was a man who did not see very far, but as far as he did see he saw very clearly. When the crisis burst on the French and British armies . . . he was quick to see that there was only one end to it: the French would crack and he must get as much of the British army as he could back to England . . . He saved the men of the BEF. And being saved, they were able to fight again another day.

And that was the point.*

In many ways, although it was hardly apparent at the time, Dunkirk was something of a boon. The army had lost almost every piece of equipment, but it was good riddance to the useless tanks, the obsolescent artillery, the butchers' vans and bakers' lorries – 2,472 guns and 63,879 vehicles according to the official returns. If the Germans had succeeded in landing in England in any strength the army would not have been able to put up much of a fight; but there was now no possible doubt about the urgency and priority of re-equipping the army, and the losses at Dunkirk were an extra spur to bringing into service better equipment which might otherwise have remained longer on the drawing board.

The BEF had fought better than might reasonably have

---

* It is curious, to say the least, that one of the lessons of the First World War in the higher organization of command was ignored: in 1940, GHQ BEF was the campaign headquarters – looking back to London and to left and right to allies – as well as having executive command of the fighting. By 1915 the original BEF had subordinate army HQs; the BEF of 1940 needed that level of command too – a point made clearly in the report of the Bartholomew Committee of Inquiry which took evidence from senior commanders and selected commanding officers after the evacuation.

been expected of an organization brought into being less than a year before. It had been poorly equipped materially, doctrinally and morally to deal with modern war, especially *Blitzkrieg*, but there had been some singular examples of fighting spirit. For the regular regiments this might have been par for the course, but the new battalions of the second citizens' army to be raised in a quarter of a century had had even less time to prepare than Kitchener's. The 6th Battalion of the King's Own, for example, composed of older reservists and conscripts alike, was formed only four months before Dunkirk as a pioneer (labour) unit to improve the fixed defences on the Franco-Belgian frontier. Yet wearing the lion of England as their cap badge would have made them at least a little conscious of their forebears' reputation – the 'old sweats' would have made sure of that – and this must in part account for the stand they made on 27 May at the village of Merville: not having expected to wield more than picks and shovels, the battalion nevertheless held out all day against attacks on three sides.

And in Parliament Churchill singled out the selfless defence of Calais, where, as at some of the other smaller embarkation points, the rearguard could not break clean:

> The Rifle Brigade, the 60th Rifles, and the Queen Victoria's Rifles [TA], with a battalion of British tanks and 1,000 Frenchmen, in all about four thousand strong, defended Calais to the last. The British brigadier was given an hour to surrender. He spurned the offer, and four days of intense street fighting passed before silence reigned over Calais, which marked the end of a memorable resistance. Only thirty unwounded survivors were brought off by the Navy, and we do not know the fate of their comrades. Their sacrifice, however, was not in vain. At least two armoured divisions, which otherwise would have been turned against the British Expeditionary Force, had to be sent to overcome them. They have added another page to the glories

of the light divisions, and the time gained enabled the Graveline water lines to be flooded and to be held by the French troops.*

The brigadier, like many others, was to die in captivity.

The BEF's fighting spirit had indeed both impressed and taken a heavy toll of the Germans. After Dunkirk, General von Reichenau (6th Army) reported that

> The English soldier was in excellent physical condition. He bore his own wounds with stoical calm; the losses of his own troops he discussed with complete equanimity. He did not complain of hardships. In battle he was tough and dogged . . . Certainly the Territorial divisions are inferior to the Regular troops in training, but where morale is concerned they are their equal.

Churchill was also very clear in his speech to Parliament, however, that 'We must be very careful not to assign to this deliverance the attributes of a victory. Wars are not won by evacuations. But there was a victory inside this deliverance, which should be noted.'

But if 200,000 men had been recovered safely to Britain, there were 68,000 casualties to be counted – and 40,000 taken prisoner, the majority of them cut off in the race for the sea, or as rearguards. The job of reconstruction was urgent, and it would have to be undertaken against a background of failure that looked very much like defeat. When Sir John Moore's army had limped into barracks in England to lick its wounds it had not been many months before another, and greater, general was to take the regiments back to the Peninsula to gain victory after victory over their erstwhile vanquishers. The British army of 1940, however, would have much longer to wait.

* Speech to House of Commons, 4 June 1940.

25

Never a Victory

1942

IN THE FOURTH VOLUME OF HIS WAR MEMOIRS, *THE HINGE OF Fate*, Churchill draws a line in the sand – almost literally – at a fly-blown railway halt in the desert 60 miles west of Cairo: 'It may almost be said, "Before Alamein we never had a victory. After Alamein we never had a defeat". It was rather more symbolically true than literally. Besides several successful actions at sea, not least the sinking of the *Bismarck*, the battle of Britain had seen the RAF winning fighter superiority over southern England and thereby dashing all German hopes of invading. And in December 1940, 36,000 men of the Western Desert Force under Lieutenant-General Richard O'Connor had counter-attacked an Italian army which had advanced into Egypt from Libya and driven them back 500 miles, destroying ten divisions, taking 130,000 prisoners and leaving Mussolini with only the most precarious foothold in North Africa.*

* Italy had declared war – 'jackal-like' – after the Fall of France. Mussolini must have secretly rued his decision soon afterwards, for besides the humiliation in Libya, a month before at Tarranto the Fleet Air Arm had

This, however, brought down vengeance in the shape of the Deutsches Afrika Korps, which Hitler despatched under General Erwin Rommel to bolster his lame-duck ally. There followed eighteen months of see-saw war in the desert, with great advances by both sides followed by hasty and equally long withdrawals. In all of this the reputation of Rommel – 'the Desert Fox' – grew, while that of one British general after another was destroyed. Notable among these were Wavell and then Auchinleck, men who, as Tacitus wrote of an early Roman emperor, would have been thought capable of ruling if only they had not actually been called upon to rule. (Each, having failed to match Rommel, was in turn 'promoted' and sent to be CinC India.) The nadir of this phase was the fall of 'Fortress' Tobruk, the strategically critical port in northeast Libya, of which Churchill learned to his chagrin during his meeting with President Roosevelt on 21 June 1942. It was this disaster which would bring Montgomery out to Egypt to command the 8th Army in Auchinleck's place.

The fall of Tobruk even made Churchill begin to doubt whether the army had any fight left in it, only four months after the terrible shock of the fall of Singapore. The Japanese, whose simultaneous attacks on Pearl Harbor, Malaya and Hong Kong had made fighting allies at last of Britain and the United States, had swept down the Malay Peninsula, outflanking British, Indian and Australian troops by superior jungle craft, and taken the great island naval base from landwards: the opposite direction from that on which the entire concept of its defence rested.

'The prime defence of Singapore is the Fleet,' Churchill had minuted the war cabinet in September 1940 when the CinC Far East Command (Air Chief Marshal Sir Henry

Footnote continued from p451
badly mauled the Italian fleet; and in May 1941, after a two-pronged offensive from Kenya and the Sudan by British, Indian and South African troops, all Italian forces in East Africa surrendered.

Brooke-Popham) had requested reinforcements in case of Japanese landings in Malaya. 'The fact that the Japanese had made landings in Malaya and even begun the siege of the fortress would not deprive a superior relieving fleet of its power . . . The defence of Singapore must therefore be based upon a strong *local* garrison and the general potentialities of sea-power.'

Two days after the Japanese landed in Malaya, fighter-bombers of the imperial navy sank the battleship *Prince of Wales* and the battlecruiser *Repulse* off the east coast. There were no other capital ships available for the defence of Singapore, and from then on, although the Japanese were numerically inferior to the British, Indian and Australian troops available, there was something of an awful inevitability to the mass surrender which would follow on the island – 80,000 men (including 15,000 Australians), bringing the total to 130,000 captured since the first landings.

Singapore was the biggest capitulation in the army's history. And it was all the more dismaying for the lack of any determined defence. As the Japanese were about to launch their attack across the straits Churchill had frantically signalled Wavell, whose command by then included the whole of South-East Asia,

> There must at this stage be no thought of saving the troops or sparing the population. The battle must be fought to the bitter end at all costs. The 18th Division has a chance to make its name in history. Commanders and senior officers should die with their troops. The honour of the British Empire and of the British Army is at stake. I rely on you to show no mercy to weakness in any form. With the Russians fighting as they are and the Americans so stubborn at Luzon [Philippines], the whole reputation of our country and our race is involved. It is expected that every unit will be brought into close contact with the enemy and fight it out.

They were not, and they did not. And so on 15 February 1942 the honour of the army was forfeited – to be redeemed at some as yet unforeseeable place and time. And if the fall of Singapore would not have quite the far-reaching consequences of the surrender at Yorktown 161 years earlier, it had the same stupefying effect in London. Combined with the news which soon began to follow from Burma, where the Japanese, invading through Thailand (and French Indo-China), were pushing back all before them, it made the new CIGS, Brooke (appointed in December), confide to his diary: 'Burma news now bad. Cannot work out why troops are not fighting better. If the army cannot fight better than it is doing at present we shall deserve to lose our Empire.'

One man, at least, knew why the troops were not fighting better. Lieutenant-General William (Bill) Slim, who in the First World War had worn the same cap badge as Montgomery (the Royal Warwickshire Regiment) before transferring to the Gurkhas, had been placed in command of the scratch Burma Corps as it reeled before the Japanese. Later he would command 14th Army (the 'Forgotten Army', as its men took a rum pride in calling themselves) which finally halted the Japanese on the Indian border and in turn took the fighting back to them in the jungles of Burma – and beat them. In his account of the campaign, *Defeat into Victory*, Slim describes the tactical lessons of the long retreat from Burma, which he used as the foundations for retraining the army. First, he wrote, the jungle itself was neither impenetrable nor unfriendly, and the Japanese had no innate superiority. He told his logistical troops that there were no non-combatants in jungle warfare: they, too, must fight in their own defence. He told his infantry and supporting arms that offensive patrolling was the key to mastery in the jungle, and that having Japanese parties to the rear of their positions did not mean that they were surrounded: it was the Japanese who were surrounded. Trying to hold long, continuous lines was futile, he concluded, just as frontal attacks ought rarely to be made, and never on narrow fronts. He was certain that tanks could be used in any country except swamp, but that they should not be scattered about in penny-packets. 'When the Japanese have the initiative they are formidable,' he warned, 'but their cohesion is seriously threatened when they lose it: mobility off-road, surprise and offensive action seizes the initiative.'

Slim was soon training 14th Army along these lines. It was a predominantly Indian army, with an increasing number of East and West African units, but there were British regiments in every brigade, and the Forgotten Army's doctrine would have a profound effect on shaping that of the post-war army, not least in the emphasis on preparing for dispersed, in-dependent operations, on resilience and initiative in commanders, and on physical toughness and field discipline in their troops. Although Slim was regarded by some as a 'sepoy general', as Wellington had at first been, his success in the hard-fought defensive battles of Imphal and Kohima on the Indian border and his conduct of the successful counter-offensive brought him in the end to Whitehall, where in 1949 he would take over as CIGS from Montgomery for four crucial years of post-war retrench-ment and the beginnings of withdrawal from empire.* His views, disseminated both during his time as CIGS and through his subsequent lectures and essays, had enormous influence on the post-war generation of officers. *Defeat into Victory*, published in 1956, is still one of the most widely read books of any on the prescribed list of the Defence Academy, speaking as it does to every rank from lieutenant to four-star general. Neither victory in the Falklands in 1982

---

* He had actually retired in 1948, but the prime minister Clement Attlee rejected Montgomery's proposal of General Sir John Crocker (a former RTR officer, and one of Monty's corps commanders in Normandy) as his successor, and brought Slim back to the Active List in the rank of field marshal.

nor the narrow margin of success achieved in Afghanistan in 2006–7 would have been possible without the realization of Slim's vision of the army, based on 'the high quality of the individual soldier, his morale, toughness, and discipline, his acceptance of hardship, and his ability to move on his own feet and to look after himself'. And, indeed, Slim's own style of command – including as it did so many of Wellington's best qualities – became one of the two models for commanders of the late twentieth century (and continues to inspire in the early years of the twenty-first). In *Retribution: The Battle for Japan, 1944–45* (2008) Sir Max Hastings describes him thus:

> In contrast to almost every other outstanding commander of the war, Slim was a disarmingly normal human being, possessed of notable self-knowledge. He was without pretension, devoted to his wife, Aileen, their family and the Indian Army. His calm, robust style of leadership and concern for the interests of his men won the admiration of all who served under him ... His blunt honesty, lack of bombast and unwillingness to play courtier did him few favours in the corridors of power. Only his soldiers never wavered in their devotion.

The alternative model of command was – and remains – of course Montgomery, although he is not nearly as popular a model now as he was in the 1970s and 1980s at the height of the Cold War, when the British Army of the Rhine was everything. 'Monty', as he became (semi-) affectionately known, was not the first choice of general to take over command of 8th Army when Auchinleck was sacked in August 1942 (he was, in fact, earmarked for command in Operation Torch, the Anglo-US landings in Vichy North Africa later that year). Churchill, despite Brooke's misgivings, had wanted 'Strafer' Gott, the 45-year-old commander of XIII Corps and veteran of the fighting in North Africa, but

Gott was killed when the aircraft in which he was travelling back from Cairo to his desert headquarters was shot down.

The trouble with Montgomery was that he upset almost everyone he met. He had been very outspoken after Dunkirk and could hardly be civil to Auchinleck when the latter was GOC Southern Command (and his superior) during the invasion scare. But Montgomery's mastery of training, which had paid such dividends during the retreat to Dunkirk, continued to impress, and so he was sent to the desert in the middle of August 1942 although he had not seen action in two years – something which, on top of the resentment felt by the old desert hands at having 'the Auk' removed, did not seem to augur well.*

Montgomery's impact was, however, immediate and profound. In its see-saw of advances and retreats 8th Army had come to rest near El Alamein, long recognized as a good place for positional defence.† There were a number of promontories which offered good observation and, crucially, the front was narrow (only some 35 miles) and could not be outflanked either in the north, where it met the sea, or in the south, where the Qattara depression was impassable to tanks. Indeed, Auchinleck had already stopped Rommel in his tracks here in July. Desperately short of fuel at the end of an over-extended supply line, Rommel's German and Italian forces had then dug in.

Montgomery had already grasped the operational reality by the time he took command. Knowing from Ultra‡

---

* During those 'inactive' years Montgomery had, however, studied the problem of fighting the Germans more than any other commander had been willing or able to.

†In 'positional defence' the defending troops occupy positions which dominate the ground in between, 'fortifying' them as strongly as possible, and fighting *in situ* rather than the 'hit and then move' of mobile defence.

‡ 'Ultra' was the generic term for intelligence gained by intercept of German strategic communications (encoded by the 'Enigma' machines) and decrypted at the code-breaking centre at Bletchley Park.

intercepts that Rommel would renew the attack, and where, he immediately reinforced the Alam el Halfa ridge to the south-east, brought the hitherto separate headquarters of 8th Army and the Desert Air Force together to create unity of effort – one of the fundamental principles of war – and, in a stroke of high drama to impress on the army his character and intent, ordered all plans for further withdrawal to be burned, telling his staff that 'If we cannot stay here alive then let us stay here dead.' But this of itself might not have been enough to turn the tide of defeat in the minds of the men of 8th Army if he had not also held out the promise – indeed, the cocksure certainty – of victory. Immediately on taking command he gathered together the officers of his headquarters and addressed them from the steps of his caravan. He told them of the need to change, and what he wanted; what he would not tolerate, what he would do. 'The great point to remember,' he concluded, 'is that we are going to finish with this chap Rommel once and for all. It will be quite easy. There is no doubt about it.'

Within a fortnight Rommel struck, and 8th Army sent him reeling back from Alam el Halfa ridge as easily as Montgomery had said they would. Churchill wondered why he did not follow up at once in a counter-attack, but in his own mind Montgomery still had work to do. For one thing, his promised new tanks (250 American Shermans, superior to anything Rommel possessed) had not yet arrived; and Montgomery was not a man to take needless risks – certainly not at this stage of his reputation-making. Despite Churchill's increasing urging – almost goading – he stood his ground until he was ready. And he was not ready for a month and more.

The extraordinary thing about Alamein was that in essentials it was a battle that looked like those of 1917. Rommel had known that Montgomery would attack at some stage – Alam el Halfa had been meant as a 'spoiling attack' –

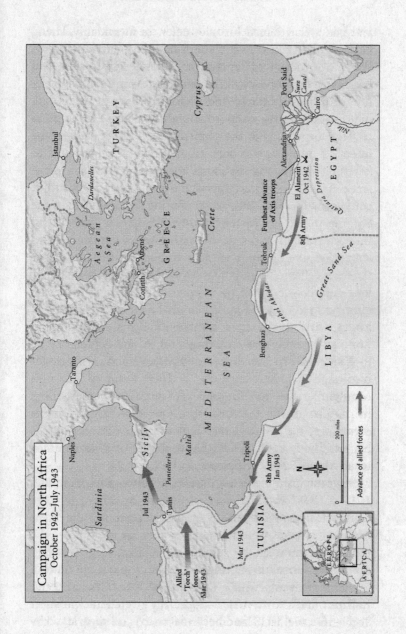

Campaign in North Africa
— October 1942–July 1943 —

TURKEY

Istanbul

Dardanelles

Aegean Sea

GREECE

Corinth Athens

Taranto

Naples

Sardinia

Sicily

Pantelleria

Malta

Tunis

TUNISIA

Jul 1943

Allied 'Torch'
forces
Mar 1943

Mar 1943

MEDITERRANEAN SEA

Cyprus

Crete

Tripoli

8th Army
Jan 1943

Port Said
Suez Canal
Cairo

Alexandria

El Alamein
Oct 1942

EGYPT

Qattara Depression

Nile

Furthest advance
of Axis troops

8th Army

Tobruk

Jebel Akhdar

Benghazi

LIBYA

Great Sand Sea

N

200 miles

0

EUROPE

AFRICA

Advance of allied forces

and had strengthened his own defences formidably, laying half a million mines, mainly anti-tank but with some anti-personnel mines to deter clearing, and deep belts of barbed wire. With no scope for outflanking, 8th Army would have to mount a frontal attack. And with a frontage of a mere 35 miles Montgomery would be able to achieve only limited tactical surprise as to the point of main effort – where he intended breaking through – and very little surprise as to anything else beyond the actual hour and day. In fact, as the weeks passed, so the certainty of the imminence of the attack could only increase. But, rather like the preliminaries to the Somme, the delay gave the British and imperial forces useful time for training and rest, while Rommel's men laboured to strengthen the defences without respite, for the Desert Air Force was able to pound them day and night and harass their long supply line to the point where the prodigious logistic effort yielded only the most marginal benefit. Nor was their leader with them: after eighteen months' campaigning Rommel's health was suffering, and in late September he took sick leave in Italy and Germany, leaving the army under command of Lieutenant-General Georg Stumme. The new commander was not in good spirits, having undergone a court martial after the Russians captured secret documents at Stalingrad, a loss for which Hitler blamed him personally. Nor was he in good health; he would die of a heart attack when Montgomery's offensive opened.

But perhaps the greatest difference between Alamein and 1916 or 1917 was that tactical victory by 8th Army here could clearly precipitate a strategic defeat for the Axis: Rommel's troops were numerically inferior, his allies – the Italians – were weak, his supply was inadequate, and he had no reserves. Breakthrough at El Alamein could set in motion a chain of events which would throw the Germans and Italians out of North Africa – especially given that Operation Torch, the Anglo-US landings in Morocco and Algeria, Vichy territory, were to begin in November. That the Germans *could* indeed be evicted from the Western Desert 8th Army understood because Montgomery went round every unit before Alamein and told them so, in no uncertain terms. And for the most part they believed him.

The dominating factor in the coming battle would be the minefields, up to 5 miles deep. Getting through them was the equivalent of crossing no-man's-land in 1916–17. And then there were the German tanks, waiting to learn where 8th Army's tanks would break through, in order to swoop like raptors to pick them off as they emerged from the cleared minefield lanes. Montgomery knew that until enough panzers were destroyed Rommel would remain a force to be reckoned with, especially in the open country of the break-out which would strongly favour the German armour – and especially the much-feared 88 mm anti-tank gun.* And he knew exactly what the coming battle would entail: a frontal attack and a duel of armour – in the words of General Sir David Fraser, writing from direct experience of the war, 'a battle of attrition, needing perfect preparation, moral force and persistence unto death. It could never be cheap in materiel or human life.'†

Montgomery had three priceless assets in the planning and execution of the battle of El Alamein, however.‡ First was the moral courage and authority which came of his fighting experience in the Great War and the retreat to Dunkirk: he knew the battle would be exceptionally hard

---

* The most formidable German gun of the war, the '88' started life as an anti-aircraft gun; and indeed some Luftwaffe anti-aircraft units were re-directed to tank-killing at various times. The 88 would also be mounted later on the Tiger tank and the Panther tank destroyer.
† David Fraser, *And We Shall Shock Them: The British Army in the Second World War.*
‡ The 'El' is sometimes dropped: the name derives from the ridge between the railway and the sea, Tel el Alamein – the hill of twin cairns.

fought, and he was prepared to fight it as hard as necessary. Second was his experience in the planning of the counter-offensives of 1918, with their careful coordination of all arms, together with intense professional study and application in the inter-war years (in part a reaction to the death of his wife) and a total immersion in the lessons of the fighting in 1940–1, which he had applied with the greatest zeal in his home commands. Third, linked of course to the first two, was unbounded self-confidence.

The plan he devised was simple enough, though it would be hard to carry through even if things went well. First would be the break-in to the German positions; then the dog-fight, in which his tanks would destroy the German armour while his infantry 'crumbled', as he called it, the German and Italian defensive positions; and finally, with the breaking of resistance, would come the break-out into open country. In all, he calculated the battle would take twelve days. And for this he reorganized 8th Army (now consisting of some 220,000 men and 1,100 tanks) into three corps: XXX Corps, composed of five infantry divisions (51st Highland, 9th Australian, 2nd New Zealand, 1st South African and 4th Indian) and an armoured brigade; XIII Corps, made up of two infantry divisions (44th and 50th) plus 7th Armoured Division and a Free French brigade; and X Corps, comprising 1st and 10th Armoured Divisions. His point of main effort would be in the north, in XXX Corps area, which was why he had strengthened the corps to five infantry divisions: initial penetration – 'capture' – of the minefields was a task only the infantry could take on. These five divisions would attack in line on a 6-mile frontage after a thorough but short artillery preparation, and then X Corps (with the bulk of the armour), massively reinforced with sappers to clear lanes through the minefields, would follow up on two divisional axes, debouching from the minefield lanes and passing through the infantry to an objective line 2 miles beyond.

From here they would halt the German tank counter-attack which Rommel would have to make in the attempt to avoid envelopment and destruction in detail. Either way, Montgomery would achieve his aim of destroying the critical mass of German armour.

A deception plan would mask the point of main effort. At H-Hour (the time at which the leading troops cross the start line) XIII Corps was to mount a 'noisy' but cautious armoured attack to deter the southern group of panzer divisions from moving north. But the artillery fire plan was cunning itself: at 9.40 p.m., just after dark, there would be a whirlwind bombardment of all the known Axis artillery positions, followed by five minutes' silence during which – standard procedure – the whole German and Italian line would stand to for the expected attack. There would then follow a simultaneous and extremely violent bombardment of the Axis infantry defences along the entire front – 'over a thousand guns' (Montgomery's own figure), the greatest fire plan since the Somme – confusing the enemy as to the intended point of attack. And then there would be the selective, surreptitious shift of the weight of fire on to the real targets in XXX Corps area.

The Royal Artillery, said Montgomery after the war, was the arm that had done the most to secure victory. It was not on the whole a tactful thing to say: the infantry had lived in discomfort in every theatre of war and had died in droves, and the Royal Armoured Corps had lost thousands of tank crews in the horrifying 'brew-ups' which followed the detonation of mines or the penetration of high-velocity shot; but the remark echoed Napoleon's maxim that 'it is with artillery that war is made'. Certainly the gunners in the Second World War achieved a far greater proficiency than their predecessors had in 1914–18 (and they had done pretty well in the circumstances), especially in predicted fire, where no previous shells had been fired to enable correction for range and line, and also in rapid correction once firing began. More easily portable radios allowed forward observation officers (FOOs) at the sharp end to respond to an infantry commander's needs rather than merely to implement an inflexible fire plan. Later in the war FOOs would fly light aircraft to spot the fall of shot (with the Glider Pilot Regiment, the forerunners of today's Army Air Corps). Improved communications also meant that fire across a wide front could be switched, intensified or suddenly concentrated on a particular target, and 'lifted' (called off), as the situation required – one of the most effective and immediate ways in which the ground commander could (and still does) influence the battle. And the Gunners were equipped with one of the all-time greats of British design: the Royal Ordnance QF 25-pounder, towed by the four-wheel drive Morris 'Quad', an enclosed truck which also carried the detachment (crew).* The 25-pounder was very accurate, with reliable high-explosive and shrapnel ammunition (considerably more lethal than its German equivalent), rapid-effect smoke and brilliant illuminant.

There would have been no point to the fire plan, however, without the Royal Engineers to clear minefield lanes. The sappers had emerged from the Great War a much strengthened and diversified corps, though still in many ways siege-minded and concentrated on construction and demolition, particularly of bridges. As the Second World War progressed, however, they developed the role which defines them today: mobility and counter-mobility – helping the combat arms to move on the battlefield, while denying the same to the enemy. Their particular focus became the anti-tank mine – its laying, and especially its lifting (which

---

* Some 25-pounders were later mounted on the Grant-Sherman chassis as self-propelled artillery – known as the 'Sexton' – which considerably speeded up the response to calls for fire.

involved first detecting the buried mine, then removing the fuse and taking both to a safe place for detonation). In October 1942 there was no effective technical means of allowing the tanks to do their own clearing: there were rudimentary flails fitted to tanks, but they were not reliable, and were meant only to deal with 'nuisance' mines, not whole minefields which could be several thousand yards deep. Later, specialist equipment – giant rollers, ploughs, heavier flails and explosive hoses – would be developed for the Normandy landings and the breaching of 'the Atlantic Wall', but at Alamein detection and lifting devolved on the sapper – on foot, equipped with a mine detector (the same device the treasure-hunter walks the fields with today), a probe, and his bare hands to unearth the mine and carry it to the rear. The pressure required to detonate the mine was that of the tank track, not the human foot, so the job was theoretically safe – except that anti-personnel mines were sown randomly to slow them down, and the clearing parties were of course vulnerable to small arms and artillery. Nor could the mine-detector headphones entirely mask the sound of the artillery: catching the tell-tale change of frequency in the constant hum of the headphones was rather like listening for birdsong at the side of a motorway – and was never going to be as easy for real as in training.

At ten o'clock on the evening of 23 October, in desert moonlight brightened by the continuous muzzle flashes of 500 guns, the infantry advanced with bayonets fixed and, in 51st Highland Division, to the skirl of the pipes.

It began well, the gunners having done a good job in neutralizing the Axis forward positions, but the defences in depth had not been subdued, and mines now began slowing down the infantry – just as the barbed wire had done in 1916 – although the Australians in the north managed to reach their objectives (as planned, however, no armour was following their diversionary attack). It was all taking longer than Montgomery had hoped, so that by daylight the tanks had not been able to break out of the minefields. Indeed, a great armoured traffic jam was building up in the cleared lanes, not helped by the swirl of dust created by artillery fire and vehicle movement. And at the sharpest end the situation was far from clear-cut, as the sappers of 7th Field Squadron discovered. Having found his clearing party ahead of the infantry, and under fire, one of the troop commanders, Lieutenant John Chartres, had just managed to get his men to make a U-turn without mishap when

> Somehow a provost 3-tonner loaded with MPs [military police] in smart white belts and red caps with mine-gap torches already lit also managed a U-turn with its tailboard towards the enemy ... It had all the appearance of an illuminated float in a carnival procession! Arthur Bakewell and I watched fascinated as tracer rounds of every imaginable calibre spattered around it – and somehow missed. I distinctly remember that Arthur and I both laughed hilariously.*

Laughter in such circumstances, if not always comprehensible to the outsider, is usually an indicator of good morale. But the job of the sappers required more than good humour: 'the sight of sappers lining up and going over a ridge to probe for mines with bayonets was terrible and awe-inspiring to watch,' recorded one tank man. 'Every one of them deserved a medal, as they seemed to go to certain death. They no sooner "went over" than bursts of enemy machine-gun fire seemed to wipe them out; then another line would form up, stub their cigarettes out and move over the top.' Forty years on, in the Falklands, sappers would be doing the same thing, and again during the First Gulf War –

---

* Chartres, '7th Field Squadron in World War II', *Royal Engineers Journal*, 1981.

and often in the same primitive way, for every sapper knows what his forebears did at Alamein: when technology cannot find the mine, there is nothing left but the sapper's probe and his cool courage.

Later in the morning 10th Armoured Division, in the southern part of XXX Corps sector, at last managed to debouch; but their leading regiments found themselves on the forward slopes of the Miteirya Ridge and outgunned by the waiting panzers and 88s, and so withdrew behind the crest. Montgomery was furious and ordered the infantry to continue the 'crumbling' fight while he pressed X Corps commander – Lieutenant-General Herbert Lumsden, a former Lancer – to get his armour to attack with, as Monty saw it, more aggression.

And attack they did – all day and the following night, before Montgomery called off the attempts to break out in the New Zealand division's sector, reinforcing 1st Armoured Division instead and ordering Lumsden to switch his main effort to the Highlanders' sector. Rommel, who had raced back from leave and had only just rejoined his headquarters,* brought this new line of attack to an abrupt halt by hurling all his available tanks at it, losing a good many of them in the process. That suited Montgomery, even if there was no progress on the ground. Like Wellington, he was happy to let the enemy come on to an unshakable firing line.

But Montgomery eventually called off this attempt to break through in the Highlanders' sector, deciding instead to exploit the Australians' success on the extreme right. He would make one more effort south of the Australians, however, in the New Zealand sector, with three British brigades – one armoured and two infantry (Durhams and Highlanders) – led by the prodigiously brave commander of the New Zealand division, Bernard Freyberg (VC and three,

---

* He would be similarly caught out on D-Day.

later four, DSOs). This attack, launched on the night of 1 November, almost broke right through, so that the following evening Rommel decided to withdraw. As he was trying to do so, 1st Armoured Division at last broke out of the New Zealand salient and threw the Germans' orderly withdrawal into disarray.

Montgomery now tried to get all three armoured divisions out (he had brought 7th Armoured up from the south several days before) to cut off the Axis retreat, but the congestion was too great, and although 8th Army took 30,000 prisoners, a third of them German, Rommel was able to get his Afrika Korps and the remains of the Italians away by 4 November. With only twenty tanks, however, he dared not pause until reaching Mersa Brega on the twentieth, having abandoned the whole of Cyrenaica.

On 8 November, therefore, the church bells, silent for so long (waiting to warn of invasion), rang out across England for the first time since the war had begun. Montgomery had lost 13,500 men killed and wounded, two-thirds of them British – half as many again as had Rommel. He had lost heavily in tanks, too, but it did not matter: he had tanks to lose; the Axis had not. Indeed, he had more tanks at the end of the battle than Rommel had started with. The game was now all but up for the Axis in North Africa; there could be no possible recovery.

Some historians are reluctant to acknowledge Alamein as quite so famous a victory, however. Correlli Barnett, for example, in his ever acerbic *Britain and her Army*, writes:

In Britain they rang the bells for Second Alamein. The Prime Minister had wished for a resounding victory before the British war effort was absorbed into the American alliance, and the resolute Montgomery and his dogged troops had granted the wish. Nevertheless Alamein (even First and Second Alamein taken together) cannot rank with Blenheim or with Waterloo, nor even with Salamanca or Vitoria. Those were triumphs over principal field armies of the great European enemy, while Alamein was a victory over a minor German expeditionary force.

Leaving aside the unfavourable comparison of Nazi Germany with Napoleonic France, this judgement makes little allowance for strategic risk. The 'minor German expeditionary force' had been a very real threat to the Suez Canal and the oil supplies of the Middle East on which the war effort depended. There had been no certainty of stopping Rommel at 'First Alamein' in July: Auchinleck's army had been looking over their shoulders for too long. But most of all the Barnett view fails to recognize the transformation in the army's fighting power which the battle – a bruising, hard-fought, ten-day match with plans changed on the hoof – brought about. D-Day and the slog from 'Normandy to the Baltic' (the title of 21st Army Group's record of the fighting in 1944–5) could not have gone as it did without Alamein. The battle quite simply put fight back into the army, redeeming the honour which, by Churchill's own reckoning besides that of enemies and allies alike, had been lost at Singapore. Indeed, without so resounding and hard-fought a victory as Alamein, Churchill and Brooke could hardly have advanced forcefully or credibly any counsel on war strategy with Washington.

In the making of the British army, Alamein was and remains one of the most important battles in its history, perhaps *the* most important. It was an object lesson in all-arms and inter-service cooperation (if by no means a smooth one), in just how hard men could and needed to fight, and in how command before and during battle should be exercised. And if it did not entirely lay to rest the ghost of the Somme, it put paid to the haunting. For Montgomery had known that this battle would exact far heavier casualties than 8th

Army had sustained in any of its previous engagements – indeed, heavier than the British army had yet sustained anywhere in this war – and he had been mentally prepared for it. When one of the armoured commanders protested that he would lose too many tanks as the plan stood, Montgomery icily assured him that he would accept 100 per cent casualties. As for the human cost, his chief personnel staff officer asked for a figure on which to base the medical plan. Montgomery went to his caravan and emerged two hours later with the answer – 13,000, which turned out to be within 4 per cent of the actual figure. But somehow Montgomery seemed to instil a confidence in his troops (if not in every one of his corps and divisional commanders) that all would be well, that he would not throw their lives away. They understood that war could not be made without casualties; they only wanted to be confident that their army commander knew his business. And thus the ghost was banished: the British army could after all plan and successfully execute a large offensive battle.

The lessons of Alamein served generations of 'Cold War warriors' in Germany (in the British Army of the Rhine), and in the First Gulf War; and even today they shape the approach to the all-arms battle, for all that the environment of war may have changed. Montgomery fought his battles according to strictly thought-out principles which he was able to turn into thorough and crystal-clear operational plans, driving these through with unrelenting determination and authority – though he did not, as is sometimes claimed, stifle initiative. Indeed, his 'master plans' set out what was to be achieved and with what; the 'how' was the business of the subordinate commander, though Montgomery did not always help in this perspective by his tendency afterwards to insist that 'everything went to plan'. War cannot be scripted; but neither can it be entirely improvised – or, at least, not for long. This, with some modification and reinterpretation, has remained the army's approach.

In terms of leadership, 'Monty' stands with Slim in the pantheon of British greats – if not as well loved as 'Uncle Bill', then certainly as fervently admired. Both men created an 'atmosphere' (the word which Montgomery himself used) of confidence in their commands. Slim said that the four best levels of command were the platoon, battalion, division and army; and in explaining why the platoon, he said simply, 'because, if you are any good you will know your men as well as their mothers do, and love them as much'. The notion remains at the heart of Sandhurst training today; and it connects with an even longer tradition, stretching back to Marlborough and even to Monck. And 'atmosphere' is everything. After Sir John Moore's army had been ejected from Spain, and Sir Arthur Wellesley had arrived with his own expeditionary force, Marshal Soult was quick to sense the change in this 'atmosphere' – and very soon to feel its effect. 'In truth, a new actor had appeared upon the scene,' wrote Sir William Napier, who had fought at Corunna, and subsequently with Wellington: 'The whole country was in commotion: and Soult, suddenly checked in his career, was pushed backward by a strong and eager hand.'*

Both Slim and Montgomery, as new actors on the scene, had dramatically changed the atmosphere in their respective theatres of war; and the enemy in Burma and North Africa was now being pushed back with hands as strong and as eager as those of Wellington.

* Napier, *History of the Peninsular War* (published in six volumes between 1828 and 1840).

26

Never a Defeat

1942–5

CHURCHILL'S IMAGE OF ALAMEIN AS A WATERSHED – 'BEFORE Alamein never a victory; after Alamein never a defeat' – was no truer in absolute terms after the battle than it had been before. But what is true is that after Alamein the trajectory of the fighting was unstoppably towards victory, even in the Far East, for as Churchill explained in the fourth volume of his memoirs, *The Grand Alliance*, with the United States now in the war 'all the rest was merely the proper application of overwhelming force'.

After Alamein, 8th Army cleared Libya (February) in a drawn-out affair which led to American criticism of Montgomery's caution, finally in May 1943 joining with the Anglo-American 'Torch' forces, which had advanced east from Algeria, to clear Tunisia. With the southern shores of the Mediterranean thus in allied hands, the invasion of Sicily (Operation Husky) in July went generally well, as did the subsequent landings on the mainland of Italy. But progress thereafter up the long leg of Italy, with its mountainous interior and narrow coastal plains, was painfully slow and costly. After the Italians changed sides in September the Germans quickly occupied the whole of the country, and Rome did not fall to the Allies until June 1944. Italy was not good 'tank country', and in the winter around Cassino and Anzio the infantry found their war was remarkably similar to that on the Western Front twenty-five years earlier. After the fall of Rome, the fighting in Italy became if anything even harder, and the campaign remains controversial: some historians argue that it was a critical diversion of resources from the main effort in north-west Europe – a symptom of the residual British infection, the indirect approach. Brooke remained convinced, however, that it was a crucial diversion of German resources that might otherwise oppose the allied advance through the Low Countries.

And that axis of advance in north-west Europe had itself been hard won. Alamein was the first battle of the war that the British army had fought *à l'outrance*, but the Normandy landings – 'D-Day' – were a far bigger affair, planned with an even greater sense of 'whatever it takes'. Indeed, British (and allied) troops were effectively committed to a fight to the death, for such were the huge resources dedicated to gaining a foothold on the first day that there could be no 'Plan B'; nor could there have been much evacuation from the beaches. The only alternative was strategic failure; and in the context of the commitment to Stalin to open a true 'second front', the failure would have been far-reaching.

In large measure the planning for Operation Overlord was Montgomery's triumph,* just as Alamein had been. He had

* 'Overlord' referred to the allied invasion of north-west Europe. The assault phase was known as 'Neptune' – the landing on the beaches, and the supporting operations required to establish a beachhead in France. 'Neptune' began on D-Day (6 June 1944) and ended on 30 June 1944. 'Overlord' began on D-Day, and continued until allied forces crossed the River Seine on 19 August 1944. The 'battle of Normandy' is the name given to the fighting between D-Day and the end of August.

been recalled from Italy in December 1943 to take command of the allied ground forces which would spearhead the landings under the overall command of the US general Dwight Eisenhower (Supreme Allied Commander Allied Expeditionary Force) and to remain in command until the build-up of forces ashore permitted the formation of two separate 'army groups'* – Montgomery's 21st (largely British and Canadian, and latterly Poles) and Omar Bradley's US 12th – at which point Eisenhower himself would take the reins on the Continent.

The landings went better than expected, with casualties, though heavy (particularly on the American beaches), far fewer than most commanders had feared. For the allies had achieved both strategic and tactical surprise. They had through a sustained and imaginative deception plan convinced Hitler and the German high command that the assault would be in the Pas de Calais, and that landings elsewhere (in Normandy) would be a diversion. And so German reserves, which might have counter-attacked successfully on D-Day (6 June) when the allies were at their most vulnerable trying to get ashore and establish a decent bridgehead, were held back. Tactical surprise was achieved with, among other things, the help of the weather. A huge storm had forced Eisenhower to postpone the invasion for twenty-four hours, and the Germans, whose meteorology was not as good as the allies', had concluded that no landings could take place for several days and had reduced their alert state (Rommel, by

---

* 'Army group' was a new concept for the British (one of sorts had been formed very briefly for operations in Tunisia in February 1943): a response to the requirement to have an intermediate headquarters below that of Supreme Headquarters Allied Expeditionary Force to direct and coordinate the various self-contained armies, not least in logistic and air support. Operations in 1940 had shown the need, as well as recognition of the sheer scale of the enterprise in north-west Europe. The Germans had army groups from the early days of the war; the Russians too, though they called theirs 'fronts'.

then in command of the counter-invasion forces, had even taken the opportunity to slip away to Germany for his wife's birthday). And during the night, before the invasion fleet appeared off the Normandy coast, airborne forces (parachute and glider troops) seized key bridges, gun emplacements and dominating ground, further confusing and then confounding the defenders.

But equally dramatic was the surprise gained by innovative equipment. The Germans had placed much reliance on their formidable beach obstacle belt – 'dragons' teeth', wire and mines, steel-reinforced concrete gun emplacements – but the allies (in truth, largely the British) had been looking at how to breach the belt as a technical rather than as solely a tactical challenge. As a result, armoured vehicles had been adapted for all sorts of specialist tasks by one of the army's earliest and most respected tank experts, Major-General Percy Hobart – who, having been sacked in 1940 after a row about the use of armour, had now been brought out of retirement at Churchill's insistence.* No matter what the problem identified by the planners, 'Hobo' always had a solution. The sand was too soft for wheeled vehicles: 'bobbin tanks' unrolled flexible trackway like a carpet salesman. The promenade-wall was too high for tanks to climb: 'ark tanks' drove against the wall and extended their ramps to allow other tanks to drive over them like a slow-motion game of leapfrog. Whatever the problem, there was one of Hobo's 'funnies' to deal with it. Never has so much technical energy been applied to a single day's fighting in the history of warfare. For the stakes on D-Day could not have been higher.

The most astonishing of the 'funnies' began its work well before reaching the shore, however. The infantry needed intimate fire support as soon as they hit the beach, but

* For more on Hobart, see 'Notes and Further Reading'.

getting tanks in with them was always going to be difficult since their landing craft, being bigger than those of the infantry, were vulnerable both to fire from the gun emplacements and to the explosive obstacles with which the Germans had studded the low-water line. The answer was a swimming tank – the DD (duplex drive) Sherman.

The DD was a standard gun tank fitted with a collapsible rubberized canvas screen which enabled it to float, and with a propeller driven by simple off-take shafts geared from the track-drive (hence 'DD'). The Americans, to their great cost on Omaha beach (each of the landing beaches, of varying widths, was given a code name), lost the majority of theirs in the swim in, but some of the British regiments, notably the 13th/18th Hussars, who had lost every one of their vehicles at Dunkirk, were able to get theirs ashore in the crucial minutes before the infantry landed. They launched their DDs (two squadrons, thirty-four tanks in all) some 5,000 yards out, and in waves whipped up by force 5 winds headed for the shore at an agonizing 100 yards per minute – almost an hour's swim, in which all of the crew except the commander, who had to stand on top of the turret to direct the driver, were below the water line. Miraculously – thanks to their training in the Moray Firth in January – only three of the Hussars' tanks foundered, with the loss of only four crew.

The corporal in command of one of the DDs recalled the moment the tracks made contact with the shingle of Queen beach on the far left of the allied landing zone. Except for the odd frogman-sapper, they were perhaps the first British troops on that stretch of the Normandy shore:

'75, HE, Action, Traverse right, steady, on. 300 – white fronted-house – first floor window, centre.'

'On.'

'Fire!'

Within a minute of dropping our screen we had fired our

first shot in anger. There was a puff of smoke and brick dust from the house we had aimed at, and we continued to engage our targets. Other DD tanks were coming in on both sides of us and by now we were under enemy fire from several positions which we identified and to which we replied with 75mm and Browning machine-gun fire.

The beach, which had been practically deserted when we had arrived, was beginning to fill up fast. The infantry were wading through the surf and advancing against a hail of small arms fire and mortar bombs. We gave covering fire wherever we could, and all the time the build-up of men and vehicles continued. [Driver] Harry Bone's voice came over the intercom:

'Let's move up the beach a bit – I'm getting bloody wet down here!'

They were landing on an incoming high tide, and with the canvas flotation screen lowered so that the gun could be used, water was coming through the driver's hatch. But the mines on the beach had not yet been cleared, so driving on would likely have had but one result. Then a wave suddenly broke over the rear of the tank, and the engine spluttered to a halt and obstinately refused to restart. The crew continued to give covering fire, but with all power gone and the tide fast overwhelming them, there was soon no alternative to baling out:

We took out the Browning machine guns and several cases of .3-inch belted ammunition, inflated the rubber dinghy [which each DD tank carried in case of foundering during the run-in] and, using the map boards as paddles, began to make our way to the beach. We had not gone far when a burst of machine gun fire hit us. Gallagher [co-driver: the Sherman had a crew of five] received a bullet in the ankle, the dinghy collapsed and turned over, and we were all tumbled into the sea, losing our guns and ammunition. The water was quite deep and flecked with bullets

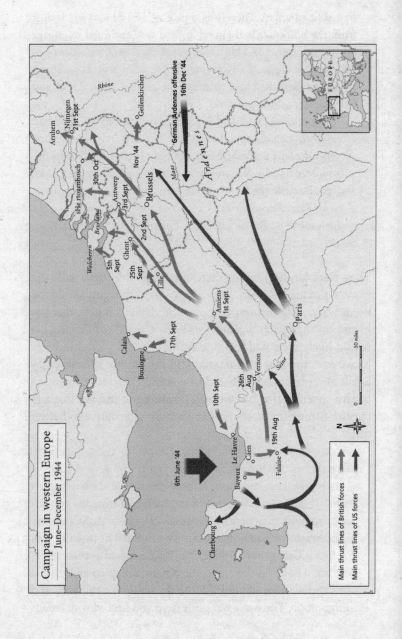

Campaign in western Europe
June–December 1944

Rhine

Geilenkirchen

German Ardennes offensive
16th Dec '44

Arnhem
21st Sept

Nijmegen

Nov '44

Ardennes

sHe rtogenbosch
30th Oct

Maas

Antwerp
3rd Sept

Brussels

Walcheren
5th Sept

Beveland

Ghent
25th Sept

2nd Sept

Lille

Amiens
1st Sept

Paris

Calais

Boulogne

17th Sept

50 miles

10th Sept

Vernon

26th
Aug

Seine

Le Havre

Cherbourg

Bayeux

Caen

6th June '44

Falaise

19th Aug

N

Main thrust lines of British forces

Main thrust lines of US forces

all around us. We caught hold of Gallagher, who must have been in some pain from his wound because he was swearing like a trooper [trooper was, indeed, his rank], and we set out to swim and splash our way to the beach. About half way there, I grabbed hold of an iron stake which was jutting out of the water to stop for a minute to take a breather. Glancing up I saw the menacing shape of a Teller mine attached to it; I rapidly swam on and urged the others to do so too.

Somehow, we managed to drag Gallagher and ourselves ashore. We got clear of the water and collapsed onto the sand, soaking wet, cold and shivering. A DD tank drove up and stopped beside us with Sergeant Hepper grinning at us out of the turret. 'Can't stop!' he said, and threw us a tin can. It was a self-heating tin of soup, one of the emergency rations with which we had been issued. One pulled a ring on top of the tin, and miraculously it started to heat itself up. We were very grateful for this as we lay there on the sand, in the middle of the battle taking turns to swig down the hot soup. We were approached by an irate captain of Royal Engineers who said to me:

'Get up, Corporal – that is no way to win the Second Front!'*

With appropriate cavalry nonchalance, after the war the 13th/18th Hussars, who on D-Day won five MCs and twelve MMs, a record for a single regiment in one day, added a few bars of 'A Life on the Ocean Wave' to their regimental march.

'D-Day' stands as one of the army's finest hours, as well as a model of what inter-service and inter-allied cooperation could achieve. The sheer ingenuity of every branch, combat and logistical, in preparing to overcome the redoubtable

---

* Patrick Hennessey, *Young Man in a Tank* (self-published, 1988). Corporal Pat Hennessey was subsequently commissioned into the RAF and rose to the rank of group captain before retiring in 1984.

defences of the 'Atlantic Wall' – *Festung Europa* (Fortress Europe) – and the problems of supply over the beaches until a major port could be captured has never been surpassed. The daring of the glider troops who swooped out of the moonlit sky to take Pegasus Bridge on the Caen Canal would be hailed as a masterstroke even today, when troops have helicopters and satellite navigation instead of the flimsy airframes flown by serjeant-pilots with their maps on their knees. The intrepidity of the 'amphibious' tank-men in their long swim, the cool courage of the sappers among the mines and booby-traps in the rising tide, the sheer determination of the infantrymen to get across the beaches and inland to grips with the depth defences – these and exploits like them were the common currency of 6 June 1944. D-Day was a battle in itself, fought largely by men who up to that day had not fired a shot in anger – even the regulars, many of them, not since 1940. And those who went ashore that day knew in their hearts that they might have to fight to the last to gain a beachhead for those who would come ashore on D+1 and in the days and weeks afterwards – the men who would take over the business of fighting the Germans out of Normandy. It was unspoken, in the main, and some learned of it later indirectly: the 13th/18th Hussars, for example, were slow to get replacements for the tanks they lost on 6 June, for the planners had simply not expected them to survive the day as a formed unit, and had therefore not allocated any. It was the same across the entire first wave.*

The fighting inland was a grim affair, the Germans seeming to contest every house and field. The countryside – *le bocage* – was so close, offering easy concealment for anti-tank weapons, that the tank had few advantages, and the

---

* On D-Day, the allies landed around 156,000 troops in Normandy. For details, see 'Notes and Further Reading'.

infantry's casualties mounted as they cleared the hedgerows and hamlets yard by yard. It was the middle of August before the Germans cracked – but when they did it looked as if their hasty retreat might turn into rout as the remains of the German 7th Army – perhaps 100,000 men and a prodigious number of tanks, guns and transports – struggled to escape what became known as the 'Falaise Pocket', the area south of Caen which an allied pincer movement had turned into a shooting gallery for the fighter-bombers. The RAF's rocket-firing Typhoons and the USAAF's Mustangs and Thunderbolts pounded the Germans remorselessly: between 20,000 and 50,000 – the estimates are still imprecise even today – managed to slip out of the noose, but 50,000 surrendered, and the roads and fields around Falaise were littered with the dead – 10,000 at least.

The battle for Normandy was at last over, and with it the uncertainty of the price that the British army would have to pay in the defeat of Germany. For by now the Russian offensive in the east was making progress, and henceforth the road from Normandy to the Baltic, although it would be a hard one, could never be completely closed to the allies for more than a few weeks, perhaps only days, at a time. They had lost 200,000 men in Normandy: 37,000 of these had been killed, 16,000 of them British, Canadian or Polish, a slightly higher percentage than the US. The Germans had lost 300,000.

But inter-allied rivalry, mistrust and straightforward differences of strategic opinion dogged the subsequent advance through France and the Low Countries. In early September Montgomery over-reached himself in a serious miscalculation – or, as Field Marshal Lord Carver, then a 28-year-old armoured brigade commander, put it in his *Seven Ages*: 'Montgomery now let success go to his head.'

'Monty', by now a field marshal but having ceded overall command of the allied ground forces to Eisenhower at the end of the battle of Normandy,* had been advocating for some time that one of the two army groups, preferably his own, be given the logistic priority so that a decisive effort could be made on a narrow front instead of continuing with Eisenhower's 'broad front' policy (which Montgomery dismissed as 'everyone attacks everywhere'). The principle was sound, but the difficulty lay in identifying the object against which to concentrate the effort. In other words, what was of decisive importance to the enemy's capacity to resist?

By September the logistical problems had also increased to the point where even concentrating support on one army group was not going to be enough to sustain it in a major thrust. Not since Marlborough's 'scarlet caterpillar' had the British army been so stretched on the Continent. Day in day out the trucks of the magnificent Royal Army Service Corps hauled the 'C Sups' – combat supplies (ammunition, rations and 'POL' – petrol, oil and lubricants) – from Cherbourg, the only major port in full operation, and the few captured Channel ports sufficiently cleared of demolitions and mines, along 400 miles of frequently indifferent roads which the sappers struggled constantly to maintain, marshalled throughout by the immaculate red-capped, blanco-belted (later 'Royal') Military Police.† It was indeed a continuous khaki conveyor belt, the image of Cicero's notion of the 'sinews of war'.

* Later than he might have done, for the build-up of US troops had placed them in the majority. It made sense, however, to wait until the battle for Normandy was won.
† Boulogne and Calais were not taken until 22 and 30 September respectively, and Dunkirk did not surrender until the end of the war. The US army transport battalions were also supplying 21st Army Group, and supply routes were an allied concern: roads were designated one-way to and from the front line, and the whole operation was known by the Americans as the 'Red Ball Express'. Even by September only a very limited amount of supplies could be moved by rail, such was the effectiveness of the preliminary allied bombing campaign, and air transport could move only a small proportion of daily requirements.

The need to shorten that conveyor belt was all too evident; but, instead of directing a major part of 21st Army Group's effort towards capturing Antwerp, which the Germans were still holding stubbornly, Montgomery now proposed an uncharacteristically bold plan to thrust north-east over the Rhine, envelop the Ruhr and thereby bring the war to an end. But to get across the Rhine (his troops were still only at the Albert Canal on the Belgian–Dutch border) he would have to seize a number of bridges on the way, including the enormous span over the Waal at Nijmegen, and cross the Lower Rhine itself at Arnhem. Eisenhower was not prepared to call a halt to the US 12th Army Group's easterly advance and put all supplies Montgomery's way (for Patton, commanding US 3rd Army, was anyway moving fast and 'Ike' did not want to slow him down), but he reckoned that extending the allied line as far as the Rhine would anyway protect Antwerp once captured, so he gave the plan the green light and allocated Montgomery the 1st Allied Airborne Army – formed only the previous month – with which to seize the bridges.*

The commander of the Allied Airborne Army was General Lewis Brereton, who had commanded the US 9th (tactical) Air Force in Normandy, his deputy the British lieutenant-general Frederick ('Boy') Browning.† Brereton had initially been sceptical about the need for a separate army-level command for glider and parachute troops. He had not been confident that the First Airborne would be used en bloc as an

---

* In fact Montgomery had first lobbied Eisenhower for a concentrated drive on Berlin, so convinced was he that German fighting power was broken. 'Ike' ruled this out at once: there was simply not the means of maintaining such a thrust, even with Antwerp open. The thrust would in any case be vulnerable to counter-attack along its entire length, and he was correct in estimating German fighting capacity to be greater than Montgomery supposed, as the Ardennes offensive in December proved.
† Browning was married to the writer Daphne du Maurier. Their daughter, Tessa, married Montgomery's only son in 1967.

army, for a number of operations had already been planned in detail and then abandoned at the last minute as ground troops over-ran the airborne objectives. Now, with the German collapse, it even looked as if its divisions would never jump again. Browning especially was anxious for them to be committed lest they be grounded and used as normal infantry: the Parachute Regiment had only been formed in 1940, and its status was therefore precarious. It is hard to fault a commander with such fighting spirit, but this was scarcely the military judgement expected of a lieutenant-general in command of a strategic asset. Browning's eagerness to use his own 1st (British) Airborne Corps, together with Montgomery's over-optimism, now led to the last of the British army's true battlefield defeats – albeit a magnificent defeat, and one which ironically has had a greater moral effect than if the operation had been a resounding success.

Operation Market Garden consisted of two separate but dependent elements: 'Market', under Browning's tactical command, was the airborne operation to seize the bridges along the main road through Nijmegen and beyond to Arnhem, while 'Garden' was the follow-up ground forces plan. The advance of 2nd (British) Army north over the secured bridges and beyond the Rhine would be spearheaded by XXX Corps under the ebullient Lieutenant-General Brian Horrocks. There were eight major water obstacles between XXX Corps' start line and the ultimate objective, the north bank of the Rhine beyond Arnhem some 60 miles distant, and Browning judged that bridges over all these obstacles would have to be seized simultaneously or else the Germans, guessing what was happening, would reinforce or demolish those that remained. And although the smaller canals and rivers could be bridged by XXX Corps' sappers without too much loss of time, the larger rivers – the Waal at Nijmegen and the Rhine at Arnhem – were too wide for 'combat bridging' (in other words, they could not be bridged from one bank only, and under fire). To make matters even harder, Highway 69, astride which XXX Corps would advance, lay like a dike or levee across the flat terrain of the Dutch polder, single carriageway for most of its length, and the ground on either side was too soft to support tactical vehicle movement.

'Market' would need three of the five divisions of the Airborne Army. The US 101st under Major-General Maxwell Taylor would drop in two places just north of XXX Corps' start line to take the bridges north of Eindhoven at Son and Veghel. The 82nd under Brigadier-General James Gavin would drop north-east of the 101st to take the bridges at Grave and Nijmegen, while the British 1st Airborne Division, under Major-General Roy Urquhart, and the Polish 1st Independent Parachute Brigade, commanded by the redoubtable veteran Major-General Stanislas Sosabowski, would take the road bridge at Arnhem and the rail bridge at Oosterbeek and hold on for forty-eight hours – the time it was calculated that XXX Corps would need to reach them. For all his urgent desire to get his airborne corps into action, Browning said memorably to Montgomery beforehand, 'I think we may be going a bridge too far.' But without a bridgehead beyond the Rhine the operation was unlikely to be worth the effort, for all its secondary purpose in securing Antwerp (once, that is, Antwerp had fallen to the allies – which did not happen until November). *A Bridge Too Far* (1974) was the title of the third of former war correspondent Cornelius Ryan's Second World War trilogy of battle accounts, and also of Richard Attenborough's epic 1977 film, parts of which veterans say are accurate.

Although there had been anticipatory planning, the actual operation was put together in not much more than a week. The plan had a fatal flaw, however: it wholly underestimated German strength on the ground, intelligence having failed to recognize definitively the presence of two panzer divisions (albeit much depleted) recuperating near Arnhem.* On the other hand, RAF intelligence wholly overestimated the strength of the German anti-aircraft defences at Arnhem, forcing Urquhart to break a fundamental rule of airborne warfare – such as they were in those early days – and choose a drop and landing zone too far from the objective: 6 miles, indeed. At best marching speed (1st Airborne had next to no wheels), unmolested, this would give the Germans a little over an hour to reinforce or destroy the bridge. The only chance lay in there being no Germans for miles – though this was unlikely on a road of such importance – or in the airborne recce squadron's jeeps with their heavy machine guns contriving to race from the landing zone to seize Arnhem Bridge and hold on until the Paras could relieve them.†

Market Garden would be the largest airborne operation in history. Beginning on the morning of 17 September, 20,000 troops were dropped by parachute and almost 15,000 flown in by glider, together with over 1,700 vehicles and 263 artillery pieces; and over the nine days of the fighting – not the three anticipated – some 3,500 tons of combat supplies were delivered by glider and parachute drop. But although

* One of the Airborne Corps' intelligence officers had his suspicions, but he was ignored in the general desire to get the operation mounted.
† The previous commander of 1st Airborne had been killed in Italy, and Urquhart took command in January 1944, aged 43. He was new to airborne forces. Indeed, until the year before he had been commanding an infantry battalion, but he was fresh from the fight in Italy where he had won a DSO. He had an uphill struggle gaining the confidence of the division, not least of his brigadiers, one of whom, Gerald Lathbury, had been told unofficially that he was to command the division. Urquhart had been commissioned into the Highland Light Infantry, and a fellow subaltern had been the future Hollywood star David Niven. In Niven's outrageous autobiography *The Moon's a Balloon*, he describes Urquhart as 'a serious soldier of great charm and warmth'. Urquhart's daughter is married to former Liberal Democrat leader Sir Menzies Campbell.

the US divisions eventually took their objectives, Urquhart's division could do no more than get a single battalion (2nd Battalion, the Parachute Regiment) to the great road bridge over the Rhine – and even these could not take its southern end.

Over the next three days 2 Para, under their remarkable commanding officer Lieutenant-Colonel John Frost, were whittled down under ferocious attacks from German infantry (including Waffen-SS) and armour from both north and south of the river, Frost himself becoming a casualty and taken prisoner. A ceasefire was arranged at one point to evacuate the wounded of both sides, and then the fighting began again in earnest, SS troops pressing their attacks with suicidal courage. Yet even during the bitterest fighting a certain wry, if black, humour prevailed among the Paras. Father Bernard Egan, 2 Para's Catholic chaplain, recalled how he met Frost coming out of a lavatory in one of the shattered buildings near the bridge looking, like the rest of the battalion, 'tired, grimy and wearing a stubble of beard'. 'Father,' said Frost, his face suddenly lighting up, 'the window is shattered, there's a hole in the wall, and the roof's gone. But it has a chain and it works.'

Some time later, Father Egan was trying to make his way to the wounded in a house on the other side of one of the streets leading to the bridge: it was being shelled and he feared he could not cross. But then he caught sight of 'A' Company commander, the languidly tall Major Digby Tatham-Warter, whom Frost described as 'a Prince Rupert of a man', calmly strolling up the middle of the street carrying his trademark umbrella (which he claimed was a recognition symbol since he could never remember the password). Tatham-Warter saw the sheltering padre, made his way over and beckoned him across. Father Egan pointed out the mortar barrage, to which Tatham-Warter replied disarmingly, 'Don't worry, Padre; I've got an umbrella.'

On getting back to battalion headquarters, which had been shelled relentlessly all day and was now on fire, Father Egan did the rounds of the wounded in the cellar, where the battalion joker, Sergeant Jack (inevitably) Spratt, chirped, 'Well, Padre, they're throwing everything at us but the kitchen sink.' Spratt had barely finished his sentence before the building took a direct hit, the ceiling fell in, and in with the shower of plaster and assorted debris crashed the sink from the kitchen upstairs. 'I knew the bastards were close,' coughed Spratt, 'but I didn't think they could hear us talking!'*

With this spirit 2 Para held on until at last, soon after dawn on 21 September, with no ammunition left and the houses all about them in flames, the remnants of the battalion were finally overwhelmed. The last radio message from the bridge – 'Out of ammo, God save the King' – was heard only by German intercept. Everyone else was still out of signal range.

Faulty communications had in fact bedevilled the operation from the beginning, leaving Urquhart and 1st Parachute Brigade's commander pinned down and out of contact with anyone for crucial hours on the first day when they went forward to find out what was happening. And while the fighting at the bridge was running its isolated course, elsewhere in Arnhem itself, in its suburbs and on the landing ground, the casualties mounted as the Germans threw in everything they could against the scattered battalions. On the ninth day, XXX Corps having been held up every mile of the way (not even managing to get across Nijmegen bridge until the evening of the fourth day, by which time they were meant to have been at Arnhem for forty-eight hours), Montgomery ordered 1st Airborne to break off the fight and get back across the Rhine as best they could. But of Urquhart's 10,000 who had dropped by

* Fr Egan was awarded the MC, and Tatham-Warter the DSO.

parachute or landed by glider, less than a third now made it back. Of the other two-thirds and more, 1,200 had been killed; the rest were taken prisoner, a great many of them wounded. Their fight was, in the words of David Fraser, who had been with the Grenadiers in the Guards Armoured Division spearheading XXX Corps' advance, 'one of the noblest fought by the British Army in the Second World War, and its glory will last as long as the British Army's story is remembered'. Eisenhower himself wrote: 'There has been no single performance by any unit that has more greatly inspired me or more excited my admiration than the nine day action by the lst British Parachute [*sic*] Division between September 17 and 25.' The division won five VCs on Market Garden.

Arnhem set the bar for determined fighting even higher. When Lieutenant-Colonel 'H' Jones was killed leading 2 Para against the odds at Goose Green in the Falklands nearly forty years later (a battle in which he too won the VC), he and all his battalion knew their operational heritage: giving in, while there was ammunition left, was not an option. Arnhem was a powerful if unquantifiable factor in the fighting that day in 1982, and it would continue to exert its force as the Paras marched and fought their way across the Falklands, uncertain of what lay ahead. It was something usually unstated, in the background, but just occasionally the Arnhem legend could be invoked explicitly to screw an extra dose of courage to the sticking post: 'Look, we've done bloody well today,' Major John Crossland told 'B' Company as evening fell at Goose Green in that desolate corner of East Falkland in the depths of the South Atlantic winter:

> Okay, we've lost some lads; we've lost the CO. Now we've really got to show our mettle. It's not over yet, we haven't got the place. We're about 1,000 metres from D Company; we're on our

own and the enemy has landed to our south and there's a con-siderable force at Goose Green, so we could be in a fairly sticky position. It's going to be like Arnhem!

They knew what he meant and understood what he expected.

And when a fortnight later 2 Para's sister battalion, 3 Para, attacked the Argentinians entrenched on Mount Longdon, their losses mounting steadily in a fierce night battle in which Serjeant Ian McKay would also win a posthumous VC, there was not a man who did not know of the battalion's near-annihilation at Arnhem and the death of its commanding officer there. Certainly 3 Para's commander in the Falklands, Lieutenant-Colonel Hew Pike, was very conscious of his battalion's reputation when he wrote to his wife after the battle: 'We finally assaulted Mount Longdon on night 11/12 June, and I suppose it was as fine a feat of arms as 3PARA has ever under-taken, surpassing anything since World War 2 and I reckon at least the equal of any World War 2 battles fought by this battalion.' Earlier, on hearing the news of 2 Para's casualties at Goose Green, he had confided to his diary the words of Lord Moran, a regimental medical officer in the First World War and Churchill's doctor in the Second: 'The individual shrinks to nothing. He has no right of an opinion. Only the regiment matters', adding his own observation that 'Small wonder, in this all pervading atmosphere of uncertainty and loneliness, that comradeship grew even tighter. It was the stuff of survival.'*

Among the British army's many highs and lows, the Second

---

* The reference is to Lord Moran's *The Anatomy of Courage* (1945), quoted by Hew Pike in *From the Front Line*. Although Arnhem is primarily a Parachute Regiment battle honour, it is shared by several infantry regiments which served in the (glider-borne) Airlanding Brigade, as well as by the supporting arms. But the army as a whole has been conscious of the example, and the raised height of the bar.

World War was perhaps its ultimate turning point: from 1945 onwards, although there has been many a bloody nose, there has been a consistent success on operations which none of the earlier periods of reform was able quite to sustain. The army of 1945, though it was led at battalion level and above by the men of 1939, was an entirely different affair from that which had been chased out of Norway, France, Libya, Greece, Malaya and Burma. The general sense of 'ignorant poverty', in David Fraser's words, that prevailed before Dunkirk had been replaced by a well-resourced professionalism: by 1945 officers and NCOs had become, in the main, thoroughly skilled in their jobs, especially those in the burgeoning technical arms. The Royal Corps of Signals, for example, which had been formed from the Royal Engineers signals branch in 1920, had over 150,000 officers and men wearing the 'winged Mercury' (the messenger of the gods) as their cap badge by the war's end. They were providing battle-field and strategic communications in every theatre of the war, and intercept and jamming of the enemy's communications (now known as electronic warfare), including the handling of Ultra. Entirely new corps were forged in the course of the war too – such as the Royal Electrical and Mechanical Engineers (REME), formed in the very month of Alamein for the maintenance and recovery of the army's growing inventory of equipment, from the smallest pistol to the biggest howitzer, from radio sets to radial engines, from typewriters to tanks. As the army grew – and grows – ever more dependent on equipment, so the REME has increased in both range and size. To Benjamin Franklin's often-quoted assurance that 'In this world nothing can be said to be certain, except death and taxes' the British soldier today would add '—and more REME'.

Perhaps the most disappointing arm – ironically, for it had seemed to promise so much in 1917–18 – was the armoured, which by the end of the war comprised twenty-four 'battalions'* of the Royal Tank Regiment and all the cavalry (except the Household Cavalry, who although mechanized remained technically part of the Household Division, along with the Foot Guards), the Reconnaissance Corps, which was disbanded in 1946, and those yeomanry regiments in the armoured and reconnaissance role.† The handling of armoured forces had never been truly inspired in North Africa, and despite some dashing regimental actions during the break-out from Normandy, and no lack of courage, by and large British armour had remained far less handy than German panzer troops in north-west Europe, and less thrusting than the Americans. It is certainly difficult to imagine that a Rommel or a Patton would not have reached Arnhem somehow.‡ In part this was because armoured doctrine had never been 'nailed', and so remained a business of local opinion (in effect, that of the senior commander, who was not usually an armoured man), which was frequently pedestrian and often downright suspicious. Wavell's sacking of Hobart exemplified the quarrels. 'Hobo' was not an easy man. His gaunt, bespectacled appearance was more that of an irascible schoolmaster than a soldier, and his belief in the tank as a means of forcing a decision – independently, almost, of other arms – won him no friends among the community of senior officers at the time. In fact no RTR officers had achieved high enough rank

* The Royal Tank Corps, as the RTR was known until 1939, had comprised eight battalions. After 1939 its constituent units were styled consecutively 1st Royal Tank Regiment (1RTR), 2RTR etc.

† Other elements of cavalry were re-roled as gunners and even signals, but there were none with horses still, for the horse had finally disappeared from the army's regular order of battle (though there were mules and the odd horse in 14th Army in the Far East).

‡ An opinion stated baldly by General Sir David Fraser in his assessment of Market Garden (*And We Shall Shock Them*). The armoured cars of the Household Cavalry had actually got up to the Rhine, by-passing resistance, but it made no difference to the follow-up. In fairness, however, none of the divisional commanders was RAC.

to make a real mark on tactics on behalf of the armoured arm, and the cavalry on the whole came too late to the top table – although towards the end of the war there was accelerated promotion for some of the younger stars, of whom perhaps the most brilliant were Michael Carver, commanding an independent armoured brigade at the age of 28, and 'Pip' Roberts, who commanded 11th Armoured Division in the heavy fighting in Normandy at 37.* Likewise, cavalry officers, who had exchanged their horses for tanks, frequently found themselves at odds with the predominantly infantry corps commanders who were suspicious of the independence with which they wished to employ the armour, or else (as at Alamein) resentful of what they perceived as hanging back in the face of an 'occupational hazard' – anti-tank fire. Conversely, from time to time cavalry commanders were sacked for an excess of *élan* – notable among them Brigadier 'Looney' Hinde (his nickname gained in the hunting field for fearless riding), whom Montgomery fired along with the divisional and corps commanders after setbacks in Normandy in 1944.

In many ways, however, the problem had been the tank itself: because no clear idea had emerged early enough about how best to use it, no very good design appeared until the closing months of the war when the Comet, and then the magnificent Centurion, entered service (though the latter was too late to see action). As Max Hastings observes in *Overlord*, his study of the fighting in Normandy, it remains an extraordinary feature of the war that despite the vast weight of technology available to the allies, British (and American) soldiers 'were called upon to fight the German army in 1944–45 with weapons inferior in every category save that of artillery'. Nevertheless, one man did emerge with

---

* Roberts retired not long after the end of the war, but Carver became a chief of the defence staff, and field marshal.

the experience to drive home the lessons after the war: General Sir John Crocker, an RTR man from the earliest days who would write the definitive armoured corps doctrine for the Cold War.*

But above all by 1945 it was the organization *for* war – the whole staff system from War Office to brigade headquarters, which had advanced so much during the First World War and then withered so perilously in the following two decades, that had become so impressively, and permanently, tuned. No army, no matter how good its fighting men and its supporting arms and services, could prevail against a first-class enemy such as the Germans and Japanese if its staffwork was not equally first rate; and if that lesson of 1914–18 had been half-forgotten in the prevailing atmosphere of 'never again' in the 1920s and 1930s, from Alamein onwards the 'never again' applied more to the army's resolve not to allow itself to forget that lesson a second time.

One thing had not been resolved, however: the position and role of what would become known as 'Special Forces' (SF). In Burma, Major-General Orde Wingate's 'Chindits' (from *chinthe*, the mythical, elusive beasts of the jungle whose statues guard the Burmese temples) had operated deep behind the Japanese lines. Resupplied by air, they had kept alive the offensive spirit when the rest of Slim's 14th Army had been either on the defensive or regrouping and retraining. They were not SF in today's sense of small, élite teams

---

* Crocker had commanded 6th Armoured Division after the fall of France, and then a corps, but did not fight in the desert – his active service in North Africa being later in Tunisia, where he was wounded. Montgomery rated him highly, and he commanded the predominantly infantry 1st Corps in Normandy. Though futile, it is nevertheless interesting to speculate what might have been had Crocker commanded the armour under Auchinleck in the desert – or even, for that matter, under Montgomery at Alamein.

working on missions of strategic effect (the Special Air Service); rather, they were selected units, specially equipped and trained, fighting the enemy conventionally but beyond the front line.*

There were two Chindit expeditions: the first, in brigade strength (some 3,000 men) marched over 1,000 miles during its three-month jungle sojourn beginning in February 1943, losing over a quarter of its number in the process; the second, the following year, consisted of 20,000 British, Indian and Commonwealth troops and was even more controversial in its conception; Wingate himself was killed during its course. Slim had been far from convinced, even before the first expedition, that the Chindits' tactical or even strategic gains would be worth the diversion of men and materiel from the main effort, and in *Defeat into Victory* he gives his decided opinion: 'They did not give, militarily, a worth-while return for the resources in men, materiel, and the time that they absorbed,' he wrote, concluding that 'The rush to form special forces arose from confused thinking on what were, or were not, normal operations of war.' Indeed, Slim believed that 'any well-trained infantry battalion should be able to do what a commando could do; in the Fourteenth Army they could and did'. And this indeed became the basis of the army's post-war development, with units moving relatively seamlessly between roles every half-dozen years or so – parachuting excepted, which in fact Slim specifically excluded from his 'normal operations' on practical grounds of training expense.

But Slim's dislike of 'private armies', as many of the ad hoc forces had come to be known – a dislike shared by most

---

* One of the best accounts of Chindit operations, with their sense of fighting the Japanese beyond the front line, is *Beyond the Chindwin* (1945) by one of the Chindit commanders, Brigadier Bernard Fergusson (the River Chindwin was in effect the front line in north-west Burma).

senior officers, including Montgomery, who would take over from Brooke as CIGS in 1946 – extended only to the 'over-classification' of the normal operations of war and the consequent large-scale diversion of resources. In the same paragraph on 'misnamed special forces' Slim argued for a more acute capability:

> There is, however, one kind of special unit which should be retained – that designed to be employed in small parties, usually behind the enemy, on tasks beyond the normal scope of warfare in the field . . . to sabotage vital installations, to spread rumours, to misdirect the enemy, to transmit intelligence, to kill or kidnap individuals, and to inspire resistance movements. They will be troops, though they will require many qualities and skills not to be expected of the ordinary soldier . . . and should operate under the control of the Higher Command . . . [and] may, if handled with imaginative ruthlessness, achieve strategic results.

This specification is almost exactly that of the SAS today. Founded originally by the then Lieutenant David Stirling, a Scots Guards officer, the SAS were originally intended for raiding and sabotage deep behind enemy lines in North Africa. The name 'Special Air Service' was meant as a deception, and their first mission, by parachute in November 1941, was a disaster, only twenty-two out of the sixty-two troopers getting away – and these only with the help of the Long Range Desert Group (LRDG), whose job was vehicle-borne reconnaissance. Shortly afterwards, however, Stirling redeemed his concept by using the LRDG to take his men into an attack on three German airfields, destroying sixty-one aircraft without a single casualty. The SAS had almost by accident discovered its modus operandi, and thereafter quickly gained regimental status, with a second regiment formed by Stirling's brother Bill. By 1945 they were operating extensively in Italy with the partisans, having also helped train the French resistance before D-Day. Soon after the war ended, nevertheless, the SAS was disbanded – only to be re-formed in 1947 for the Malayan Emergency.

After his electoral defeat at the end of July 1945, with Hitler dead and the capitulation of the Japanese only a fortnight away, Winston Churchill wrote a final note of thanks to Field Marshal Sir Alan Brooke: 'Our story in this war is a good one, and this will be recognized as time goes on.'

The British army's story in the Second World War was in fact by no means a brilliant one, but in the end it was indeed good enough – and at times it was glorious. It had fought the war with the burden of the Somme weighing on the shoulders of senior officers and politicians alike; for eighteen months it had shouldered that burden without allies save for the imperial and dominion forces; and it had had to over-come its initial gruelling setbacks and humiliations while fighting in three continents. In his compendium of the army's experience from the end of the Boer War to its involvement in the Balkans, *Britain's Army in the Twentieth Century* (1999), Field Marshal Lord Carver writes of the balancing act entailed in trying to be ready for multifarious commitments:

Using the experience of the past as a guide to the balance required to meet future demands has . . . often proved un-reliable; but imaginative visions of how to meet them have also been, if not false, at least premature. The army has generally been distrustful of and slow to change, except under the stress of a major war.

Like Wellington himself, the army has always been a conservative creature; even in the Second World War its collective instincts had remained strategically, tactically and

497

for the most part technically cautious – although there was some brilliant improvisation at times.* But as in 1918, in the Second World War the army had in the end prevailed: it had beaten the Germans (and the Japanese). And it had done so without the debilitating losses of the First. Although 144,000 killed was a bad enough count, it was less than a fifth of the 1914–18 figure. In part this was because only on rare occasions – Alamein and D-Day the prime examples – were commanders at every level prepared to drive home attacks almost regardless of cost; in part it was greater skill; and in part it was luck.

The British soldier could not match the fanaticism of the Japanese or the indoctrinated aggression of the German in battle, but he would, in the words of one historian, 'fight with unflinching doggedness while absolutely necessary and then break off for tea'. However, side by side with the Americans (and Canadians), who did not of course drink tea, the British army had fought its way from Normandy to the Baltic – and side by side with troops from the Empire, who did drink tea, from Imphal to Rangoon. It was, after all, in Montgomery's tent on Lüneburg Heath that on 4 May 1945 the 'Instrument of Surrender of all German Armed Forces in Holland, in Northwest Germany including all Islands, and in Denmark' was signed. There can be no arguing with victory.

How, though, in the demobilization that was sure to follow victory, would the British army's hard-won capability be maintained?

---

* Hobart's 'funnies' were the shining example, and they continued to serve until the day of the German capitulation. Perhaps the most succinct judgement is war correspondent John North's in his *North-West Europe 1944–5* (1953): 'The war as fought by Britain's 79th Armoured Division when supporting an infantry assault was the ultimate in mechanized fire power, and the conception and its execution had been British throughout.'

27

Only the Regiment Matters

The 'austerity' years, 1946–53

SINCE 1949 LEIGHTON HOUSE, A HANDSOME NINETEENTH-century mansion in 50 acres of parkland on the edge of the little Wiltshire town of Westbury, has been home to the Regular Commissions Board, which was set up to select officers commissioned in wartime for full-time (regular) commissions, and entrants for Sandhurst, in the post-war army. But not all the 'War Office selection boards' that had been hastily established to select those wartime temporary officers could be disbanded immediately, for although the army in August 1945 was not much smaller at 2.9 million men than it had been in November 1918, the political situation militated against the same wholesale demobilization. The new Labour government under the premiership of Clement Attlee (Churchill's deputy in the wartime coalition) had expected that it would soon be 'business as usual' for the army, with a return to voluntary recruiting; but the peace of 1945 was quite unlike that of 1918, or 1815, or even 1715. The imperial situation was far shakier, in part because of a surge in nationalism, in India especially, and in part because of the spread of Communism in Asia – allied to the uncertainty of how difficult it would be to recolonize territory lost to the Japanese in 1941. And in Europe the enemy had been replaced by an even more powerful threat: the Red Army.

The wartime conscription act could not be prolonged indefinitely, however, and so in July 1947 Attlee's government took the unprecedented step of passing an Act of Parliament providing for conscription in peacetime.* There was to be one year's compulsory military service for all fit males aged 18, with a call-out liability for the following six years. This, it was calculated, would produce an active army of 305,000. However, Montgomery, who took over from Brooke as CIGS in the middle of 1946, wanted to plan for the contingency of having to field an army which mirrored that of 1944 against the new Soviet threat, and so the period of 'National Service', as it was comfortingly called, was soon extended to eighteen months, with the reserve liability reduced to three and a half years during which the reservist would be on the paper strength of a TA unit. But even this increase was not enough, and in 1950, by which time Montgomery had in turn been replaced as CIGS by Slim, the period with the colours was increased to two years, generating an army of nearly 400,000, a little over half of which was regular. By 1952 it had increased again to 442,000, including eight battalions of Gurkhas transferred to the British establishment after Indian independence in 1948.

Yet even with this unprecedentedly high number of troops in 'peacetime' the army was soon overstretched. With the breakdown of negotiations over the future of Germany, Stalin's encouragement of a Communist coup in

---

* Unprecedented, that is, excepting the Military Training Act of April 1939, which was passed in the expectation of war.

Czechoslovakia and the Soviet blockade of Berlin in 1948, the British Army of the Rhine (BAOR) was hastily converted from a non-operational occupation force of roughly two divisions to an operational army of five divisions under the improvised direction of the new Western European Union comprising Britain, France and the Benelux countries. After 1949 and the foundation of the North Atlantic Treaty Organization (NATO) there was a more formal military structure resembling that of 1944–5, with Eisenhower as the Supreme Allied Commander Europe (SACEUR) and Montgomery as his deputy. NATO's object, as its first secretary-general, Lord Ismay (who as General Sir Hastings 'Pug' Ismay had been military secretary to Churchill's war cabinet), memorably put it, was 'to keep the Russians out, the Americans in, and the Germans down'.

But war with Communism broke out further away than anticipated – in Korea, on 25 June 1950, where the withdrawal of the US occupation force the year before had emboldened the North Korean Communists, inspired by the success in China of Mao Tse-tung's 'People's Liberation Army' against Chiang Kai-shek's Nationalist troops, to launch a surprise attack across the border between North and South Korea, the 38th parallel.* The UN Security Council, in its first test of the Cold War and in the absence of a Russian veto (the Russians were boycotting the Council in protest at its inclusion of Chinese Nationalist represent-ation), passed a resolution authorizing intervention, and a

---

* The allies had been urging Russia to declare war on Japan since the Tehran conference in 1943, and earlier, but Stalin had bided his time until the Soviets could enter the war against Japan with minimal loss. It was not until after the US dropped the first atomic bomb (6 August 1945) that Stalin revoked the Soviet–Japanese non-aggression pact and invaded Manchuria and North Korea. President Truman, in some alarm, proposed a joint occupation of Korea by the two powers. The Soviets occupied the territory north of the 38th parallel, quitting in 1948 after installing a Communist regime.

US-led force under General Douglas MacArthur, command-ing the occupation forces in Japan, was hastily assembled. A Commonwealth brigade, the 27th (later redesignated 28th), consisting of two British battalions from Hong Kong together with an Australian and a Canadian battalion and a New Zealand artillery regiment, was sent at once; it would eventually be incorporated into a Commonwealth Division that also included the 29th Brigade of three battalions and an armoured regiment, sent from Britain in December, and a third (Canadian) brigade. In October China threw its weight behind the North Koreans, its army supported by Russian pilots and jet aircraft in Chinese colours. The fight-ing, always fierce and almost always confused, ebbed and flowed along the whole length of the peninsula for three years.

The Korean terrain was not entirely alien to those with experience of the North-West Frontier or Italy – a bare and mountainous landscape of scattered farms and villages – but to the National Servicemen who made up the bulk of the army it might as well have been the moon. In the summer it was hot and humid; in the winter it was bitterly cold. British units were relieved every twelve to fifteen months (and sometimes after only six) during the three years of the fight-ing, so that by the end of the war a good many soldiers – in the infantry especially – were wearing the British Korea Medal and that of the United Nations, the UN's first cam-paign medal. But one action in particular, at the Imjin River between 22 and 25 April 1951, has passed into army legend to stand with those of earlier wars – including Arnhem and Rorke's Drift – as an example of defiant defence against the odds, demonstrating yet again how much rested on the fighting spirit of the infantry.

The Chinese had launched their spring counter-offensive against UN forces on the lower Imjin, close to the border, with the aim of breaking through to recapture the South

Korean capital, Seoul. The British 29th Infantry Brigade, consisting of the first battalions of the Royal Northumberland Fusiliers, the Gloucestershire Regiment ('Glosters') and the Royal Ulster Rifles, together with a Belgian battalion and its company of Luxembourgers, supported by tanks and artillery, held the westernmost part of the UN line, astride the main road to Seoul, with the 1st Republic of Korea (ROK) Division to their west and the Americans to their east (the Commonwealth Division had not yet been formed, so the brigade answered direct to the US corps HQ). The brigade's frontage was 12 miles long in mountainous country – far too long for the four battalions to cover while also giving each other support. Nor were the actual positions prepared as strongly as they might have been. Digging had not been extensive, and wiring and mining were almost perfunctory, for the UN forces had been on the offensive, pushing the Chinese and North Koreans back across the border, and they regarded the Imjin as a pause-line from which to continue the advance as soon as the situation allowed. A Chinese counter-attack had been talked about, but few were taking it seriously, especially since patrols north of the river found little evidence of a build-up of troops. Despite the lessons of history, scant heed was given to the old adage that 'sweat saves blood'.

In making the best of a difficult hand, the brigade commander, Brigadier Tom Brodie, had also gambled on a compromise to achieve both coverage of his entire front and mutual support between battalions. He sited the battalions on the best of the high ground, aligned north-east to south-west, with the Belgians on the right, north of the river, the Fusiliers a mile to their left but south of the river, and the Glosters a mile and a half further left of the Fusiliers, with the Ulster Rifles and C Squadron 8th Hussars (sixteen Centurion tanks) in reserve 2 miles to the rear. In turn, the four companies of each battalion were sited four-square but not as tight as the commanding officers would have preferred (as was customary at that time, defensive positions were 'sited two down' – the brigade commander siting the companies, the battalion's commanding officer the platoons, and so on). It would have been better, as both the Fusiliers' and Glosters' commanding officers argued, to site the brigade more tightly, denying penetration *between* battalions, and dealing with any outflanking movement by concentrated artillery fire and the reserve, although the brigade had no guns heavier than 25-pounders, and no means to call on heavier calibres from the neighbouring US division (though that problem could have been fixed relatively easily). So far in Korea the infantry had been copying the tactics of 8th Army in the mountains of Italy, but Brodie had commanded a brigade column in the second Chindit campaign and had therefore lived with the idea of being encircled; it is just possible that he saw no more reason to be dismayed at the prospect of Chinese penetration in mountainous terrain than he had been at Japanese encirclement in the jungle.

Late in the day on the twenty-second, the brigade learned that there were Chinese troops moving towards them in strength, and so the battalions ordered 50 per cent stand-to-arms throughout the night. Darkness fell, and the first encounter was a brush with a Chinese patrol that had slipped undetected round the Belgians towards the two bridges across the Imjin to their rear. An Ulster Rifles platoon deploying (late) to secure the crossings for the night was driven back with heavy losses. Chinese follow-up forces now attacked the Belgian positions and took the bridges, while others forded the river and attacked the neighbouring Fusiliers' right rear ('Z') company. Further downstream, more Chinese troops managed to get across and attacked the Fusiliers' left forward ('X') company. Too far from the other Fusilier companies for support, 'X' Company withdrew closer into the centre of the battalion position before first light, which in turn exposed the right forward ('Y') company. Pressure increased as the Chinese encirclement progressed, each company compelled to fight by itself rather than in a concerted battalion action, and all but 'W' Company in the left rear, with the guns of 45 Field Regiment Royal Artillery close by, began steadily giving ground.

It was St George's Day, the Fusiliers' regimental day. Red and white roses had been flown in from Hong Kong for each man to wear with the red and white hackle in his beret. Instead of the customary rose presentation parade, however, the battalion spent the whole day fighting – repelling attack after attack, and making costly counter-attacks to recover lost ground. If it would not go down in history as one of the regiment's finest days (for 'X' Company might have held on longer, keeping the battalion position more intact) there was raw courage enough – the Fusiliers, the old 5th Regiment of Foot, were not nicknamed 'The Fighting Fifth' for nothing ('Z' Company commander would receive one of the brigade's three DSOs for his leadership, which says enough). But faulty deployment did for them in the end, and by evening they had been pushed off their hills, though they managed to keep cohesion as they pulled back, and in turn helped cover the withdrawal of the Belgians (who had also fought a good action).

On the far left of the brigade, on Hill 314, were the Glosters, who had earned their distinctive 'back-badge' (a miniature of their Sphinx badge but worn on the back of the beret) by fighting back-to-back at Alexandria in 1801.* Their position was even further out on a limb, though the companies were sited rather more tightly than the Fusiliers' had been, and in their commanding officer and adjutant they had two men of extraordinary and complementary

* Described in ch. 11.

character – Lieutenant-Colonel James Power Carne and Captain Anthony Farrar-Hockley. Carne was 45, old for a commanding officer even in 1951 when the army had not yet settled back into its promotion routine. Pipe-smoking, taciturn to the point of seeming inarticulate, even with his DSO from Burma 'Fred' Carne had never in his service been described as a high-flier; but in such a predicament as the Glosters' now, he was probably the best sort of commanding officer they could have had. Farrar-Hockley, on the other hand, was as fiery as Carne was stolid. He had enlisted in the regiment under-age during the Second World War, been commissioned into the Parachute Regiment and had won an MC at 20. Pugnacious and uncompromising, he was the perfect adjutant for a man like Carne.

The battle began well for the battalion. Their standing patrol forward on the Imjin was able to throw back the first Chinese attempt to cross soon after dark, alerting the companies in good time for them to stand to properly – and then repelled a further three attempts before they ran out of ammunition and had to withdraw. Not long after this attacks began in earnest on the left- and centre-forward companies ('A' and 'D') and continued all night, with the right ('B') and rear ('C') companies coming under attack from Chinese troops outflanking the position to the east. By morning 'A' and 'D' Companies had suffered severe casualties and were giving ground – critically at 'Castle Site' which overlooked the left-forward company. Farrar-Hockley, who had gone forward to 'A' Company soon after first light, recounts the desperate business of trying to retake the hill:

Phil [Lieutenant Philip Curtis, commander 1st Platoon] is called to the field telephone: Pat's [Major Pat Angier, 'A' Company commander] voice sounds in his ear. 'Phil, at the present rate of casualties we can't hold on unless we get the Castle Site back. Their machine-guns up there completely dominate your platoon and most of Terry's. We shall never stop their advance until we hold that ground again.' Phil looks over the edge of the trench at the Castle Site, two hundred yards away, as Pat continues talking, giving him the instructions for the counter attack. They talk for a minute or so; there is not much more to be said when an instruction is given to assault with a handful of tired men across open ground. Everyone knows it is vital: everyone knows it is appallingly dangerous. The only details to be fixed are the arrangements for support-ing fire; and, though A Company's machine-gunners are dead, D Company [to their right] will support. Phil gathers his tiny assault party together. It is time; they rise from the ground and move forward to the barbed wire that once protected the rear of the forward platoon. Already two men are hit and Papworth, the Medical Corporal, is attending to them. They are through the wire safely – safely! – when the machine-gun in the bunker begins to fire. Phil is badly wounded: he drops to the ground. They drag him back through the wire somehow and seek what little cover there is as it creeps across their front. The machine-gun stops, content now it has driven them back – waiting for a better target when they move into the open again. 'It's all right, sir,' says someone to Phil. 'The Medical Corporal's been sent for. He'll be here any minute.' Phil raises himself from the ground, rests on a friendly shoulder, then climbs by a great effort on to one knee. 'We must take the Castle Site,' he says; and gets up to take it. The others beg him to wait until his wounds are tended. One man places a hand on his side. 'Just wait until Papworth has seen you, sir.'

But Phil has gone: gone to the wire, gone through the wire, gone towards the bunker. The others come out behind him, their eyes all on him. And suddenly it seems as if, for a few breathless moments, the whole of the remainder of that field of battle is still and silent, watching, amazed, the lone figure that runs so painfully forward to the bunker holding the approach to the Castle Site: one tiny figure, throwing grenades, firing a pistol, set to take Castle Hill. Perhaps he will make it – in spite of his wounds, in spite of the odds – perhaps this act of supreme gallantry may, by its sheer audacity, succeed. But the machine-gun in the bunker fires into him: he staggers, falls, and is dead instantly; the grenade he threw a second before his death explodes after it in the mouth of the bunker. The machine-gun does not fire on three of Phil's platoon who run forward to pick him up; it does not fire again through the battle: it is destroyed; the muzzle blown away, the crew dead.*

Philip Curtis was 24. He had been seconded to the Glosters from the Duke of Cornwall's Light Infantry, and his action at Castle Site was recognized by the posthumous award of the VC.

But it was not going to be possible to retake the dominating ground. Major Angier spoke by radio to Colonel Carne. 'I'm afraid we've lost Castle Site. I want to know whether I am to stay here indefinitely or not. If I am to stay I must be reinforced as my numbers are getting very low.'

Carne told him in his quiet, measured way that the company must stay put.

Angier acknowledged, and then added 'Don't worry about us. We'll be all right.' He was killed fifteen minutes later.

There was now only one officer alive and unwounded on 'A' Company's position.

Fighting continued all day until towards last light Carne, fearing they would be overwhelmed in the dark, decided to pull back the forward companies to form a tighter defensive position on Hill 235.

The remnants of 'A' and 'D' did manage to withdraw but 'B' Company, on the right, could not disengage, fighting off seven separate attacks during the night. Daylight brought

* From *Eye-Witness History* edited by Jon E. Lewis (1998), and in conversation with author, 2005.

respite at last and they were able to get back, but only seventeen of the company remained in action.

A relief force – a Philippine combat team with US support – was hastily despatched by the divisional commander to relieve the Glosters, but their progress was hampered by Chinese who had got into the rear of the brigade position. Throughout the twenty-fourth, therefore, the Glosters, now almost literally back-to-back, Alexandria-like, on Hill 235 – 'Gloster Hill' as it remains known even today – fought off wave after wave of Chinese infantry with the support of a single battery of 25-pounders still in range.

In the afternoon Carne received a radio message from brigade headquarters. Acknowledging it without emotion, he turned to Farrar-Hockley and said, 'You know that armour-infantry column that's coming to relieve us?'

'Yes, sir?'

'Well. It isn't coming.'

'Right, sir.'*

The attacks continued unabated throughout the night – and always to the unnerving call of bugles and fanatical screams. Farrar-Hockley, who as adjutant was the titular commander of the Corps of Drums, ordered the Glosters' drum-major to answer each Chinese call with a defiant one of his own. And so everything from 'Cookhouse' to 'Officers Dress for Dinner' sounded over the darkened hillsides, adding to the growing legend that was the 'Glorious Glosters' at Imjin. By morning, however, the game was up. At 9.30 Carne received a message from the brigade commander that he would soon lose his already limited artillery support since the remaining battery within range was having to withdraw, and left it to his discretion whether – in truth, when – to withdraw.

* From *The Edge of the Sword*, Farrar-Hockley's account of the battle and his subsequent experience as a PoW (including his several escape attempts).

Once an infantry battalion is without artillery support it can only fight the direct battle – what it sees to its front. Even with its own mortars it cannot much influence what the enemy is doing out of range or sight of its riflemen and machine-gunners, and the enemy then has freedom to manœuvre and concentrate against the weakest point. And then, when small-arms ammunition, grenades and mortar bombs are expended, the bayonet buys only a few seconds more. The Glosters were down to three rounds a man, and the Bren guns to one and a half magazines (40–50 rounds). Carne, who had fired every round in his own pistol, ordered the companies to break out as best they could, and to go it alone.

But the Chinese were swarming so deep in the brigade position that evasion proved all but impossible. Only the remnants of 'D' Company – just forty men – made it.

Farrar-Hockley, who was taken prisoner after the battle, would later rise to four-star general. For his conduct on this occasion he was awarded the DSO – an exceptional decoration for a captain at this time, for since the First World War the DSO had come to be bestowed principally in recognition of leadership at battalion-command level.* Colonel Carne, likewise captured, would endure eighteen months' solitary confinement and drug-assisted brainwashing; he won the battalion's second VC. He had 'moved among the whole battalion under very heavy mortar and machine-gun fire, inspiring the utmost confidence and the will to resist among his troops,' ran the citation: 'On two separate occasions, armed with rifle and grenades, he personally led assault parties which drove back the enemy and saved important situations. His courage, coolness and leadership was felt not only in his own battalion but throughout the whole brigade.'

* His son would win the MC commanding a company of 2 Para at Goose Green.

There had been courage across the whole brigade, indeed – not least in the Hussar squadron's sterling efforts to cover the withdrawal of the battalions. At one stage, with Chinese infantrymen crawling all over the Centurions, trying to prise open the hatches, the tanks were reduced to hosing each other with their machine guns. The squadron leader, Major Henry Huth, received the DSO for what he described as 'one long bloody ambush', and his second-in-command the MC – an uncommon allocation of medals to the armoured corps. It would be the last time that British tanks fired in anger, save for a few rounds in the Suez intervention of 1956, until the Gulf War of 1991.

In all, 29th Brigade suffered 1,100 casualties in the battle of the Imjin River, including 34 officers and 808 other ranks missing – a quarter of the brigade's fighting strength on the eve of battle. Of these, 620 were from the Glosters, of whom 522 became prisoners of war,* 180 of them wounded. Fifty-nine Glosters had been killed in action, and a further thirty-four would die in captivity. But the Chinese had bought their victory dearly, for their casualties were estimated at around 10,000. As a result, the Chinese 63rd Army, which had begun the offensive with three divisions and approximately 27,000 men, was pulled out of the front line.

Seoul did not fall to the spring offensive. But when the armistice was signed in 1953 the conflict was frozen rather than resolved: it was, said the UN commander, General Maxwell Taylor, who had commanded 101st Airborne on Market Garden, 'a suspension of hostilities – an interruption of the shooting'. The 38th parallel became a 'demilitarized zone', and to this day 25,000 American troops remain in

* Including the legendary Padre Sam Davies (at whose theological college the author once studied), who had stayed behind with the wounded, and who was to be particularly savagely treated in captivity. His utterly un-self-serving account of that time, *In Spite of Dungeons* (1954), powerfully illustrates the role of the regimental chaplain.

South Korea – more than a quarter of the British army's entire strength.

The legend of 'the Glorious Glosters' is still a powerful one. It is probably the more powerful because although the regiment was known throughout the army from their days as the 28th Foot and for their unique back-badge, they were in no way special. Gloucestershire did not have the reputation, say, of Liverpool or Tyneside as a brawling sort of recruiting area; the regiment did not attract particularly 'smart' officers (indeed, it had more than its share of officers on attachment, a sign that it was not attracting enough regulars); and in its ranks was the same mix of National Servicemen, reservists recalled to the colours and regular soldiers, not all of whom would have been entirely willing volunteers. Its leadership was solid, however, from top to bottom – and in some cases inspired. As Major Pat Angier was buried with hasty rites within the company position his batman wept; and his citation, like that for Carne's VC, stated that 'His courage, coolness and leadership was felt not only in his own battalion but throughout the whole brigade.' What made them glorious was not so much what the Glosters were but the way they had fought – though it was the way they were that had made them fight as they had – and the action at Imjin was felt throughout the army. Once more the conduct of a workaday infantry battalion had demonstrated fighting spirit to every other regiment in a way that could only encourage emulation.*

Imjin is a battle to study not for its tactical lessons – other than, in the duke of Wellington's old phrase, in how not to

---

* And the legend of Imjin lives on. When the headquarters of the (NATO) Rapid Reaction Corps moves from Germany to the UK in 2010, its new home, formerly RAF Innsworth in Gloucestershire, will be renamed 'Imjin Lines'.

do it – but for a shining example of how a battalion of 700 men can conduct themselves when they have been let down. Lord Moran's words again ring true – if not perhaps the whole truth, then sufficient unto the day: 'The individual shrinks to nothing . . . Only the regiment matters.'

28

Recessional

East of Suez, 1948–68

God of our fathers, known of old—
Lord of our far-flung battle line—
Beneath whose awful hand we hold
Dominion over palm and pine—
Lord God of Hosts, be with us yet,
Lest we forget—lest we forget!

The tumult and the shouting dies—
The Captains and the Kings depart . . .

Rudyard Kipling, 'Recessional'

Shortly after breakfast on a fine tropical June morning in 1948, Arthur Harris, the manager of one of the Anglo-Malayan Rubber Corporation's plantations in the northern Malay state of Perak, set off on his rounds. As he left his bungalow three ethnic Chinese gunmen shot him dead. It was the beginning of the 'War of the Running Dogs' – as the insurgents called the collaborators with British rule.

In the following weeks there were more attacks. Unarmed British officials, planters and managers were shot, as well as Malays, Chinese and Indians who worked with them in the tin mines and rubber plantations – the 'running dogs'. The governor-general reluctantly declared a state of emergency. The counter-insurgency campaign that followed would last until 1960 when the final defeat of the Communist terrorists ('CTs') would pave the way for Malayan independence – the second great step on the long withdrawal from empire which Kipling's poem 'Recessional' had portended sixty years earlier.

In the book of the army's finest hours the 'Malayan Emergency' stands as one of the finest. The strategy for this sustained campaign with huge manpower and logistical demands – much greater than those of the Korean War – was above all an imaginative and well-executed strategy that brought victory, but it was also one applied with superb tactical skill. And this was a campaign fought against an enemy whose political creed and experience of guerrilla warfare made them perhaps the most formidable of any in the 'small wars' which – despite the interruption of the bigger conflicts – have been the staple of the army's operational experience for the past century and a half.

Malaya in 1948 was an explosive mix of ethnic groups. The largest segment – 50 per cent – of the population of four and a quarter million were the indigenous Muslim Malays, who accepted overall British rule but were strongly loyal to their sultans, the sovereign heads of the nine states of the Malay Federation (Johore, Pahang, Negri Sembilan, Selangor, Perak, Kedah, Perlis, Trengganu and Kelantan). There were around two million Chinese, whose numbers had grown rapidly in the preceding ten years, in part as a result of the Japanese occupation when they had been brought in as slave labour. Many were second-generation Malaya-born but their loyalty, culturally at least, was to

China, whether Communist or Nationalist. Crucially, half a million of the Chinese in Malaya were squatters who had fled the towns during the Japanese occupation, and who had no title to the land on which they eked out a living. There were half a million Indians, mainly Tamils working in the plantations, and the same number of other non-indigenous residents and transient workers, of whom the British (some 12,000) were significant both as administrators and in running the country's businesses. And of key importance, though the least significant when the Emergency began, were the aboriginal tribes living deep in the jungle, stubbornly refusing to recognize the authority of the sultans. No one knew quite how many they were, for no one had any dealings with them; estimates ranged between 50,000 and 100,000.

British rule in Malaya stretched back to 1874. When the Japanese over-ran the peninsula in 1942 the Malayan Communist Party (MCP) organized the Malayan Peoples' Anti-Japanese Army (MPAJA), a largely Chinese force despite its name, which with clandestine British support and implied promises of equality for the Chinese after the war had helped speed liberation, although the country was not in the end retaken militarily.* This, on top of the débâcle of defeat in 1941, made re-establishing British rule tricky, and after the majority Malay opposition forced the Attlee government to drop plans for equal rights the MCP initiated a campaign of terrorism to oust the British and install a Communist regime. With exquisite oriental irony, it was led by the 24-year-old Chin Peng, who had been an MPAJA liaison officer with the British army and had been awarded an OBE and two mentions in despatches for his service. The

---

* Perhaps the best known of the British officers who remained in Malaya after the fall of Singapore was Spencer Chapman, who wrote a classic account of the guerrilla war, and thereby of jungle tactics, *The Jungle is Neutral* (1949).

MPAJA now transformed itself into an insurgent army, simply substituting 'British' for 'Japanese', though prudently changing its name soon after to the Malayan Races' Liberation Army.

By the end of 1948 eight MRLA 'regiments' were operating throughout the peninsula in large cells on traditional Maoist guerrilla warfare lines (i.e. in rural areas, as opposed to the Leninist approach of focusing on towns and cities), financed in good measure through extortion from the local squatter population. Chin Peng's plan was to raid isolated estates, tin mines, and police and government buildings to drive the security forces into the urban areas. He would then set up guerrilla bases in the 'liberated areas' to train new recruits from the Min Chung Yuen Thong (usually simply 'Min Yuen'), the urban-based 'mass revolutionary movement'. In the third phase the expanded army would move from the rural areas to attack the towns, villages and railways, with the Min Yuen acting as auxiliary saboteurs to cripple the economy. From this would come the climactic phase: with the country on its knees, the British army would be defeated in conventional battle. The concept was succeeding in China itself, and would succeed in French Indo-China (and to a large extent in US-supported Vietnam); its first real test against the West, however, in Malaya, would prove an instructive failure.

Attacks on British administrators and 'running dogs' were stepped up, but at first the army was not keen to get embroiled. There were only nine infantry battalions in Malaya Command – six of them Gurkha – plus three Malay and a field artillery regiment. The GOC, Major-General Charles Boucher, himself a Gurkha officer, asked the commander-in-chief of Far East Land Forces (in Singapore), Lieutenant-General Sir Neil Ritchie, for reinforcements, but Ritchie had different strategic priorities, believing that a conventional threat to the region from a Russo-Chinese pact claimed all his resources. It took the new commissioner-general for South-east Asia, Ramsay MacDonald's son Malcolm, to persuade him to help, and by then the insurgency had gained a dangerous momentum which the police, even with greater powers and now armed, could not contain. By the end of 1948 Boucher had six more battalions, including three in 2nd Guards Brigade – the first time the Guards had served so far from London in what was officially peacetime.

Boucher's plan was to break up the MRLA concentrations and drive them deeper into the jungle, cutting them off from active Min Yuen support and isolating them from the squatters from whom they might extort money, food and intelligence. But the overall campaign – military, police and civil affairs – lacked any real unifying drive, which was hardly surprising since the high commissioner, Sir Henry Gurney, lacked any useful experience (a deficiency already manifest in his tardy declaration of a state of emergency). Matters improved somewhat when Sir John Harding relieved Ritchie as commander-in-chief the following year, and when two more brigades including 3rd Commando Brigade Royal Marines were sent. A true winning move, however, came with the creation of the post of director of operations, and the choice of the man for the job. The new CIGS, Slim, called out of retirement a former 14th Army officer (who had also distinguished himself in East Africa and the Western Desert), Lieutenant-General Sir Harold Briggs, of whom he was to write: 'I know of few commanders who made as many immediate and critical decisions on every step of the ladder of promotion, and I know of none who made so few mistakes.'

Briggs's assessment of the situation was as immediate, critical and unmistaken as his earlier decisions had been. In addition to Boucher's aim of breaking up the CT concentrations and driving them deep into the jungle, he made a priority of protecting the squatter population. But since the squatters were scattered about the country on the jungle edge, he devised the 'New Villages' scheme – forced resettlement: some half a million Chinese would be rehoused in fortified villages, guarded round the clock by armed police. The difference between this forcible removal from farming land and the equally draconian scheme of the Boer War was significant, however (although protection from the 'kaffir' threat had also been a factor in South Africa): the squatters were eventually to be given legal title to the land on which they were resettled, and citizenship. Briggs had calculated that all but the most dedicated collectivist could therefore see that he had an interest in defeating the insurgency.

The other element of the 'Briggs Plan' was offensive action. Once the defensive measures were in place, supervised by the local joint (police, civil and military) executive committees, the army would take the fight to the CTs in the jungle, clearing the peninsula from south to north, with police and civil consolidation (education, health services, water and electricity) following up the military success. It was the reverse, indeed, of the MCP's Maoist insurgency strategy. But the plan depended on absolute civil–police–military cooperation, and this in turn depended on the right man in overall command. Before handing over as commander-in-chief in May 1951 (to succeed Slim as CIGS), Harding recommended that the high commissioner be replaced by a military officer with full powers to direct the campaign. When in October Sir Henry Gurney was killed in an ambush, Churchill, who was back at No. 10 for one last term, appointed General Sir Gerald Templer in his place. At last the single-minded direction of the counter-insurgency could begin.

Immediately troop strength was raised to 45,000 (before the fall of the Berlin Wall the whole of the BAOR would stand at only 11,000 more), including twenty-four battalions of infantry, many of them Gurkha or Commonwealth, thus beginning a continuing association with Fiji as a source of manpower. And at Harding's request the Special Air Service was resuscitated. Operating initially as 'the Malayan Scouts' before the present-day organization of a single regular regiment (22 SAS) and two TA regiments (21 and 23 SAS), and working with SAS squadrons of the Australian, New Zealand and Rhodesian armies, their priority was to locate the CT bases deep in the jungle. In this the aboriginal tribes were a key source of intelligence – sometimes actively as trackers* – which a vigorous 'hearts and minds' approach by the SAS (medical attention especially) helped secure. 'Hearts and minds' in the context of counter-insurgency was Templer's term, but it was originally coined by the second president of the United States, John Adams, writing in 1818 of the struggle for independence: 'The Revolution was effected before the War commenced. The Revolution was in the minds and hearts of the people.'†

Perhaps an extreme example of the British army's ability to learn from its previous mistakes (no matter how tardily), 'hearts and minds' would be an approach that would define all its subsequent 'low-intensity operations'.

The success of both the defensive and offensive schemes of the Briggs Plan was ultimately attributable also to unified command, patience and innovative tactics. Helicopters were

---

* The best known of the local trackers, who saw service as a formed unit, were the Ibans brought from North Borneo.

†The Briggs Plan, and Templer's application of it (hand in glove with his permanent secretary, Sir Robert Thompson) was studied by the US army in Iraq under General David Petraeus. There is, indeed, further irony in this: Briggs was, until the age of 20, an American citizen, and Thompson, after leaving Malaya, was a special adviser to President Kennedy in the Vietnam War. The US 'strategic hamlets' plan for Vietnam had some similarities with the British Malaya experience, and Briggs was confident that the counter-insurgency strategy would work. However, in its execution US forces diverged from the principles, and the rest is history.

used extensively for the first time – an enormous 'force multiplier' – and the Malayan police and the country's own army, which had fought well alongside the British army during the Japanese invasion, dramatically increased in numbers and effectiveness through the secondment of British officers and NCOs. The Gurkha battalions, though the Gurkhas themselves were hillsmen not jungle-dwellers, somehow took naturally to the environment, as they had in Burma, and provided a solid core of continuity and expertise in jungle fighting, remaining in theatre longer than the British battalions. But the British infantry battalions adapted remarkably well, too. For although in Burma they had shown an aptitude for conventional war in the jungle, the altogether stealthier techniques of sudden, leafy close-quarter action were hardly native to them. The Suffolk Regiment – from a county of open fields rather than forest – were in action in the primary jungle of the southern Malay states continuously from mid-1949 to January 1953 and claimed the highest CT head-count (literally, for in the early stages only the heads were brought out of the jungle for identification). But other regiments, some of them full of city-dwellers, were not far behind them.

It was certainly warfare needing determination and cold courage. Lieutenant Greville Charrington's platoon of Suffolks had not quite finished laying their speculative ambush on a logging trail at the forest's edge in August 1950 when a small CT party appeared. They killed two but the third, though wounded, managed to get away. Although night was fast approaching, Charrington pressed his men to the follow-up. Writing of events in the third person and referring to himself as 'the officer', he found:

a heavily bloodstained track leading off into the lalang [high grass], which he followed with his torch. He came across a rifle and a little later, where the bandit had entered the baluka [tree covered bracken] on his belly, he found a one star bandit cap covered in blood.

The officer immediately threw his two grenades and ordered the other men to do the same so that a pattern of grenade bursts was formed like a naval vessel searching for a submarine with a pattern of depth charges. They then spread out and tried to sweep through. The blackness of the night and the denseness of the undergrowth soon made it apparent that they were not getting anywhere. The officer spread his men out and round the edge of the bracken and crawled down the bandit's track, following it by the light of his torch on the blood stains. It was eerie in this tunnel in the bracken with the possibility that the bandit might be waiting for him round every bend. Pig tracks, mere tunnels in the almost solid undergrowth, ran parallel to and across his bandit track, but he was able to keep to the wounded Chinese by the blood which coated the stems of the bracken on three sides of the narrow pig run. As he went he passed two places where his grenades had exploded. After a bit the blood trail started to diminish and it became more and more difficult to follow the track which was only marked every two yards with just a spot of blood. Ten yards after this the signs gave out ...

Charrington and his men continued the search but could find nothing, and instead moved on to the logging camp to interrogate the workers. In the morning the rest of his company came up with their Iban trackers to follow the trail back to the main CT camp, while on the road half a mile from the logging camp the wounded 'bandit' gave himself up to a passing police vehicle – and three weeks later was showing the Suffolks the tracks and camps used by the CTs. It had been classic counter-insurgency work: routine military action based on sketchy information which in turn yielded hard intelligence – and more dead insurgents.

The Malayan Emergency would become the model for the

army's subsequent counter-insurgency campaigns in Kenya, Cyprus and South Arabia. When, for example, in 1955 Georgios Grivas's EOKA (Ethniki Organosis Kuprion Agoniston – National Organization of Cypriot Fighters) took up arms against the colonial government in Cyprus, Churchill's successor as prime minister, Anthony Eden, lost no time in appointing a soldier – Harding, at the end of his time as CIGS – as governor, with full powers over both the military and civil effort. In turn, Harding lost no time in formulating his own 'Briggs Plan'.

And so the army began, as never quite before in its history, to cleave into two parts. In Europe, principally Germany, was that part which had grown out of Montgomery's 21st Army Group – the heavy metal, the 'petrol feet' as they were known by the other half, fighting the Cold War (though no one could know that it would stay 'cold') with tanks, armoured personnel carriers and self-propelled artillery. And beyond Europe were the 'brown knees', as the petrol feet called them – the men who wore shorts in barracks, fighting in the long retreat from empire, for the most part on foot in desert, mountain or jungle, more the heirs of Slim's 14th Army. From time to time the regiments changed places, and the transition would be painful for many months until they got the measure of their new environment, but the build-up of operational nous in the collective mind of the army would be enormous. Significantly, the SAS established both their ethos and their strategic purpose in these 'recessional' campaigns. In Kenya, in the 'Mau Mau rebellion'* – a murderous affair of dissident, mainly Kikuyu, tribesmen encouraged by Jomo Kenyatta's Kenya African Union – they would operate

---

* There are so many explanations of the term 'Mau Mau' that it is pointless listing them, especially since the rebels themselves – largely of the Kikuyu clan – called themselves by several names but not this.

against the movement's hard core, using 'turned' terrorists as guides. The killing of one of the principal Mau Mau leaders, Dedan Kimathi, late in 1956 proved pivotal, and four years later the Kenyan Emergency was officially at an end – an example of strategic effect that would have been at once recognizable to officers on the North-West Frontier a century before. The SAS are strategic troops.

In 1964, when trouble erupted in Aden and its hinterland, which had recently been united in an uncertain federation of sheikh rulers, British troops were sent to reinforce both the old colony of Aden itself and the Yemeni border, and to range in the mountainous Radfan area against Egyptian-backed insurgents trying to incorporate the new polity into the Yemen. While up-country the pattern of operations was familiar enough, in the teeming streets of Aden port fighting the terrorists, as opposed simply to dealing with civil disorder, proved a largely new – and very vicious – business. It was the first time, too, that television was able to paint any immediate picture of what was happening, and public opinion in Britain would become a significant factor in both political and operational decision-making. One regiment in particular, the Argyll and Sutherland Highlanders, gripped the attention of *Guardian* and *Telegraph* readers alike, if rather differently. After an Argylls patrol had been ambushed and three soldiers killed, the high commissioner and the senior military commanders ordered the troops to draw back from the troublesome district of Crater – to the utter dismay of the Argylls' commanding officer Lieutenant-Colonel Colin Mitchell (whom the press had already dubbed 'Mad Mitch'). A fortnight later the Argylls defiantly re-occupied Crater with scarcely a shot fired but much skirling of pipes. In so doing, Mitchell had disobeyed orders, Nelson-like. But unlike Nelson, who was promoted for the consequent victory, Mitchell was effectively sacked. The British public rather approved of the Argylls' devotion to regimental honour, though: Mitchell was subsequently elected to Parliament, and when defence cuts were announced in 1968 – and the Argylls were slated for disbandment – a nationwide campaign to 'Save the Argylls' was mounted, and the regiment was indeed saved.

The Argylls had not come green to the Aden insurgency. There had been another formative campaign earlier in which they had played a prominent if unseen part. Indeed, the entire campaign was and remains little known, and at the time was so hedged about with political sensitivity that a euphemism was coined for the conflict: 'Confrontation'. In part a legacy of the Malayan Emergency, it began with a rebellion in 1962 against the sultan of the tiny oil-rich state of Brunei on the northern coast of Borneo, the largest island of the Indonesian archipelago. The rebellion was supported from across the border by Indonesia, whose pro-Communist President Sukarno was opposed to the incorporation of Sarawak and Sabah (former British North Borneo) into the new Federation of Malaysia which was due to form in September 1963. Perversely, Brunei's – or rather, the sultan's – refusal to join the federation was the cause of the rebellion, for Sukarno wanted to incorporate all three of the Borneo states into Indonesia and believed it would be easier if they were already federated. The rebellion was quickly put down with the help of two battalions (one of them Gurkha) and Royal Marines from Hong Kong. But Sukarno then significantly stepped up the pressure on the other two states by sending regular troops across the border (hitherto they had been disguised as Indonesian volunteers supporting local insurgents).

The border between Indonesian Borneo – Kalimantan – and the states along the north coast is 1,000 miles long, densely forested even now, and in parts mountainous, rising to 8,000 feet. The Indonesian strategy was straightforward:

troops would cross the border in company strength, sometimes double company, set up a jungle base and from there intimidate the aboriginal (mostly Dyak) tribesmen, with the aim of bringing about a gradual de facto extension of Indonesian territory. The man in command of British troops in northern Borneo, working to the newly created Malaysian National Operations Committee, was Major-General Walter Walker, an immensely forthright Gurkha officer who had been born on a tea plantation in India, had fought in Burma and had seen a good deal of the Malayan Emergency.* At first, for political reasons, he had had to keep his troops back from the border, relying on the SAS to monitor what was happening, largely through contact with the Dyaks; but when it became clear what the Indonesians' strategy was he set up jungle bases himself to counter the penetration, and resupplied them by helicopter.

By the end of 1964, with penetration increasing and growing Communist subversion of the Chinese population, Walker's command was increased to 14,000 men – three Malay battalions, eight Gurkha, ten British and two Royal Marine commandos, supported by five batteries of light (helicopter-portable) artillery, two squadrons of armoured cars, sixty RAF and Royal Navy (mainly troop-lift) helicopters, and forty observation and liaison helicopters from the Army Air Corps, formed seven years earlier. The numbers would rise to 17,000 the following year with Australian and New Zealand troops sent from the Commonwealth Brigade in Singapore. Crossing the Indonesian border had initially been permitted only in hot pursuit, though an exception was made for the SAS whose incursions, based in large measure on 'sigint' (signals intelligence, i.e. radio intercept), were highly secret, and for a few

---

* Ever undaunted, his final command, as a four-star general, would be of NATO troops in the Arctic ('a good horse runs on any going').

'trusted' regiments. For the rest of the 'green army', as the SAS called them, the campaign was largely one of patrolling and ambushing – and occasionally being ambushed, when the intensity of the fighting could be as great for the patrol involved as any of the larger-scale actions in Korea had been. The VC citation for Lance-Corporal Rambahadur Limbu of the 10th Princess Mary's Gurkha Rifles is apt testimony:

On 21st November 1965 in the Bau District of Sarawak Lance Corporal Rambahadur Limbu was with his Company when they discovered and attacked a strong enemy force located in the Border area. The enemy were strongly entrenched in Platoon strength, on top of a sheer sided hill the only approach to which was along a knife edge ridge allowing only three men to move abreast. Leading his support group in the van of the attack he could see the nearest trench and in it a sentry manning a machine gun.

Determined to gain first blood he inched himself forward until, still ten yards from his enemy, he was seen and the sentry opened fire, immediately wounding a man to his right. Rushing forward he reached the enemy trench in seconds and killed the sentry, thereby gaining for the attacking force a first but firm foothold on the objective. The enemy were now fully alerted and, from their positions in depth, brought down heavy automatic fire on the attacking force, concentrating this onto the area of the trench held alone by Lance Corporal Rambahadur. Appreciating that he could not carry out his task of supporting his platoon from this position he courageously left the comparative safety of his trench and, with a complete disregard for the hail of fire being directed at him, he got together and led his fire group to a better fire position some yards ahead.

He now attempted to indicate his intentions to his Platoon Commander by shouting and hand signals but failing to do so in the deafening noise of exploding grenades and continuous automatic fire he again moved out into the open and reported personally, despite the extreme dangers of being hit by the fire not only from the enemy but by his own comrades. It was at the moment of reporting that he saw both men of his own group seriously wounded.

Knowing that their only hope of survival was immediate first aid and that evacuation from their very exposed position so close to the enemy was vital he immediately commenced the first of his three supremely gallant attempts to rescue his comrades. Using what little ground cover he could find he crawled forward, in full view of at least two enemy machine gun posts who concentrated their fire on him and which, at this stage of the battle, could not be effectively subdued by the rest of his platoon.

For three full minutes he continued to move forward but when almost able to touch the nearest casualty he was driven back by the accurate and intense weight of fire covering his line of approach. After a pause he again started to crawl forward but he soon realised that only speed would give him the cover which the ground could not. Rushing forward he hurled himself on the ground beside one of the wounded and calling for support from two light machine guns which had now come up to his right in support he picked up the man and carried him to safety out of the line of fire.

Without hesitation he immediately returned to the top of the hill determined to complete his self imposed task of saving those for whom he felt personally responsible. It was now clear from the increased weight of fire being concentrated on the approaches to and in the immediate vicinity of the remaining casualty the enemy were doing all they could to prevent any further attempts at rescue. However, despite this Lance Corporal Rambahadur again moved out into the open for his final effort.

In a series of short forward rushes and once being pinned down for some minutes by the intense and accurate automatic fire which could be seen striking the ground all round him he eventually reached the wounded man. Picking him up and unable now to seek cover he carried him back as fast as he could through the hail of enemy bullets.

It had taken twenty minutes to complete this gallant action and the events leading up to it. For all but a few seconds this young Non-Commissioned Officer had been moving alone in full view of the enemy and under the continuous aimed fire of their automatic weapons. That he was able to achieve what he did against such overwhelming odds without being hit is miraculous. His outstanding personal bravery, selfless conduct, complete contempt of the enemy and determination to save the lives of the men of his fire group set an incomparable example and inspired all who saw him.

Finally rejoining his section on the left flank of the attack Lance-Corporal Rambahadur was able to recover the light machine gun abandoned by the wounded and with it won his revenge, initially giving support during the later stages of the prolonged assault and finally being responsible for killing four more enemy as they attempted to escape across the border. This hour long battle which had throughout been fought at point blank range and with the utmost ferocity by both sides was finally won.

At least twenty-four enemy are known to have died at a cost to the attacking force of three killed and two wounded. In scale and in achievement this engagement stands out as one of the first importance and there is no doubt that, but for the inspired conduct and example set by Lance-Corporal Rambahadur at the most vital stage of the battle, much less would have been achieved and greater casualties caused. He displayed heroism, self sacrifice and a devotion to duty and to his men of the very highest order. His actions on this day reached a zenith of determined, premeditated valour which must count amongst the most notable on record and is deserving of the greatest admiration and the highest praise.

By the time Walker's tenure of command was over in 1965 the Indonesian army's spirit had been severely weakened. In October a power struggle began, eventually bringing the anti-Communist General Suharto to power, and cross-border 'Confrontation' fizzled out. In all, the campaign had lasted four years, and, given the size of the operational area, Indonesian resources and the fragility of the new Malaysian Federation, victory had been gained with remarkable economy of effort. It proved to be a formative experience not only for Walker but for a number of other officers who would go on to high rank, significantly in Northern Ireland but also in the Falklands and the First Gulf War.* Indeed, some of the appreciation and planning for the recapture of the Falklands – on raiding and other techniques – was done by a general staff colonel who as a 23-year-old company commander in Borneo had won an MC with the 2nd Goorkhas (as the regiment always spelled it): Lieutenant (later Lieutenant-General Sir) Peter Duffell. The 2nd were one of the regiments trusted to operate on the Indonesian side of the border, and Duffell's company, acting on intelligence gained by one of the long-serving Intelligence Corps officers who liaised with the local tribes, had destroyed a fifty-strong company of marines in a stealthy ten-day operation.†

Duffell was at that time a National Service officer, his service somewhat delayed. Going on to become a regular officer, after Staff College he was posted as a brigade major in Northern Ireland in the early days of the Troubles, and later commanded the Gurkha Brigade. As Major-General of the Gurkhas in the early 1990s he would take the decision

---

* The CGS at the time of the Falklands, Field Marshal Lord ('Dwin) Bramall, commanded a battalion of the Green Jackets in Borneo.
† Another of the 2nd's officers was the future CGS and field marshal Sir John Chapple.

finally to abandon the old Indian Army organization and method and align the Gurkhas instead with British practice, so that they could properly integrate with British brigades on operations.*

The Gurkhas in fact bore much of the burden of Confrontation, with a high proportion of the 114 killed and 181 wounded (an exceptionally high ratio of killed to wounded, which from the Second World War onwards had been roughly ten wounded to every one man killed). But the British soldier's apt remark – 'Them's Gurkhas, Miss. Them's us!' – has applied increasingly since their incorporation in the British army's order of battle rather than the Indian Army's. And Rambahadur Limbu's VC has been a further benchmark for its courage.

Throughout the period of post-war emergencies, rebellions and confrontations the army continued to be reduced. In 1956 the political defeat of the Suez intervention† brought the ex-Guardsman Harold Macmillan to No. 10 in place of the ex-Rifleman Anthony Eden (who had won the MC in the First World War).‡ Macmillan had been both minister of defence (at that time still a coordinating post, the three separate service ministers sitting in cabinet), and chancellor of the exchequer, and he was determined – in the way of all political flesh – to reduce defence expenditure. Not only that, he was also determined to abolish National Service, for the

* See ch. 32 and epilogue.
† Following President Nasser's nationalization of the British- and French-owned Suez Canal in 1956, an Anglo-French force seized the canal zone in an airborne and maritime operation in which the last British operational parachute drop took place. US pressure compelled a humiliating withdrawal soon after.
‡ Macmillan served with the Grenadiers in the First World War and was wounded three times. At the Somme he had quietly lain in a trench with a bullet in his pelvis, waiting to be evacuated, reading Aeschylus in the original Greek.

recall of 23,000 largely ex-NS reservists for Suez had not been popular. Call-up of conscripts was therefore finally ended in 1960. To achieve savings and meet commitments with a smaller army, large overseas bases were to be closed and reliance placed instead on a home-based strategic reserve and long-range air transport. Tactical nuclear weapons – small-yield bombs and artillery rounds – designed for use on the battlefield in support of conventional operations would also make it possible to reduce numbers in BAOR; and more locally raised troops would be the mainstay of colonial defence. Barring the obvious technological differences, it was an approach not dissimilar to that envisaged in the military retrenchments of the eighteenth and nineteenth centuries.

Arguments now began to rage in the War Office over the size of the new all-professional army. Templer, the CIGS, believed the minimum needed was 200,000; the war minister, Duncan Sandys (who had been wounded in Norway in 1940 serving with the TA), believed that only 165,000 could be recruited. In the end, because overseas garrisons could not be reduced as much as hoped, for the operational demands were just too great and the impracticalities of relying on strategic airlift too many,* by 1960 the figure had crept up to a heady 196,000.

There had been amalgamations and disbandments in the meantime. The second battalions of infantry regiments had gone in the immediate post-war cuts (except for those of the Guards), but the number of battalions had stood at eighty-five even in 1951. Under the new arrangements they would be reduced from seventy-seven to sixty. The Royal Armoured Corps would lose almost a quarter of its strength too, cut

---

* There were never enough serviceable aircraft, they did not have the range to reach the Indian Ocean without refuelling on the ground, and the airspace through which they flew belonged to other people.

from thirty regiments to twenty-three; and the Royal Artillery would shrink to twenty regiments, though their motto *Ubique* ('Everywhere') for the most part still held good. The regimental depots, the great feature of the Cardwell reforms, disappeared, the regiments instead brigading their training on a regional or 'functional' basis, so that, for example, the regiments of the north-west formed the Lancastrian Brigade depot at Preston – except for the Lancashire Fusiliers, whose recruits trained with the other English fusilier regiments at their new depot in Sutton Coldfield in Warwickshire. There were always just such historic flies in the ointment of infantry reform, as there continue to be. The mistake has frequently been to try to remove them.

This brigading to save money on the manning and infrastructure of recruit training and administration then began to develop another momentum. The regiments of each brigade wore a common cap badge – more or less – and some of the groupings began operating in effect as a single multi-battalion regiment. The battalions of the Green Jackets Brigade, which trained at the Rifle Depot in Winchester, now began referring to themselves not just as the Oxfordshire and Buckinghamshire Light Infantry (who, although light infantry and therefore historically red-jacketed, wore 'virtual green' because of their early association with Sir John Moore), the King's Royal Rifle Corps or the Rifle Brigade, but as 1st, 2nd and 3rd Green Jackets. Some of the 'heavy infantry', the regiments of the East Anglian Brigade, followed suit. In 1964, therefore, the Army Council expressed its 'wish and intention' (it felt it could not compel) that the infantry reorganize itself formally along these lines to permit flexibility when increasing or decreasing its strength without the upheaval of further amalgamations. Cynics might have seen this as merely a way of cutting the infantry by stealth, as indeed seemed to be the case when the newly formed Light Infantry had its fourth battalion cut less than a year after the regiment was formed, along with the fourth battalions of other four-battalion regiments. There was, alas, no safety in numbers. Indeed, the large regiments would be cut to two battalions in 1993. The Army Dress Committee, which authorizes all changes in distinctions of dress, emblems and 'accoutrements', almost certainly has the greatest number of files of any committee of the MoD.

After the Borneo Confrontation, then, the trend in man-power was ever downwards – in the TA, too – with both Labour and Conservative governments trying to find more and more economies in expensive volunteers. National Servicemen, coming and going all the time, may have been a training burden, and they may have been 'bolshie', but they had been cheap. Manpower was expensive, the defence budget was inevitably finite, and military equipment was costlier by the year. BAOR was an equipment-intensive organization, and the human element was consequently squeezed ever harder. To some extent the TA could come to the rescue, no longer fielding formed brigades and divisions but instead filling in the increasing gaps in the regular army's order of battle for general war. But it was not 'tidy'.

By the end of the 1960s Britain had all but withdrawn from east of Suez. There remained only the garrison in Hong Kong (the contingent of the Commonwealth Brigade was soon to leave Singapore) and a Gurkha battalion in Brunei, maintained at the expense of the sultan. In the Mediterranean, British troops quit Malta, thinned out in Gibraltar and withdrew into the handful of sovereign base areas in Cyprus (with a few more wearing UN berets to keep the peace between Greek and Turkish Cypriots following independence). And in Germany they trained incessantly over the rolling plains of Lower Saxony to be ready for the Soviet army.

In 1968 not a single British soldier was killed in action. It was the first year of which this could be said in the lifetime of any serving man – the first, in all probability, in the army's history.*

* Though even 1968 is questionable, for an SAS soldier was killed in Ethiopia in ambiguous circumstances.

29

The Troubles

Northern Ireland, 1969–2007

'THE CHIEF CONSTABLE BELIEVES IT IS A LAW AND ORDER situation; the GOC *knows* it is counter-insurgency.'

So said the staff officer briefing the company commanders of a newly arrived infantry battalion in South Armagh – 'bandit country', as it was known – in 1980: eleven long years after the 'Troubles' began.* Six months later when the battalion left, five of them were dead, and the Royal Ulster Constabulary (RUC) and the army were no nearer defining what it was they were doing. Notwithstanding – perhaps even because of – this strategic failure, Operation Banner, the longest operation in the army's history, was one of its most formative experiences, a 38-year continuous commitment with a final bill of more than 300,000 troops, 763 of whom had been killed and 6,116 wounded as a direct result of terrorist action.

'Op Banner' taught three generations of officers and

---

* Author's diary.

NCOs the habit of tactical decision-making, of taking operational responsibility, often under fire and always under the threat of ambush by bullet or bomb. And it did so within a political context so demanding that each tactical decision might have an immediate strategic consequence. Indeed, the IRA would time their attacks with an eye to the six o'clock news. The reaction of the commander on the ground, whether lance-corporal or captain, was as much a battle for the headlines – for the route to hearts and minds on both sides of the water and both sides of the border – as it was for the streets of Belfast or the roads of South Armagh. *Minimum force*: the words were etched on the tactical consciousness of every officer and NCO, a principle developed for counter-insurgency 'east of Suez' where a miscalculation might not be so catastrophic (there were no 'Bloody Sunday' inquiries for Malaya, Cyprus, Kenya or Aden).* The experience east of Suez together with that of Operation Banner begat a doctrine for 'operations other than war' that would make the British army's intervention in the chaotic disintegration of Yugoslavia in the 1990s both possible and beneficial at a time when other armies held back in ignorance of what to do. But did too much 'peace-keeping' damage the army's health, as the US army was wont to ask? Did it blunt its fighting edge?

Ireland had not troubled the army much since 1922, but in Ireland, perhaps as in no other place, old sins cast long shadows. The Catholic Irish had worn red coats for centuries; even after 1922 they continued to do so, though not in those regiments that had gained the tough fighting reputations of Wellington's day, or on the North-West

---

* Minimum (necessary) force is defined by the army as 'the measured and proportionate application of violence or coercion, sufficient only to achieve a specific objective and confined in effect to the legitimate target intended'. The contrasting principle in war is *overwhelming* force.

Frontier, or in France and Flanders – the Dublin Fusiliers, the Munster Fusiliers, the Leinsters, the Royal Irish, and the incomparable Connaught Rangers. They joined instead the Irish Guards or, increasingly, the Ulster regiments, some of which had previously been – on the surface at least – predominantly Protestant. These southern, almost all Catholic, Irishmen were in the minds of their fellow soldiers the loyal Irish, 'West Britons',* like those who in 1916 were fighting on the Western Front.

On the other hand, the Irish Republican Army (IRA),† which sprang seemingly inexplicably from the same community, would murder any man or woman in or out of uniform – any civilian, indeed – who opposed their vision of a united socialist republic of Ireland, just as they had done during the First World War, in a campaign of which the 1916 'Easter Rising' was merely the most spectacular but incompetent action. The British army, both in 1916 and after 1918 during the fighting which led to Irish independence, had had little time for 'hearts and minds'; and the excesses of the 'Black and Tans' – as the frequently ex-service police auxiliaries were known for their mixture of khaki and blue or green clothing and brown and black leather (alluding to the Scarteen Hunt's 'black and tan' foxhounds) – compounded the resentment of British troops in the minds of the Catholic population.

'The Troubles', as with poetic Irish understatement the conflict became known north and south of the border, began

---

* The Irish Nationalist leader Daniel O'Connell had used the term in the House of Commons in 1832: 'The people of Ireland are ready to become a portion of the Empire, provided they be made so in reality and not in name alone; they are ready to become a kind of West Briton if made so in benefits and justice; but if not, we are Irishmen again.'
† The IRA fractured in 1969–70. The 'Official IRA' (OIRA) soon declared a ceasefire; the Provisional IRA (PIRA – the 'Provos') continued their war with the RUC and army – and, indeed, with the OIRA.

in the summer of 1969 towards the end of a period of Europe-wide protest which ranged from the courageous and focused anti-Russian demonstrations in Czechoslovakia (the 'Prague Spring' of 1968) to the left-wing, nihilist, student–worker riots in Paris and West Germany. The Northern Ireland Civil Rights Association (NICRA, which was not solely Catholic) had been turning up the heat in their protest marches against discrimination in employment and political representation, and a series of running battles in Belfast and Londonderry, the second city of the province, in which (Protestant) Loyalists joined against the NICRA, had left the RUC not only exhausted but reviled: by the (Catholic) Nationalists and also by many Loyalists who had been on the receiving end of RUC even-(if heavy-)handedness. Whatever the reality, the appearance was of a police force intent on enforcing Loyalist dominance. In consequence, Harold Wilson's Labour government ordered in the army as a 'peacekeeping' force more acceptable than the RUC to the Nationalists.

It is interesting to speculate what would have been the outcome (assuming the Troubles, at that stage at least, were indeed a case of enforcing law and order and not a true insurgency) had the RUC been reinforced by the Metropolitan Police and other constabularies, as those forces in the Yorkshire and Midland coalfields were during the miners' strike of 1984–5. But Ireland was, after all, over the water, and the province had a long tradition of 'militarism'. And the RUC – in particular its auxiliary part-time constabulary, the 'B Specials' – was not merely a police force but an anti-guerrilla force. As well as police stations it had 'barracks', armoured cars and medium machine guns. To an Ulsterman, the arrival of soldiers on the streets did not perhaps have the same impact it would have had on the inhabitants of a mainland British city. To the Loyalists they were another echelon of the civil power, which was in effect the RUC; to the Nationalists they were likewise another echelon, but more impartial than the RUC.

To the IRA, however, British troops posed a very significant threat, though not in the way that might first be supposed. For the troops were welcomed initially in the Bogside and Creggan areas of Londonderry, the scenes of the worst violence later. The Prince of Wales's Own Regiment of Yorkshire, not long home from Aden and sent from Colchester with little notice, were given mugs of tea by Catholic women who hailed them as protectors from both the Loyalists and the RUC. It was the same in the Falls Road area of Belfast, later notorious for its gun battles (it was here in 1973 that Lieutenant, now General Sir, Richard Dannatt won the MC). But this role of the British army as protector of the Catholic–Nationalist population was inimical to the IRA's own strategy of replacing the RUC in the 'Green' areas.* The IRA therefore began direct action – attacks on police stations and RUC men, particularly Special Branch – and continued to foment disorder through NICRA, provoking more sectarian violence.

Troops began surging in and out of Northern Ireland. Usually, there were just two battalions resident in Belfast and County Down; during the 1970 'marching season', the summer months when the various Protestant associations took to the streets to mark King William's victories against Catholic James, there were fifteen. Both civil disorder and attacks on the security forces, which included the new, locally raised Ulster Defence Regiment, increased. Internment was introduced, which brought a temporary

---

* The term derived from the maps issued by HQ Northern Ireland, with their colour overlays showing the preponderant sectarian make-up of an area: green for Catholic/Nationalist/Republican, orange for Protestant/Loyalist, and white for 'non-sectarian' – business areas, professional centres such as the universities, or middle-class residential districts (sectarian violence was almost exclusively a working-class phenomenon).

respite. But internment of suspected members of both the IRA and the UVF (Ulster Volunteer Force, the Loyalist para-militaries), on the certification of an RUC inspector, and robust methods of interrogation, only hardened support in the Nationalist areas, which soon turned into IRA strong-holds. By the summer of 1971, just eighteen months after Yorkshire soldiers had been drinking Catholic tea, armed gangs had all but taken over the Bogside in Londonderry, and daily clashes with youths of the Protestant lodges were rapidly making the streets look like the East End of London in the Blitz. Soon there were barricades all over the place and the IRA openly manned checkpoints, eagerly filmed by the international media. By the end of the year, seven soldiers had been killed in the city.

Although police primacy was the ostensible principle, in reality the RUC had withdrawn from the hard areas. Here the army's presence too was only transient, being confined to risky foot patrols and pointless, even provocative, mobile patrols in obsolescent armoured troop carriers – the heavily up-armoured one-ton Humber, known as a 'pig' for its snouty appearance, or the bigger Saracen with its revolving turret and machine gun. In the absence of specialist riot gear and techniques, soldiers could only respond to violence in the rough and ready way they had in places like Aden, where the general absence of the media at the point of contact meant that such tactics worked. In Northern Ireland, where almost every clash was filmed and not all were faith-fully edited, the same tactics were often counter-productive. The IRA certainly comprehended this, and had every reason therefore to provoke excess. NICRA, some of whose members were entirely well-meaning but whose leaders were – to put it mildly – manipulated or manipulating, also saw that violent protest paid, especially when initial concessions made by the Northern Ireland government at Stormont con-vinced the leadership that more violence would lead to more concessions. This was textbook insurgency: the IRA now had the means to bring on proxy battle with the army in which the military response would only further the political object. In words that South American revolutionaries of that time were using, it was about making society brittle. In this situation it was vital for the army (and RUC) to avoid taking the bait; but in Londonderry in particular there was an accident waiting to happen, and early the following year the IRA won perhaps its most significant victory of the campaign.

What exactly happened that day – 'Bloody Sunday' – has been the subject of the longest-running judicial inquiry in history, its report still not public at the time of writing this book. 'The facts that are undisputed are well known,' the prime minister, Tony Blair, told Parliament when announcing the inquiry by Lord Justice Saville almost twenty-six years to the day after the event:

On 30 January 1972, during a disturbance in Londonderry following a civil rights march, shots were fired by the British Army. Thirteen people were killed and another thirteen were wounded, one of whom subsequently died. The day after the incident, the then Prime Minister set up a public inquiry under the then Lord Chief Justice, Lord Widgery. Lord Widgery produced a report within 11 weeks of the day. His conclusions included the following: that shots had been fired at the soldiers before they started the firing that led to the casualties; that, for the most part, the soldiers acted as they did because they thought their standing orders justified it; and that although there was no proof that any of the deceased had been shot while handling a firearm or bomb, there was a strong suspicion that some had been firing weapons or handling bombs in the course of the afternoon.

Blair went on to say that the report's timescale meant that Widgery was not able to consider all the evidence that might have been available – testimony from the wounded still in hospital, for example, or substantial numbers of eye-witness accounts. New material had since come to light, including ballistic and medical evidence. But he added his appreciation for 'the way in which our security forces have responded over the years to terrorism in Northern Ireland', which 'set an example to the world of restraint combined with effectiveness, given the dangerous circumstances in which they are called on to operate'. And he included the inevitable palliative that 'Lessons have, of course, been learnt over many years – in some cases, painful lessons.'

In his autobiography *Soldier*, General Sir Mike Jackson, who on that day was adjutant of 1 Para, which had been sent from Belfast to reinforce the brigade in Londonderry, describes what he saw:

> here [the Bogside] Nationalists had declared 'Free Derry', a barricaded 'no-go' area for the Security Forces, openly policed by armed and hooded IRA men. A community alert system was used to mobilize the IRA to repel incursions by the Security Forces. Women would sound the alarm by banging dustbin lids. The Irish flag flew over 'Free Derry', which to all intents and purposes was no longer part of the United Kingdom ...
>
> Londonderry lay within the responsibility of 8 Brigade, commanded by Brigadier Andrew McClelland. By comparison with 39 Brigade [in Belfast, to which 1 Para belonged] it was 8 Brigade's practice to use CS gas in large quantities, while in Belfast we hardly used it at all. Although circumstances in Londonderry were quite different from those prevailing in Belfast, there was a sense in 39 Brigade that 8 Brigade hadn't been firm enough.

Jackson goes on to explain the fears of the city's business community that the coming march, which was illegal, would end in the destruction of property as well as business. A plan was devised to hold the marchers on a 'containment line' by barriers across the roads into the commercial and civic centre, and to arrest the 'worst of the hooligans':

> It was a snatch operation, of the type we did all the time. The main difficulty was to stop people running away: to put a cork in the bottle. We had evolved a tactic of going in behind the crowd or of coming in at the flank to cut them off. Clearly this was a large operation which would require a large number of troops. We envisaged using three companies, in a pincer movement, to surround the rioters and cut off their retreat, arrest them, bundle them into the pigs and take them off for handover to the RUC.
>
> Derek Wilford [commanding officer] gave us a briefing shortly before the march. His attitude was confident and professional. There was a sense that we might be heading for some sort of set-piece confrontation: it seemed unlikely that there wouldn't be at least some aggro. But it's worth stressing that the Battalion's mission was to capture rioters in the immediate vicinity of the barricades, which of course meant going over the containment line ... Of course we anticipated that the IRA might react when we 'invaded their turf'. We had to be prepared to be attacked at any time. However, there was no sense in which we planned to use the arrest operation to 'teach the IRA a lesson'. That was not on the agenda.

The march got under way in the afternoon, and Jackson, who was with the commanding officer throughout, describes how they heard on the radio that Support Company, which in any battalion included the older, more experienced soldiers, had shot a man trying to detonate a nail bomb. Towards four o'clock the marchers came down the hill towards 'Aggro Corner', near where the Paras waited:

> There was a lot of shouting and chanting. The vast majority of

them did not linger by the barricades, but moved on beyond the Rossville Flats to 'Free Derry Corner', a well-known gathering place, where they were addressed by various speakers. But the hooligans hung back by the barricades so they could take on the soldiers. This was what we had anticipated. The thugs and the marchers were now separated, so we could get behind the thugs and isolate them.

After a frustrating delay, the Paras were given the word to go in.

Through the window I saw our Support Company's vehicles moving towards the Rossville Flats, about two hundred yards off. The barricades had been pulled to one side to allow vehicle access. Now they were debussing in the shadow of the flats. Some soldiers were carrying batons ready to make arrests, their rifles slung across their backs, but some had their weapons ready to cover the others. There was nothing unusual about this; on the contrary, it was standard drill for some soldiers to cover those making arrests. Almost immediately it became apparent that they thought they had come under fire. They began zig-zagging from side to side and looking for cover.

Derek Wilford was a commander who liked to be forward among his soldiers, to see what was happening for himself. After a few minutes, he decided to do just that. We tore down the stairs together, round the corner and out into the street, and then sprinted across the waste ground. There was a lot of noise. I saw soldiers hunched up, trying to make their bodies as small as possible. As I dashed forward with my head down I had the definite impression that someone was firing at me from a vantage point somewhere ahead. When you are used to the sound of gunfire you can estimate the range and the direction it is coming from. You first hear a crack, which is the sound of displaced air as the bullet passes close to you; then a thump, the delayed noise of the rifle being fired. The thump gives you an indication of direction, the time gap between the crack and the thump gives you a sense of range.

Trying to describe the minutes that followed was never going to be easy: 'It was a confused situation', writes Jackson. 'My next clear memory of that day is of being back in a factory building,' and it was here in the early evening that it was 'becoming clear that there had been a large number of fatalities' – considerably more indeed than in one action before or since:

> I was left with some very mixed and worrying feelings. I imagine that others in the Battalion felt the same. I hated the thought, as some commentators would state straight away, that our soldiers might have lost control. It would be very un-professional to have done that, and in the army one is very proud of professionalism. I knew these men, and I knew their quality. So far as I was concerned the Paras were tough, but they were disciplined. I found it difficult to accept that there could have been any mass breach of discipline ... But however incredible the accusations, it was a terrible thought that we might have killed innocent people, whatever the stress of the moment, and whatever the provocation. There was no doubt in my own mind that we had come under fire, and in those circumstances it was legitimate to return fire at properly identified targets. The question remains whether the response of some of our soldiers was proportionate or not, considering the nature of the threat.

The Widgery report in the immediate aftermath of the affair had in fact concluded that some of the firing had 'bordered on the reckless'. The Saville inquiry – whose costs are approaching £400 million – may be able to refine that judgement, though it is difficult to see how. As Jackson says, the fact that it has taken so long and consumed so much public money is perhaps in itself evidence of the difficulty in establishing the truth. 'As well try to write the history of a ball,' said the duke of Wellington on learning that someone wanted to write an account of Waterloo. But it is difficult to believe that the truth was Blair's object. A finding that the Paras were not to blame would be dismissed by the Nationalist community; that they *were* to blame would be a most inconvenient truth. The inquiry could only have been about 'process' – and during that process what appears to be peace has come to Northern Ireland.*

'Lessons have, of course, been learnt over many years,' Blair had told Parliament, with that Delphic addition, 'in some cases, painful lessons.' In fact the lesson-learning needed no formal process: the troops on the ground throughout the province knew they had lost what sympathy remained among the Nationalist community, and among most of the media. Crucially, it handed a propaganda coup to the Nationalist fund-raising machine in the United States; and it would be nearly three decades before US Democrat politicians in particular saw beyond the 'freedom fighter' badge that Bloody Sunday had bestowed on the IRA and recognized them for what they were. No matter what the truth, at the time perception was everything. When, for example, the coroner closed the official inquests eighteen months later – and a full year after Widgery – he felt himself moved to remark (whether properly or not is immaterial):

> It strikes me that the Army ran amok that day and shot without thinking what they were doing. They were shooting innocent people. These people may have been taking part in a march that was banned but that does not justify the troops coming in and firing live rounds indiscriminately. I would say without hesitation that it was sheer, unadulterated murder.

* See end-of-chapter note.

The coroner was retired Major Hubert O'Neill, late of the Royal Artillery.

The IRA now had a 'legitimate' *casus belli*. The British army was portrayed as the enemy of the Nationalist population, and the IRA would battle them out of Northern Ireland. And for the next thirty years they tried to do just that, with the active support of a significant part, and the approval of the greater part, of the Nationalist population. Even those who shared Nationalist aspirations but abhorred violence were in an ambivalent position: to them the IRA's means may have been wrong, but the cause was right; they could not be *wholly* condemned therefore. The Catholic Church in the province seemed at times to be particularly awash with ambivalence; certainly many of its priests were Nationalist sympathizers, and some were known to be not entirely inactive supporters. And the Provisional IRA, being merely socialist and Republican, was more palatable to the Catholic hierarchy, for all their sins of violence, than the Official IRA and their hard-line communism. In fact the OIRA declared a ceasefire soon afterwards, and became a spent force.

The 'Troubles', and attempts to deal with them, were bedevilled by this complex mix of perspectives and elements. The lack of normal policing in the Nationalist areas, urban and rural alike, especially along the border with the South, allowed crime to flourish (with much of the proceeds going to the IRA) and everyday laws to be flouted. As such the situation was one of law and order. The IRA's war with the security forces, however, was a clear insurgency – armed rebellion against the state. The Troubles were also in part civil war, a continuation of internal conflicts that had plagued the island for centuries; the sectarian violence was at times truly atavistic. And overlaying, perhaps even underlying, all this – and certainly weaving in and out of the tangled thicket of the 'Irish problem' – was a contest of political ideas which would not have been out of place in the Russia of 1917. The bombs that exploded in public places in Northern Ireland and on the mainland were more lethal than those of the Fenian outrages in London in the nineteenth century, or the anarchist bombs in France and Central Europe, but they were out of the same book of violent revolution. If anyone truly understood what all this was about, no one had a clear idea how it was to be ended – except, perhaps, by the passage of time alone.

In all this the RUC struggled bravely to reform itself and to uphold the law. And in upholding the law it had to deal with the chief challenge to the rule of law, the IRA. Its Special Branch was reinforced and its activities expanded. But since the RUC was no longer a paramilitary organization, it could not take on the IRA in direct combat. The army was therefore increasingly drawn in, its mission – variously stated – being to 'hold the ring' while a political solution was found, and to 'drive a wedge between the men of violence and the Nationalist community'. It was never formally tasked with seeking out and destroying the IRA; the gun battles occurred more often as a consequence of guarding the network of RUC stations, urban and rural, or on speculative ambushes and patrolling. Occasionally, however, there were active operations based on hard intelligence. In the early days, as in Malaya or Borneo, this was the infantry's job; but as the campaign progressed and 'hard' intelligence became even more valuable, the SAS took the lead.

And in the early days the infantry's war, in the rural areas especially, was not dissimilar to what it had been in the Far East. In February 1973 a staff-serjeant platoon commander of the King's Own Royal Border Regiment (KORBR)* was

* The regiment was an amalgamation (1959) of the King's Own, whose fortunes in earlier centuries have been chronicled in previous chapters, and the Border Regiment. Its name disappeared from the Army List in 2006 when it was amalgamated with two other infantry regiments to form the Duke of Lancaster's Regiment. The KORBR was a tight-knit battalion from the north-west of England, and served longer in Northern Ireland than any other regiment.

killed in an exchange of fire with an IRA 'active service unit' (ASU) near Strabane on the border in County Tyrone. Soon afterwards the battalion received intelligence that the ASU was based at a farm close by, and the KORBR's commanding officer, Lieutenant-Colonel (later Major-General) David Miller, who had won an MC as a company commander in Aden, ordered one of his platoons to set up an observation post (OP) ready to take offensive action if the ASU showed themselves. The platoon commander was 21-year-old Lieutenant Stephen Flanagan, who was not long out of Sandhurst and had spent most of his first year in jungle training with the battalion in Malaysia. Miller gave him a very free operational hand, as Flanagan's diary records:

0900hrs 2nd February – summoned to CO's office. Gave me 24hrs to devise a plan to sort them out. Three provisos; don't cross the border, abide by the Yellow Card [rules of engagement] and keep in communication. Spent all day studying my Jungle Warfare School manuals for inspiration (I had been trained in Malaya but not at the Platoon Commanders' Course in Wiltshire).

0900hrs 3rd February – presented my plan for an OP/ambush inserted by night with a dug-in rebroadcast station for the useless A41 radio. Deception would be the key to success and survival.

3rd–5th February – preparation and rehearsals.

2359hrs 6th February – self, carrying GPMG [general-purpose machine gun], and both company snipers dropped off by covert van 1 mile short of OP, chosen from an aerial photograph. Found a suitable hedgerow and dug in. The first comms check with the rebro station at 0200hrs failed (and was never established).

Dawn 7th February – the deception plan began with the OC's [officer commanding the company] helicopter flight along the river to wake the ASU. Two men appeared at the doors of a lone house and adjacent caravan; neither appeared armed. The range was greater than I wanted. Just before 0900 sounds of three Saracens and two Ferrets [wheeled armoured vehicles] announced the arrival of 9th Platoon and troop of 16th/5th Lancers into positions on the Border to the north of our OP. We couldn't see them. After 45 minutes, single shots fired from dead ground somewhere to our front at the platoon, who returned fire and then withdrew to our base as briefed.

1200hrs still nothing. OC's orders were to withdraw to the ERV [emergency rendez-vous] no later than Noon. We wanted to, but needed to wait.

1307hrs – A car arrived on the forecourt of the lone building – was it a pub? Two men got out both carrying rifles. I nodded assent; two shots, both men went to ground, but it was not clear if they had been hit. That's what the GPMG was for. Long killing burst. Then silence. Someone said it was time to get to the ERV.

8th February – OC agreed the Platoon should mount a clearance op to the same area, provided I took all available firepower. Overnight someone had constructed a brick wall with firing ports in front of the caravan. Ten minutes after we arrived, shots were fired from the firing ports and from two trenches beside the main road. All hell broke out. Six Brownings [heavy calibre machine guns] on the Saracens, the Ferrets and troop leader's Saladin were firing, plus the three platoon GPMGs. Sergeant H tried to grip the fire discipline [control the volume of fire] but they couldn't hear him; I watched transfixed at the impact of this weight of fire on a hurriedly constructed brick wall. Within seconds it was gone, exposing the caravan and the poor sods who had taken us on.

Special Branch reported that four of the seven-man ASU had possibly been killed (although this was never established). For his courage, planning and leadership during the operation, Flanagan was awarded the MC.

In the urban areas, however, the daily patrolling, with or without the RUC, was more a battle to keep the IRA off the streets, where the deadliest threat to foot or vehicle patrols was the single-shot sniper (IEDs – improvised explosive devices – tended to be used by the IRA as 'spectaculars', with a warning to gain the greatest disruptive effect). To deter the sniper a 'multiple' of three patrols, each of four men, usually on foot, two of them commanded by a corporal or lance-corporal, the third by a subaltern or serjeant, would coordinate their movements so as to defeat the IRA early-warning system and threaten the sniper's escape route. These 'multiples' became the basis of the long-term effort to win back the streets, and since two out of the three patrol commanders would be junior NCOs, Northern Ireland became known as a 'corporals' war' – just as Inkerman in the Crimea had been called a 'corporals' battle', the smoke and fog being so thick that no one could exercise control beyond a few yards.

In July 1976, Lance-Corporal David Harkness of the KORBR, which was on its second emergency tour of duty since leaving the province in 1973 after eighteen months as a resident battalion, was commanding a mobile patrol in West Belfast. As his Land Rover slowed to cross a 'speed bump' it came under sustained fire from a house at close range, one of the rounds slightly wounding him. 'In the face of murderous fire,' ran his medal citation,

> Corporal Harkness immediately called to his vehicle driver to halt and within a few seconds had dismounted his patrol and put in a follow-up attack on the house ... charging the house at the head of his patrol, gaining entry by kicking in the back door, at the same time deploying another member of the patrol to cut off the front of the house. Such was the speed and unexpected-ness of Corporal Harkness's action that the gunmen barely escaped through the front door as he entered through the back.

The gunmen leapt into a waiting car standing ready with the engine running . . . further shots were fired from the car as the patrol assaulted the house. The house was surrounded by a high wooden fence which prevented the 'cut off' member of the patrol opening fire on the car but due to Harkness's swift action a full description was passed on the battalion radio . . . enabling a vehicle checkpoint to stop the car a few minutes later.

All this had been done in the midst of passers-by and the usual traffic, with the split-second decisions – 'flight or fight' and fire discipline – taken by a 20-year-old lance-corporal who had left school at fifteen to join the Infantry Junior Leaders' Battalion.* For his courage under fierce fire, leadership and presence of mind, Harkness was awarded the Military Medal.

If 'Op Banner' was fundamentally a corporals' war, there were from time to time larger operations arising from hard intelligence or simply the routine of life in a corner of the United Kingdom – such as the secure movement of quarrying explosives. There was one strategic operation, however, in July 1972, which required half the infantry strength of the entire British army. After 'Bloody Sunday' the Nationalist no-go areas had become so extensive and high-profile – one gable-end sign, 'You Are Now Entering Free Derry', challenged British sovereignty on TV screens around the world almost nightly – that it looked as if the Ulster Defence Association (UDA) would set up no-go areas of their own in the Loyalist districts. The GOC and director of operations, Lieutenant-General Sir Harry Tuzo, a gunner who had commanded 51 Gurkha Brigade in Borneo, decided on a concerted effort to clear the barricades right across the

* The IJLB trained boys whose NCO potential had one way or another been recognized while at school. They entered at 15 until in 1972 the school leaving age was raised to 16.

province and re-establish a presence in all the 'green' areas. To do this he was reinforced to twenty-seven battalions (including artillery and armoured corps regiments in the infantry role): some 21,000 men, with a further 9,000 UDR mobilized for the operation. On 31 July, by the application of this overwhelming strength in 'Operation Motorman', he was able at last to get rid of the no-go areas province-wide – and with scarcely a shot fired.

The war now settled into one of steady attrition, on both sides. There were occasional highs and lows, some very low indeed. In 1979, on the same day that Lord Mountbatten was killed at his estate in southern Ireland, eighteen men of the Parachute Regiment and the Queen's Own Highlanders, including the Highlanders' commanding officer Lieutenant-Colonel David Blair (a devout Catholic), were killed by IEDs in skilfully executed attacks at Warrenpoint near the border south of Newry. Indeed, the IED threat in the border areas, South Armagh especially, soon became so great that there could be no overt movement by military vehicles: troops were moved instead by helicopter, and the military base at Bessbrook became for many years the busiest heliport in the world with the constant coming and going of RAF Pumas and Chinooks, and the Scouts and Gazelles of the Army Air Corps. In the four months (it would later be six) that a battalion spent in South Armagh at the height of the IRA offensive in the late 1970s and 1980s, it would lose on average six men. Between 1969 and 1997, the year PIRA declared an indefinite ceasefire, 763 servicemen were killed – the great majority of them army, including UDR – plus more than 300 RUC. Of that total, nearly half were killed in the first ten years – forty-eight of them in 1979 alone. The seventies were, indeed, the learning years, a decade in which the IRA suffered heavily too. Thereafter 'force protection' by both tactical and technical means became an increasing priority, so that in the early 1980s a senior commander in South Armagh would tell his battalion commanders that the real mission was not to take casualties since these were of disproportionate strategic value to the IRA compared with any tactical advantage gained in killing IRA men.*

But although these thirty-eight years of continuous operations put a severe strain in particular on the regiments of BAOR, who paid the price in erosion of their armoured warfare capability, they gave the army an incalculable edge. It developed sophisticated techniques of operational analysis and training as well as expertise in what in other armies was the preserve of Special Forces – not least in covert surveillance. Secret corners of England were given over to special instruction in urban patrolling and snap shooting, and in a host of Op-Banner-related tactics. These secret places would in turn be adapted for training for the Balkans, Iraq and Afghanistan. The growing awareness of tactical–strategic linkage – an ill-judged reaction on the streets of Belfast in the morning bringing a government minister to the despatch box in Westminster in the afternoon – taught the army to be media-wise. On top of this, in their contact with some of the most deprived parts of the United Kingdom and the other world of sectarianism, its officers got strong doses of 'social reality' – a distinct advantage in the foreign interventions that were to follow, and an insight which kept the army from becoming once more, in the younger Moltke's words, 'that perfect thing apart' at a time of growing social and political tension in the country as a whole.†

Op Banner taught the army yet again the value of patience

* Author's diary.
† This point was made forcefully to the author by a former brigade commander in the Province, subsequently in turn the army's director of public relations, the professional head of the infantry, and the last Commander British Forces in Hong Kong – Major-General Bryan Dutton.

in dealing with insurgencies, and especially in the acceptance of casualties. Above all, it gave the army – at every level – the confidence to operate in situations which lacked operational clarity. This is more than just a resigned acceptance of 'mission creep', or even the lack of strategic coherence: it is an acknowledgement that in any 'war among the people', as one of its most successful practitioners, General Sir Rupert Smith, has called it,* the operational situation is dynamic and the correct ultimate strategy may not be discernible at the point at which military force is committed.

Op Banner also gave the army something it had never before enjoyed: public support based on a true understanding of the circumstances in which soldiers were operating. While disasters such as 'Bloody Sunday' in the early days caused considerable recoil – not least in the press, where journalists such as Simon Winchester of the *Guardian* appeared utterly convinced that the army was a brutal, anti-Catholic organization – the IRA bombing campaign demonstrated the true nature of militant Republicanism. The army's mounting casualties and increasing self-control slowly but surely gained the respect, even admiration, of the public; indeed, it gained the grudging and unspoken respect of a large part of the Nationalist community too. This public affection for the army, though not nearly as demonstrative as in the United States, remains a strong sustaining force, and in turn shapes its own image-making.†

Far from blunting the army's fighting edge, then, the

---

* Rupert Smith, *The Utility of Force: The Art of War in the Modern World* (2005).

† It was this that in 2007 prompted the CGS, Sir Richard Dannatt, to call for public expressions of support for homecoming soldiers: the war in Iraq (and to a lesser extent that in Afghanistan) was deeply unpopular with the public, and Dannatt was intent on decoupling revulsion for the government from support for the troops themselves, without which morale – already under pressure for a number of other reasons – would have sunk ever lower. See also ch. 32.

'Troubles' actually kept the army up to the operational mark in a decade after the withdrawal from east of Suez when otherwise it would have known only training in Germany or the most benign sort of UN peacekeeping in Cyprus (or for a very few, more active service in Oman*). For Special Forces – not just the SAS but the special intelligence-gathering unit 14 Intelligence Company – Northern Ireland was a singular opportunity to develop techniques for urban operations in particular which have been reprised to real effect in both Iraq and Afghanistan. Indeed, a new 'Special Reconnaissance Regiment' was born of 14 Intelligence Company's experience, and has proved a key strategic asset in both those countries. And not only was Op Banner, the 'corporals' war', invaluable for junior leadership, it was a proving ground for many of the officers who would after-wards fill senior appointments in BAOR – and thereafter the First Gulf War – for the habit of operational planning and decision-making is a transferable asset.

But before that sweeping victory of armoured forces in the Gulf, there would come one of the army's most gruelling tests of its fighting power, ever. And at the end of the longest, most tenuous, lines of communication in its history.

---

* For a few officers and NCOs seconded to the sultan's armed forces, and Royal Engineers as well as SAS, there was the opportunity for some brisk fighting in Dhofar, in the south of the Gulf state of Oman, where a remarkably little-known war was being fought against Communist insurgents. It was won, under British direction, by 1975.

End-of-Chapter Note

'There was no doubt in my mind that we had come under fire, and in those circumstances it was legitimate to return fire at properly identified targets. The question remains whether the response of some of our soldiers was proportionate or not, considering the nature of the threat' (pp 546). Jackson's question was answered, to the extent that it ever could be, when the Saville Inquiry's report was published in June 2010: 'The firing by soldiers of 1 PARA on Bloody Sunday [an interesting, perhaps even telling, judicial lapse into the vernacular] caused the deaths of thirteen people and injury to a similar number, none of whom was posing a threat of causing death or serious injury.' Saville spoke of the Paras 'losing their self-control' and 'a serious and widespread loss of fire discipline'.

The Prime Minister, David Cameron, made an abject apology in parliament, and Jackson likewise took it on the jaw: 'The Prime Minister made a fulsome apology and I join him in so doing' (though 'fulsome' has always been a perilously ambiguous word).

The findings were, indeed, what many in the rest of the army had always believed, that in operations such as these the Paras 'didn't do restraint' – at least, not very well. Incidents continued, though nothing like as grave, but in the mid 1990s, in an echo of the disbanding of the Canadian Airborne Regiment after murderous excesses in Somalia in 1992, the CGS warned the Paras' senior officers that the regiment was rapidly becoming unusable, and to 'get a grip'.

A regiment is more than its worst mistake, however; when Saville was published, the Paras were already redeeming themselves, and with interest, in Afghanistan.

30

A Near-run Thing

A South Atlantic winter, 1982

AT THE ARMY STAFF COLLEGE IN CAMBERLEY TOWARDS THE END OF 1981, A Royal Marines officer asked a visiting government minister about the growing Argentine sabre-rattling in the South Atlantic. The minister's reply, delivered with a dismissiveness bordering on disdain, was that it would be dealt with by diplomatic means. A few months later, during the final map exercise in which the Royal Navy and RAF staff colleges also took part, the techniques for an amphibious landing on the coast of north-west Africa were studied. Throughout, the admiral directing the exercise was at pains to emphasize that this was a theoretical study only, for the purposes of practising inter-service staffwork: the United Kingdom, he repeatedly pointed out, did not have the capability to launch such an operation – even just round the corner from Gibraltar, as this one was. Barely nine months later the Argentinian officer on the course, who had by then become the military assistant to the commander of the Argentine invasion force, was taken prisoner at Port Stanley, the Falklands' capital.

* * *

At the beginning of the 1980s the Argentine military junta led by General Leopoldo Galtieri faced mounting economic problems and social unrest. In their view, acquiring either by peaceful or warlike means the Falkland Islands – or Las Malvinas as the Argentinians called them – would be a welcome distraction, a much-needed boost to the junta's popularity. Foreign military conquest had, after all, worked for many a Roman emperor. Argentina had long laid claim to the islands, and with the 150th anniversary of British rule in 1983 looming, as early as July 1981 the Foreign Office judged that in the intervening months there would be mounting Argentinian diplomatic activity; direct military force, however, it judged to be likely only if diplomacy were to fail completely. In the event, the junta's invasion of the islands came as a complete strategic and tactical surprise – only marginally less of a surprise, indeed, than the astonishing improvisation which retook them for Britain. The Foreign Office had not seen it coming;* the foreign secretary, Lord Carrington, who had won an MC with the Guards Armoured Division on Operation Market Garden, at once resigned. The Ministry of Defence had even been running down its presence in the South Atlantic and had just announced that the ice patrol ship HMS *Endurance* was to be withdrawn from service in 1982 without replacement. On the whole, joint (tri-service) staff opinion was that retaking the islands, which were beyond the range of land-based air cover (except of course from Argentina), was not feasible. The defence secretary, Sir John Nott, who had himself been

---

* The line put out by the Foreign Office at the time was that until 29 March, the eve of the invasion, there was no intention by the junta to invade, and so no intelligence could have discovered anything to report. It is, of course, ludicrous: even if the invasion had been improvised, as much of it appeared to be, it could not have been mounted in forty-eight hours from conception.

a National Service officer in the Gurkhas, had serious doubts too; and according to Admiral Sir Sandy Woodward, who would command the Falklands Task Force, the US Navy considered it downright impossible.* Fortunately, the chiefs of staff were of the opinion that unless British forces *were* able to retake the islands, the future was not worth contemplating. The first sea lord, Admiral Sir Henry Leach, who had joined the Royal Navy as a 13-year-old cadet in 1937, put on full uniform and marched to the House of Commons to tackle the Prime Minister, Margaret Thatcher. She asked him if retaking the islands was possible, to which he replied, 'Yes, we can recover the islands. And we must!' Thatcher asked why 'must', to which Leach, whose father had been captain of the *Prince of Wales* and had gone down with his ship in 1941, replied: 'Because if we don't do that, in a few months we will be living in a different country whose word will count for little!' Mrs Thatcher agreed with him, and the task force was at once cobbled together.

When the task force set sail it was in something of a carnival atmosphere, although few could have doubted that if it came to a fight – as it clearly must if there were no political settlement in the interim – it would be bloody. The atmosphere was in part due to the commandeering of the two luxury liners *QEII* and *Canberra* as troop transports. *QEII*'s departure was particularly festive,

> akin to the departure of the great troopships in their heyday – a military band in red jackets, bunting and friends, families and VIPs bidding farewell. Some wives of Royal Signallers in the Headquarters and Signal Squadron found their way to a gallery overlooking the quayside, and, together with everyone else,

* Arthur Herman, *To Rule the Waves: How the British Navy Shaped the Modern World* (2004).

were waving at their menfolk. One of the ladies saw her husband, a lance corporal. Married only a few weeks, she suddenly stripped off her blouse and bra, and, according to a witness, waving the latter in the air shouted, 'Feast your eyes while you can because you're not going to see these for a while!' Needless to say all eyes swung to the young lady, including those of an enterprising crane operator who manoeuvred the hook of his jib so that the girl could put her bra on it. He then swung the jib over to the ship where it was safely delivered to a somewhat embarrassed, but probably secretly rather proud, husband amid the cheers of those who had been watching this little cameo unfold.*

The carnival atmosphere did not last beyond the harbour moles, however, as the troops on board got down to the serious business of preparation and training. Commanded by a Royal Marine, Major-General Jeremy Moore, the landing element was in effect a (very) light – and very ad hoc – division. It comprised two brigades: 3rd Commando Brigade Royal Marines reinforced by two battalions of the Parachute Regiment; and 5th Infantry Brigade, consisting of two Guards battalions (2nd Scots and 1st Welsh) hastily recalled from ceremonial duties, and 7th Duke of Edinburgh's Gurkha Rifles.† All the combat support – principally artillery and engineers – was army. Together they would have to deal with some 6,000 Argentinians defending Port Stanley, plus various outposts at key points throughout the islands, and the long cruise south, with a stop in the Ascension Islands to

---

* From *5th Infantry Brigade in the Falklands*, by Nicholas van der Bijl and David Aldea (2002).

† Using Gurkhas was still considered 'tricky', but the CGS, General Sir Edwin ('Dwin') Bramall, who had served alongside them during the Borneo 'Confrontation', insisted that if the army did not use Gurkhas this time then someone would always find a reason not to use them. When he told Margaret Thatcher that he intended sending a Gurkha battalion she replied 'Only *one*?'

stretch legs and test weapons, did at least give the unfamiliar mix of commanders and staff officers a chance to work out what they might do.

Why battalions were pulled off public duties rather than sending others who were more combat-ready has never been entirely convincingly explained. There again, so much of the army's response was desperately improvised (and some say amateurish) that it invites comparison with the way the Crimea expedition was thrown together – not least in that the failings of commanders and logistics would in the end be redeemed by the sheer fighting courage of the field officers and soldiers. There had been no serious fighting anywhere in over a decade, but Northern Ireland had sharpened basic soldier-skills and junior leadership. Two distinct armies had developed, however: the army in Germany, and the army else-where. And the irony was that although there had been no fighting in Germany for more than *three* decades, yet because of BAOR's organization and rigorous training regime, the expertise in organizing for war (battle procedure and logistics on a large scale) lay with the 'petrol feet' not the 'brown knees'. The Falklands was simply not the sort of operation that the non-BAOR army had envisaged; yet that alone cannot account for the general dislocation. There was an unaware complacency from political top to military bottom, in part, no doubt, a result of the distraction of Northern Ireland – ironically, the very experience that had given junior leaders and soldiers their edge.*

The initial landings were to be at San Carlos Bay on East Falkland, 50 miles across the island from Stanley. With

---

* The Marines were generally far more combat-ready, and their brigade head-quarters and operating procedures much more efficient – a fact not lost in the exhaustive after-action analysis. But all that said, it was a magnificent feat of defiant improvisation, parts of which – the medical support, for example – were superb.

Argentine air superiority these landings were always going to be perilous, and indeed a number of ships were hit and sunk, but the Royal Marines were able to secure the beachhead with little opposition on the ground. With the islands now effectively isolated from Argentine reinforcement (the sinking of the *Belgrano* by HM submarine *Conqueror* had demonstrated that coming by sea was not an option) the intention was to build up strength ashore before beginning the advance on Stanley. But London wanted action quickly (as indeed did some officers, who were dismayed at the loss of momentum): despite the Falklands' being sovereign territory, with the absolute right under the UN Charter to defend itself when attacked, Mrs Thatcher feared that growing international pressure for a truce might become irresistible – or, at least, politically damaging. Off went the Commando Brigade, therefore, with its two battalions of the Parachute Regiment, to 'yomp' across the island towards Stanley through a wet, wind-blown landscape, sparse in cover and devoid of all charm.

Fifteen miles to the south, at Darwin and Goose Green on the narrow isthmus linking the northern and southern halves of the island, was an Argentine garrison of 1,400 men with close support from Pucara aircraft: the only substantial threat to either the landing area or the flank of the brigade's march across the rain-lashed moorland. The 2nd Battalion of the Parachute Regiment, commanded by Lieutenant-Colonel Herbert ('H') Jones, was ordered to capture the position. He had limited fire support – his own mortars, three 105 mm howitzers from 29 Commando Regiment Royal Artillery (with fewer than 1,000 rounds), and HMS *Arrow*'s single 4.5 inch (semi-automatic) gun – plus the battalion's few 'Milan' anti-tank missiles, useful against bunkers even if there were no armoured vehicles to fire them at. The Argentinians were rather better-armed, and though most were conscripts some of them were well trained and motivated; and they outnumbered the Paras two to one. Nor was the ground friendly to an attacker: the isthmus was barely 2 miles across at its widest point, offering sparse cover and scant room to manoeuvre.

The battalion's attack began in the early hours of 28 May and was soon running into stubborn opposition, some of the fighting hand-to-hand. Lance-Corporal Gary Bingley won a posthumous MM tackling a machine-gun trench on his own and thereby allowing his pinned-down company to resume the attack. But now the Argentine main body began a determined stand along Darwin Ridge, and, with limited fire support and no opportunity to outflank the position, plagued increasingly by Argentine artillery and worsening weather, 2 Para's attack stalled completely. What happened next is still the subject of as much conjecture as record. Soon after dawn (which was not until about ten o'clock) the commanding officer, 'H' Jones, frustrated by the stalemate, took himself and his small protection party forward to tackle an entrenched machine gun on a low spur commanding the left-hand approaches to the ridge. They managed to work forward along a little re-entrant until Argentine fire drove them to ground; then in an instant 'H' sprang up and charged the trench, firing his sub-machine gun as he went. He was cut down by fire from a supporting trench, and died before he could be evacuated to the casualty clearing station. For this action he was awarded the VC.* Some commentators have been critical of Jones's action: his job was command of the battalion – fighting the battle, not individual machine guns. Perhaps, however, in airborne units expected to land beyond the front line and then to fight surrounded, things are not so clear cut. Right or wrong, Jones was doing what his Para forebears at Arnhem had done.

*Three others were killed in this desperate action, including the adjutant.

And it was not without moral effect. 'A' Company, commanded by Major (later Major-General) Dair Farrar-Hockley,* had been repulsed three times, but towards noon they finally managed to clear the eastern end of the ridge, opening up the battle once again. By last light, and by dint of sheer determined fighting in which more gallantry medals would be won for a single battalion action than any since Korea, 2 Para had been able to close up to Goose Green.

At this point the acting commanding officer, the former second-in-command Major Chris Keeble, a deeply thoughtful man who had directed the battle with subtle skill after 'H' Jones's death, called on the Argentinians to surrender, sending two of their captured NCOs with a carefully worded – and psychologically cunning – document which deceived, flattered, and exploited the Argentine commander's pride and sense of duty. The following morning the garrison surrendered: 2 Para had forced the capitulation of more than 1,000 men and had killed a quarter of that number, in turn suffering 17 killed and 64 wounded. On his return to Argentina, the commanding officer, Colonel Italo Piaggi, was cashiered.

Writing up his thoughts as 2 Para's sister battalion tabbed across East Falkland (Marines 'yomp', Paras 'tab') towards their own bloody battle on Mount Longdon, Lieutenant-Colonel Hew Pike, hard hit by the death of his fellow commanding officer, reflected on the spirit that now animated his battalion (and had animated 2 Para at Goose Green). He had received a letter in the field (the army's postal services are ever unsung heroes in their morale-sustaining work) from a former soldier of his, reminiscing about their time 'up-country' in Aden in 1964, which contained the injunction to 'keep the old tradition up and give 'em some bloody hammer'. In his notes Pike writes,

---

*Son of General Sir Anthony – 'Para' – Farrar-Hockley of Imjin fame (see ch. 27).

Most soldiers now on these windswept mountains would only have been about two years old in 1964, and some would not even have been born for a battalion is composed mainly of very young men . . . [three of those who would be killed in action with 3 Para were not even eighteen]. But the 'old tradition' as Hughie Henderson had put it, was there all right. The same attitude of mind that had carried us deep into the Wadi Dhubsan in South Arabia would carry us across East Falkland. As the Battalion Commander, I knew this – and drew strength from it. And when I sometimes pondered how this attitude of mind could pass so silently, yet so powerfully, from one generation of men to another, there could only be one answer. It must be through their training. The attitude of far, fast and without question is bred by a certain approach to training, and by the consistency of that approach. It is an approach that generates high morale, confidence and success.*

A different approach to training, with common features nevertheless, would see the 2nd Scots Guards manage to overcome the same sort of determined opposition a fortnight later, on the final obstacle to victory – the bare ridge overlooking Stanley. Hew Pike's 3 Para had taken Mount Longdon on the night of 11 June, winning the other VC of the war (again awarded posthumously, to Serjeant Ian McKay), and two Royal Marine commandos (battalions) had taken Mounts Harriet and Two Sisters to the south. Two nights later, having passed through the captured heights, 5th Brigade took over and began what was to prove the culminating action of the war, the capture of the second chain of heights – of which Mount Tumbledown, the Guards' objective, was the most tenaciously defended – by troops of the 5th Marine Battalion.

The attack, against a well-entrenched enemy and with

* Hew Pike, *From the Front Line.*

limited artillery ammunition available, soon ran into trouble: the textbook ratio in the attack is three to one, but the situation on Tumbledown was not in the textbook, and in the culminating phase of the battle the ratio of attackers to defenders was just about evens. But one of the platoon commanders, Second Lieutenant James Stuart, not long out of Sandhurst and wondering what would now happen, spoke later of his realization that somehow all would be well: 'We were the Scots Guards. There was no way that we were going to be thrown off that mountain.' Or, as the regimental motto has it, *Nemo me impune lacessit* (No one provokes me with impunity).

One of the Argentinian platoon commanders recalled how

The British soldiers crept up on the platoon position just before midnight. Before long the platoon was completely surrounded and on the verge of being overrun so I decided that the 81 mm mortar platoon [under] Ruben Galluisi should fire on our platoon.* At that moment Argentinean mortar bombs landed in the middle of the position, we had no other choice. The British had to withdraw and we started swearing at them. That was how the British were driven out during that first attack. I had up to then lost five of my own men [out of about thirty]. The British eventually got up and started attacking again. It was now around three in the morning and we had been trading fire off and on for nearly three hours.†

'Left Flank' Company of the Scots Guards (the Guards name their companies in the old Marlburian fashion) at last

---

* Bringing down fire on your own position is the last desperate measure to hold off the attacker. All then depends on how well the trenches have been dug – and luck.

† Van der Bijl and Aldea, *5th Infantry Brigade in the Falklands*.

managed to get into position to make what with only two hours of darkness left would have to be the final attempt to take Tumbledown. It had taken a long time to get the artillery adjusted, but now the rounds were falling accurately on target. The company commander, Major John Kiszely, takes up the story:

I said to the platoon commander, Alasdair Mitchell, 'Look, the rounds are on target now, so put in a platoon attack on the first ridge two or three hundred yards ahead of us'. I thought when I told him that, 'This could be goodbye Alasdair Mitchell'. But they achieved it because the enemy's heads were down . . . We moved forward to join Mitchell's platoon which was three hundred metres up the hill. As I got to the ridge where they were busy finishing clearing the position, I saw [through a night-vision device] this next ridge about two to three hundred metres up, with no activity on it, because the Argentineans' heads were still down . . . I looked around and there was a platoon down to my right sorting out their objective so I realized that all I had to do was to get them going at right angles up the hill instead of carrying on down to the right. I also realized that only I could do that because I was the only one that could see the ridge. I started rather over-involving myself in the platoon's battle and got the ones nearest to me moving by shouting at them and grabbing them, saying 'Come on!' But only about a dozen heard me. In the meantime I had run onto the next ridge only to find myself totally alone because those who were coming on had either lost direction [not all of them had night-vision aids] or were going other places, or were kneeling down, shooting, keeping the enemy's heads down. But eventually in ones and twos they found me.

It was absolutely pitch black and we couldn't see anything. However we got to this next ridge and sorted out three or four sangars. Then again I looked up, and there was this other ridge about 200 metres ahead and I thought, 'We've done it once;

we'll do it again!' The same thing happened, except this time, of the twelve or so that were with me, one was shot dead, and another was shot in the chest and we had also taken some prisoners. So we had to leave somebody with them . . . After we had done this about three times and actually got up to the crest, there were only six people with me . . .

Just at that moment a machine gun opened up and three of the six who were with me were shot. Entirely, utterly my fault.

Three hundred years of fighting reputation had been at stake on Tumbledown, in the same way that the Parachute Regiment's driving principle, embedded in its intensive four decades' history – 'far, fast and without question' – had sustained its members in their gruelling march across East Falkland and in two bloody battles. The Guards' approach to duty, whether 'public duty' (the ceremonial in London) or on operations, was one of instant, unflinching obedience to orders, to 'the word of command'; and the words of command came from men who understood the ultimate demands of that duty. 'Our officers were determined never to yield, and the men were resolved to stand by them to the last,' Lord Saltoun had told the duke of Wellington after the Guards had held the chateau of Hougoumont against enormous odds. It was the spirit of Hougoumont and many a battle like it which through years of drill and training now got the Scots Guards to the top of the mountain – and the officers, 'determined never to yield': Major (now Lieutenant-General Sir) John Kiszely, who was awarded the MC for his bravery and leadership that night, himself shot two Argentinians and bayoneted a third. Reflecting on the Guards' determination that night, he ascribes it in no small part to 'the regimental system and the competitiveness of it', explaining how 2 Para's victory at Goose Green in the first major action of the war had set the standard for the rest to follow:

H Jones set a standard of bravery; it was like a gauntlet being thrown down to other officers. In a way, 2PARA threw down the gauntlet to the rest of us and said 'Match that'. Goose Green was a good tactical victory; hard fought, bloody good. 2PARA set an example to us and, I guess, to everybody else in the Falklands, even if they didn't actually say so at the time. But they felt it subconsciously.*

As Adam Smith had written two centuries before, 'In a long peace the generals, perhaps, may sometimes forget their skill; but, where a well-regulated standing army has been kept up, the soldiers seem never to forget their valour.'†

With the SAS already nosing about Port Stanley, and the dominating heights now in British hands, white flags started to appear in the town – much to the frustration of the Gurkhas, who were robbed of their battle of exploitation which was meant to open up the way into the town. Eleven and a half thousand Argentine prisoners 'went into the bag'; nearly 700 more had gone only to their graves, half of them to the cold waters of the Atlantic when the *Belgrano* was sunk.

But the retaking of the Falklands had cost the task force heavily too – in both men and materiel. Warships, merchantmen and aircraft, especially helicopters, had been lost in alarming, even critical, numbers. The human toll was 1,000 killed and wounded, of whom 123 dead were army. To many, it was 'Margaret Thatcher's War' – a none-too-approving epithet (her injunction to 'Rejoice!' when South Georgia had been retaken jarred especially with an increasingly defeatist minority in the country, although most people were indeed

* Conversations with author; and Max Arthur, *Above All, Courage.*
† *An Inquiry into the Nature and Causes of the Wealth of Nations.* See also ch. 5.

happy to rejoice).* But there was some justice in the attribution of victory, for besides taking the strategic gamble of fighting to recover the islands it had been her government that had rapidly restored service morale on coming to power in 1979 by, inter alia, substantial pay increases.† In a very real sense, Thatcher made the army what it was in the 1980s – and continued to give priority to defence spending until the fall of the Berlin Wall.

The Falklands War had been a very near-run thing indeed. Nevertheless it had been the finest thing, too; and in June 1982 the standing of the British army in the eyes of the nation and of the world had probably never stood higher.

---

* The archbishop of Canterbury's ambivalence over a national 'thanksgiving service' in St Paul's Cathedral was particularly puzzling – to say the least – to many; and more so because as Lieutenant Robert Runcie he had won an MC with the Scots Guards in Germany in 1945.

† The nadir had been the famous 'winter of discontent' of Prime Minister Jim Callaghan's Labour government, when during widespread strikes by public service workers and others, servicemen had kept the country going. At the same time, pay was so poor that many of its junior ranks with families were receiving income support.

31

If You Want Peace . . .

Europe, and its oilfields, 1945–2000

THE ROMAN MILITARY PHILOSOPHER VEGETIUS, WRITING AROUND the year AD 390 on the tendency to neglect defence spending, summarized his thoughts thus: 'Therefore, he who wishes peace should prepare for war; he who desires victory should carefully train his soldiers; he who wants favourable results should fight relying on skill, not on chance.' Fifteen centuries later, President Harry S. Truman would add the doleful rider: 'If you're not prepared to pay the price of peace, you'd better be prepared to pay the price of war.'

Not long after the Second World War the British army, for the first time in its history, began to organize, equip, train and locate itself in peacetime to fight a specific war – a war, in Ismay's memorable words on one of the purposes of NATO, 'to keep the Russians out' of Western Europe. The intensity of the effort to deter the 'Group of Soviet Forces Germany' from crossing the Inner German Border (IGB), as the heavily mined line between East and West Germany was known, and of being ready to fight if deterrence failed, was more than merely training, however: it was 'Cold War'.

At the Yalta Conference and at Potsdam in July 1945 the allies had agreed that Germany, like ancient Gaul under the Romans, would be divided into three parts – or 'zones' – with, in addition, a small French sector adjacent to the Franco-German border (with all the best vineyards). There were similar arrangements for Berlin, which was deep in the Russian zone (or the German Democratic Republic – 'East Germany' – a polity officially unrecognized by the West). The British zone consisted largely of the north German plain which the armies of George II had known so well – the *Länder* of Nordrhein-Westfalen, Niedersachsen and Schleswig-Holstein. And as the threat of Nazi resistance receded after the crushing defeat of Germany and the programme of 'de-Nazification', and that from the Red Army increased, the British Army of Occupation became the British Army of the Rhine and turned to look east. In 1952 BAOR became part of Northern Army Group (Northag) within NATO's new command structure along with Canadian, Dutch and Belgian troops. When West Germany became a sovereign state once more in 1955 – when it was no longer necessary, in Ismay's words again, 'to keep the Germans *down*' (indeed, their manpower was needed to keep the Russians *out*) – Northag was reinforced by a corps of the new West German army. And to complete Ismay's purpose – 'to keep the Americans *in*' – Northag's sister army group Centag (Central Army Group) was formed in the former American occupation zone with two US and two German corps.

BAOR numbered some 55,000 men for most of the Cold War, plus strong supporting elements of the RAF (a 'tactical air force') and 3,000 troops in the autonomous Berlin Infantry Brigade. With dependants and civilian staff – everything from schoolteachers to NAAFI storemen, welfare workers, broadcasters, journalists, the Salvation Army and all the other institutions of the home country – there were upwards of 200,000 people in 'Little Britain over the Rhine'. Comparatively long tours of duty became the norm: infantry regiments were stationed in Germany for five years, armoured regiments rather longer (one regiment, the Queen's Royal Irish Hussars, remained in Paderborn for nine years during the 1970s); and artillery, engineer and logistic regiments were based permanently in their garrison towns, with soldiers posted in and out individually. However, even living in their English-speaking world of barracks, married quarters, schools, hospitals, shops and forces' radio, the army took on a certain 'German' identity, just as those in India before the Second World War had taken on a distinctly Indian identity. It was not so much an embracing of German culture – from which BAOR largely kept itself apart, except from the legendary beer and *Bratwurst* – as a distinctive way of soldiering.

One of the reasons for this distinctiveness was the annual training regime. After 1945, training had taken place over German land almost as if the war had not ended, and after 1955 this military 'right to roam' continued, except that compensation was paid for damage. And since the north German plain was largely under crops, large-scale armoured manœuvres could take place after the harvest with relatively little damage. Apart from the Libyan desert (which, after Captain Gaddafi's coup in 1969, was no longer available for training) or the Canadian prairie (with its huge logistical demands), it was the only place where British armoured troops could range freely and realistically – with the added advantage of its being the very terrain on which they expected to fight. Brigade, divisional and even corps commanders and their staffs were able to practise their art with the essential element of Clausewitzian 'friction' included – something which paid off handsomely in the First Gulf War. As, too, did three decades of increasingly sophisticated equipment procurement stemming

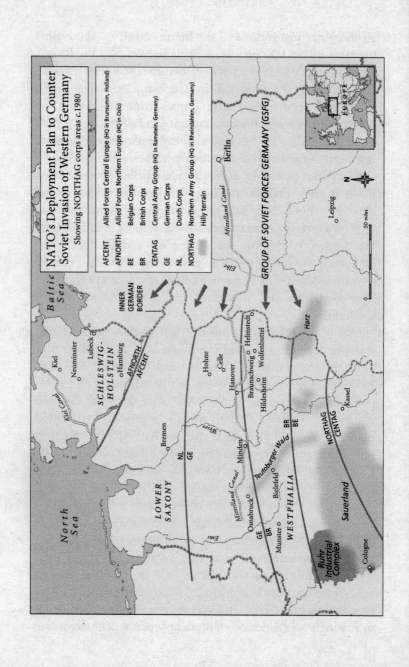

NATO's Deployment Plan to Counter
Soviet Invasion of Western Germany
Showing NORTHAG corps areas c.1980

AFCENT	Allied Forces Central Europe (HQ in Brunssum, Holland)
AFNORTH	Allied Forces Northern Europe (HQ in Oslo)
BE	Belgian Corps
BR	British Corps
CENTAG	Central Army Group (HQ in Ramstein, Germany)
GE	German Corps
NL	Dutch Corps
NORTHAG	Northern Army Group (HQ in Rheindahlen, Germany)
	Hilly terrain

GROUP OF SOVIET FORCES GERMANY (GSFG)

INNER
GERMAN
BORDER

Baltic
Sea

North
Sea

Kiel
Neumunster
Lubeck
Hamburg
Kiel Canal
SCHLESWIG-
HOLSTEIN
AFNORTH
AFCENT

Berlin

Leipzig

Mittelland Canal
Elbe

Bremen
Weser
LOWER
SAXONY
NL
GE
Mittelland Canal
Minden
Osnabruck
Ems
GE
BR
Munster
Bielefeld
Teutoburger Wald
WESTPHALIA
Ruhr
Industrial
Complex
Cologne
Sauerland

Hohne
Celle
Hanover
Hildesheim
Braunschweig
Wolfenbuttel
Helmstedt
Harz
Kassel
BR
BE
NORTHAG
CENTAG

N

50 miles

EUROPE

from BAOR's highly focused operational requirements – producing, notably, the Challenger tank and the Warrior infantry fighting vehicle, which though conceived on the north German plain at the height of the Cold War have twice battled with Iraqi forces in the desert.

Every year for the best part of four decades before the First Gulf War there were months of preliminary 'special to arm' training – firing practice for every rifle and machine gun, every tank, every artillery piece; mine-laying and bridging for the sappers, signals exercises for the headquarters and the like – preceding the annual autumn manœuvres, and always in an atmosphere of intense competition, rivalry and scrutiny. There never was a British army so primed for its task and at such a permanent peak of efficiency: the proximity of the enemy, in hailing distance across the IGB like some silent reincarnation of the Western Front, made it all both real and immediate. Indeed, the Russians monitored all radio traffic and spied on all exercises; and the troops in BAOR relished the thought of it. There were detailed deployment plans which were constantly studied, refined and rehearsed, and strict controls on manning to keep 1st (British) Corps at high readiness (no more than 10 per cent of a regiment's strength could be out of the country). Ninety per cent of armoured vehicles had to be available to move at four hours' notice, placing a constant and heavy strain on the repair and logistic chain. In the 'readiness war', the REME, RAOC and RCT were the principal warriors.* And in case the threat from across the IGB did not always seem imminent enough, there were NATO teams ready to swoop on a regiment in the middle of the night to test every detail of its fitness for the fight – from the calling in of married

---

* In 1965 the RASC had shed its non-transport functions to other corps and taken on the residual transport functions of the Royal Engineers to become the Royal Corps of Transport (RCT).

soldiers to the out-loading of ammunition. 'Satisfactory' and the commanding officer lived to prove himself another day; 'Unsatisfactory' and his promising career would be at an end.

All of this made for a remarkable cohesiveness: senior commanders knew their junior officers; soldiers knew their generals; *everybody* seemed to know everybody else (it was said that cavalry officers only married the sisters of other cavalry officers, since they were the only girls intrepid enough to leave London for a regimental dance on Lüneburg Heath in midwinter). Every weekend there were hunter trials, polo matches, sailing regattas – contests in every conceivable activity, which all added to the atmosphere of energy and competition. The army had learned how to live abroad in India and the Middle East, and it applied that experience in full measure to Germany, the only difference being that the natives were marriageable.

There was a negative side, however: a tendency to think in terms of set moves to defeat a specific part of the Soviet war machine on a particular piece of ground in a precise phase of a well thought-out plan. It was Montgomery's 'everything went to plan' taken to extremes that would in fact have dismayed him. The lessons of the Second World War in handling armour – boldly and flexibly – were also progressively abandoned because of the political requirement to hold ground as close to the IGB as possible, which resulted in a concept of positional defence not dissimilar to that of France in 1940. This state of affairs was rectified to an extent in 1984 when General Sir Nigel Bagnall became Commander Northern Army Group and persuaded the Germans to stop thinking about defending every inch of ground and embrace their tank warfare heritage instead. This in turn eased the general tactical constipation which was besetting the British army, and when Bagnall (who as an infantryman before transferring to the armoured corps had won an MC and bar

– a second MC – in Malaya) became CGS the following year,* he began a root-and-branch reform of the army's tactical doctrine – indeed, of the whole art of campaign planning. Bagnall had studied deeply the German experience on both eastern and western fronts in the Second World War, as well as Israeli tactics in the Six Day War of 1967 and the 1973 Yom Kippur War, and also contemporary American thinking about 'deep manœuvre'. By the end of the 1980s, in large measure as a result of Bagnall's single-mindedness, the British army had become a freer-thinking organization than it had been for decades, and the arrival into service of world-class equipment such as Challenger and Warrior – and, later, the Apache attack helicopter – gave wings to the new spirit of 'mission command' (what the Germans themselves had long called *Auftragstaktik*) in which a subordinate is given far greater operational latitude than in the previous regime of *Befehlstaktik* ('order command'), in which the commander supervises everything himself and imposes his view not only of what must be done but exactly how.

The fall of the Berlin Wall, and soon of the Soviet Union itself, came as a victory for the Cold Warriors of BAOR, but it was at the same time their *congé*, for the new Conservative government of John Major was determined to scavenge a hefty 'peace dividend' from the rubble. And so, as in the past after victory in war, the nation sought to reduce its army. But before the cuts could be made there would be a sort of curtain call for the victorious BAOR: Saddam Hussein, Iraq's pseudo-military dictator, had invaded Kuwait in August 1990, and the UN Security Council had authorized a US-led coalition to eject him.

Over the course of several months 1st Armoured Division,

---

* 'Imperial' was dropped from the title in 1964, there being very little left east of Suez.

consisting of two armoured brigades (including the 7th, the famed 'Desert Rats' of 8th Army) and one of artillery, moved by rail, sea and air from Germany to the Saudi desert to form the second largest of the coalition's components – with all its supporting elements, some 43,000 men (and women), and two and a half thousand armoured vehicles. The operation was named Granby after the hero of Warburg, the celebrated bald-headed commander of British troops in Germany in the Seven Years War.

After months of training, and a punishing air campaign against the entire Iraqi command and control system and its obsolescent army dug in in the desert, on 24 February 1991 US General Norman Schwarzkopf's force thrust across the Saudi–Iraq border just to the west of Kuwait (with US Marines and Saudis making straight for Kuwait City itself). After penetrating deep into Iraqi territory they turned east to attack the Republican Guard – the best of Saddam's formations – in the flank, with 1st Armoured Division protecting the right of this great turning movement.

The effect was dramatic: outgunned, outfought and totally outmanoeuvred, within forty-eight hours Iraqi troops were retreating from Kuwait. Schwarzkopf's forces pursued them to within 150 miles of Baghdad before being ordered to withdraw – perhaps one of the greatest lost opportunities of recent times (if, that is, the coalition could have got UN sanction to remove Saddam). One hundred hours after the ground campaign started, President George Bush declared a ceasefire. Iraqi casualties had been heavy; those of the coalition quite remarkably light.

Among the many lessons of the First Gulf War – in command and control, logistics, 'friendly fire' and the hundred and one other areas in which professional soldiers look to learn after any war – there was one 'geopolitical' truth that would have the profoundest effect on the shape of things to come: the only army in the world capable of operating with, as opposed to merely alongside, the US army in high-intensity war was the British. The French, who had withdrawn from the military structure of NATO decades before, were hopelessly 'national' in their equipment and doctrine, with little interoperability. The Germans were constitutionally – in both senses – unable to break loose from the *Heimat*, for the Hitler and then the Soviet years weighed as heavily on them as the Somme once had on the shoulders of British generals. And no other NATO ally possessed either the hardware or the confidence. Since the First Gulf War, the British army has thought of itself increasingly (without admitting it quite) as the US army's auxiliary, and its determination to maintain a capability for high-intensity operations of the Gulf War type has driven its restructuring and equipment programmes up to the present day. This vision was justified by the Iraq War of 2003 when again the British army was the only one capable of any integrated effort with the main US forces, although in the end its operation to take Basra and the oilfields was a discrete affair. To a very large degree, then, the British army of the early twenty-first century has been, and continues to be, shaped by its ever closer association with that of Uncle Sam.

But that close association – special military relationship, indeed – has not always encompassed the whole spectrum of conflict. After the débâcle of Vietnam and the unhappy experience of intervention in Somalia in 1992, the US army returned (mentally at least) to what it saw as its primary – indeed, some argued, its exclusive – function: defence of the Republic. In other words, 'warfighting'. The Pentagon saw 'operations other than war' not only as problematic and inconclusive (whereas the use of military force was meant to be *decisive*) but also as positively harmful to the warrior ethos. When the Balkans erupted in civil war with the collapse of the Republic of Yugoslavia in 1992, the US army was unwilling and probably doctrinally unable to join the hastily formed UN protection force (UNPROFOR) in Bosnia, for the Security Council mandate was for a peace-keeping operation only, not peace enforcement. Both the US and Britain took the view that enforcement would be tanta-mount to warfighting if either of the factions chose to resist – a relatively clear-cut operation therefore. But an inter-vention to deliver humanitarian aid while fighting continued between the factions, and with no mandate for the use of force beyond self-protection, was an altogether trickier business. The term 'peacekeeping' seemed wholly inappropriate, since there was quite evidently no peace to keep. The 'international community' decided it could not sit idly by, however; and so, despite the ambiguity, danger and double-dealing entailed, it chose to do the minimum: send peacekeepers anyway, and hope they would be able to do *something*.

As peacekeepers in Bosnia, the UNPROFOR contingents wore blue berets (or blue helmets when the situation required) and all vehicles were painted white, for peace-keepers were meant to derive their security from being recognized as UN troops. The white paint served, too, to concentrate the minds of politicians who from time to time demanded that the peacekeepers take on one or other of the warring factions when some atrocity aroused public opinion; for as the best-known of UNPROFOR's commanders General Sir Michael Rose (who had commanded the SAS in the Falklands) put it, to do so would be to make war, and 'you can't make war in white-painted vehicles'.

Into this apparently hopeless situation a British battalion, the Cheshire Regiment, stepped bravely at the outset of the mandate and began making operational doctrine on the hoof. By combining techniques learned over long years in Northern Ireland with battle skills honed in BAOR – whence they had come (in their Warriors) – they were able to do far more than had been thought possible. Other battalions which in turn relieved them stretched the boundaries of what became known as 'wider peacekeeping' even further and gave a lead to other nations who were gingerly putting on the same blue berets. Nevertheless 'peacekeeping' was a dangerous business: over 300 UN troops were killed in Bosnia – a higher casualty rate than in the Gulf War. And from time to time experience at the 'sharp end' was no different from the fighting in any number of 'small wars' east of Suez, or indeed in the South Atlantic. In March 1994 the Duke of Wellington's Regiment was one of the first units to enter the besieged Bosnian Muslim enclave of Goražde, nominally a UN 'safe haven' but under strong pressure from the Bosnian Serbs, who were trying to over-run it and hotly contesting the entry of UN troops. One of the Dukes' patrols fired 2,000 rounds of ammunition in a fifteen-minute battle with a Serb patrol in which a Dukes soldier and seven Bosnian Serbs were killed. And then the following month a young corporal, Wayne Mills, won the Conspicuous Gallantry Cross, second only to the VC. His patrol, coming under heavy fire from a Bosnian Serb patrol, returned fire, killing two, and then withdrew – as their orders required – but the Serbs followed up, still firing. As the Dukes patrol reached the edge of a clearing Mills, with limited ammunition and knowing that his orders were to fire only in self-defence, ordered his men to withdraw while he remained behind to cover them across the open ground. Not until he had killed half the patrol and its leader did the Serbs break off the attack – with Mills himself down to his last few rounds.

In late 1995 a political settlement, brokered at Dayton, Ohio, brought a fragile ceasefire between the Bosnian Serbs and Bosnian Muslims. It was to be cemented by an 'implementation force' (IFOR) based on the new NATO Allied Command Europe Rapid Reaction Corps (ARRC) with an enforcement mandate from the Security Council.

The ARRC was British-led with a predominantly British staff, and its key fighting elements were British. And there was indeed some fighting to enforce the agreement, but although intense at times it was local and renegade, and by degrees the ARRC was able to withdraw. In its place an ad hoc 'stabilization force' (SFOR) was able to oversee the ten-year process of democratization which led in December 2005 to the EU's taking responsibility for the residual military operation.

It would not be the ARRC's last encounter with the Balkans, however, for there was, as Washington called it, 'unfinished business' – unfinished from Dayton, but also from the fall of the Berlin Wall, from the Second World War, from the First World War, and even from the Congress of Vienna. After the Dayton 'humiliation' the Serbian president Slobodan Milosevic, who had backed the Bosnian Serbs both militarily and politically, turned to forcibly expelling the Muslims from the semi-autonomous region of Kosovo. In 1999, after a prolonged strategic bombing campaign by NATO air forces, he agreed to withdraw the Serbian army and paramilitaries from Kosovo and accept instead a NATO force to stabilize the ethnic conflict which had been largely of his making. Again, the commander was British – Lieutenant-General (later General) Sir Mike Jackson, the future CGS, whose famous disagreement with US General Wesley Clark over the Russian *coup de main* at Pristina airport won him plaudits on both sides of the Atlantic.* The occupation of Kosovo was, indeed, a very British combination of military diplomacy and force, both implied and explicit: at its peak there were nearly 20,000 British troops

* The Russians had sent a light-armour column from Bosnia through Serbia to sit on the runway at Pristina, announcing that they had come to join KFOR – a brilliant and well-executed tactical movement of the greatest strategic impact which could only be admired (if secretly) by most other military observers.

committed to 'KFOR'. But Kosovo has remained 'unfinished', and many soldiers wondered at the time how the government, by now headed by Tony Blair, could continue taking the 'peace dividend', further reducing the army's strength, in so patently uncertain a post-Cold-War world.

What emerged from the Balkans, however, was a renewal of military pragmatism in the British army, and a confidence to engage in operations where there is both a threat from the enemy and a battle for the support of the civil community – 'war among the people', as the former UNPROFOR commander and leader of 1st Armoured Division in the Gulf War, General Sir Rupert Smith, calls it. Dag Hammarskjöld, the UN's second and greatest Secretary-General, once said that peacekeeping was 'not a job for soldiers but only soldiers can do it'. The experience of Bosnia, especially before Dayton, and of Kosovo, underscored the word *soldier*: the peacekeeper had to be a man thoroughly expert in battle drills and equipped to fight; he had to be able to *throw* a punch if the promise of *pulling* his punches was to have any effect; he would otherwise be merely brushed aside by bruisers who did not fight by the rules.

At the turn of the twenty-first century, therefore, the British army, smaller than it had been since before Bonaparte's day, was a master of 'operations other than war'. Some officers were echoing the Americans' concern, however, that a constant diet of 'minimum force' and seeking consent was blunting the army's fighting edge. But would there ever be a need again for the all-arms, high-intensity battle of old?

The Pentagon was certain there would.

32

The Army Falters?

Iraq and Afghanistan, 2001

THE EXACT MOMENTS AT WHICH THE TWO HIJACKED PLANES crashed into the 'Twin Towers' on 11 September 2001 – '9/11' – are now erased from the mainstream media. Taste and decency demand it, and also reason: repetition risks dulling the senses. Yet although the choice of weapon and target far exceeded the evil of the surprise attack on Pearl Harbor (a day that in Roosevelt's words would 'live in infamy'), the attacks on the World Trade Center and the Pentagon, and the heroically thwarted attack on the White House, shared with Pearl Harbor the shock of sudden vulnerability – and in consequence provoked the same seismic shift of strategy. If Kipling had been alive he might have been moved to write something along the lines of 'Edgehill Fight', noting that the twisted outline of blackened steel 'changes the world today'.*

*This was not universally appreciated at the time – a dangerous disconnect between British and American strategic thinking which in part continues. In the immediate aftermath of 9/11 the author recalls listening in disbelief to an

Forty-eight hours after the attacks a defiant President George W. Bush stood amid the rubble at 'Ground Zero' (significantly, the term used of the point of impact of a nuclear weapon), his hand on the shoulder of one of the New York firemen who had borne the brunt of the courageous rescue effort, and spoke to the crowd through a microphone. When a distant voice called out, 'George, we can't hear you!' Bush smiled and answered loudly: 'I can hear you! I can hear you, the rest of the world can hear you, and the people who knocked these buildings down will hear all of us soon!'

A week later, addressing Congress, Bush expanded on his extemporary remark:

> Americans are asking, 'How will we fight and win this war?' We will direct every resource at our command – every means of diplomacy, every tool of intelligence, every instrument of law enforcement, every financial influence, and every necessary weapon of war – to the destruction and to the defeat of the global terror network. Now, this war will not be like the war against Iraq a decade ago, with a decisive liberation of territory and a swift conclusion. It will not look like the air war above Kosovo two years ago, where no ground troops were used and not a single American was lost in combat. Our response involves far more than instant retaliation and isolated strikes. Americans should not expect one battle, but a lengthy campaign unlike any other we have ever seen. It may include dramatic strikes visible on TV and covert operations secret even in success. We will starve terrorists of funding, turn them one against another, drive them from place to place until there is no refuge or no rest . . .

FCO official (today a very senior official) disputing that anything fundamental had changed.

Sitting close by the president in Congress was Tony Blair, Britain's prime minister, who had flown to Washington in a show of solidarity and to offer support (though tellingly he had not taken with him the chief of the defence staff). Bush spoke of several countries which had shown their support, then added: 'America has no truer friend than Great Britain. Once again, we are joined together in a great cause. I'm so honoured the British prime minister has crossed an ocean to show his unity with America. Thank you for coming, friend.'

Going to Washington was, of course, what Churchill had done in the aftermath of Pearl Harbor. Blair's ocean crossing would likewise define the British military response, though he played his hand with considerably less acumen than Churchill did in the Second World War.

Within a month of the attacks the United States had begun its military counter-offensive, the 'War on Terror'. Afghanistan under the Taleban had been a haven and training ground for Al-Qaeda, which had carried out the 9/11 attacks as well as bombings and assassinations over the previous decade in East Africa, the Middle East and western Asia. The US campaign to oust the Taleban from Afghanistan (Operation Enduring Freedom), endowed with legitimacy under Article 5 of the North Atlantic Charter,* was a brilliantly conceived and well-executed application of high-tech intelligence, Special Forces (including British), air power (including the RAF), and the armies of the various anti-Taleban warlords known officially as the United Islamic Front for the Salvation of Afghanistan, and commonly as the Northern Alliance.

---

* Article 5 of the North Atlantic Charter states that 'an armed attack against one or more of [the allies] in Europe or North America shall be considered an attack against them all and ... [each] will assist the Party or Parties so attacked by taking forthwith, individually and in concert with the other Parties, such action as it deems necessary, including the use of armed force, to restore and maintain the security of the North Atlantic area.'

By the middle of November, Kabul was in the hands of the Americans and their allies, and by the end of December it looked as if all that was left was merely mopping up. Indeed Sir John Keegan, writing in the *Daily Telegraph*, compared the apparent victory with one of the great turning points in the military history of the British Empire: 'The collapse of Taliban resistance in northern Afghanistan and the fall of Kabul may stand as one of the most remarkable reversals of military fortune since Kitchener's victory at Omdurman in the Sudan in 1898.'

But despite the ferocious battles in the mountains of south-east Afghanistan, in which lay the cave complex of Tora Bora to which many Taleban had bolted, the key Al-Qaeda leaders could not be found. Those Taleban who had not been killed in last stands at Tora Bora fled across the border into Pakistan.

US and British forces and their Afghan allies now began to consolidate. A *loya jirga* (grand council) of tribal leaders and former exiles, in effect an interim Afghan government, was established in Kabul under Hamid Karzai, and military attention turned to civil projects to win over the hearts and minds of those Afghans who were indifferent to who ruled in Kabul. The Taleban had not given up, however, and both American and British troops were in action throughout 2002, although far enough from Kabul as to reinforce the general impression that peace prevailed, and democratic and economic progress was being made. British efforts were in the main concentrated on 'provincial reconstruction', in which the army (with around 1,500 troops) was – in broad terms – meant to provide conditions of security within which the civilian agencies could operate. However, both London and Washington increasingly took their military eyes off the ball, while the civil agencies seemed unable or unwilling to play the game at all. The experienced British major-general in charge of the multinational 'security assistance force', John McColl (who would later become NATO's deputy supreme allied commander Europe), was replaced by a newly promoted colonel to head the British contingent. It all looked like the early symptoms of 'mission-accomplished' syndrome.*

The causes of the Iraq War of 2003, or the Second Gulf War as it is sometimes called (with either a conscious or an unconscious nod to continuity – or 'unfinished business'), were several, ranked differently depending on viewpoint: Iraq's alleged possession of weapons of mass destruction (WMD) which posed an imminent threat to US security and that of their allies in the 'War on Terror'; Saddam Hussein's harbouring and supporting Al-Qaeda in the 'nexus of terrorism'; his financial support for the families of Palestinian suicide bombers; Iraqi human rights abuses, not least on the Kurds of Northern Iraq; security of the country's oil reserves as well as wider regional security fears; and the 'neocon' doctrine of accelerating the post-Communist glide into worldwide liberal democracy, including pre-emption where necessary.† So motives might have been mixed; but mixed motives are not necessarily bad motives, and collectively the concerns clearly amounted in Washington to a strong *casus belli*.

Tony Blair, in another instinctive move that some believe

---

* So impressed with McColl was President Karzai, however, that he is reputed to have made enticing offers for him to pay back the Queen's shilling and take his instead. Blair later made McColl his special envoy to Afghanistan, and when Karzai turned down Paddy Ashdown as the UN nominee as high representative in Kabul, he asked again for McColl (by then DSACEUR) instead. Here were shades of the North-West Frontier and the 'Great Game'.
† The 'Bush Doctrine' of pre-emptive war was laid out in the National Security Council text *National Security Strategy of the United States* (September 2002): 'We must deter and defend against the threat before it is unleashed . . . even if uncertainty remains as to the time and place of the enemy's attack . . . The United States will, if necessary, act pre-emptively.'

to have been strategically correct for the 'special relationship' ('my ally right or wrong'), immediately aligned himself with President Bush in the rhetoric of the *jus ad bellum*, as he had in the immediate aftermath of 9/11. Not only was this Britain's *only* possible geopolitical option, the apologists argue, Blair also believed that by nailing his colours to Bush's mast he would be able to influence events before, during and after the war. Only the long view of history can judge Blair's strategic wisdom, but his 'tactics' proved wholly wishful: Britain had no meaningful influence in the planning or conduct of the war, nor any in the concept for post-conflict nation-building. Indeed, there seems to have been precious little planning for what would happen after Saddam was toppled. General Sir Richard Dannatt, former chief of the general staff, has put it simply: 'I think history will show that the planning for what happened after the initial successful war fighting phase was poor, probably based more on optimism than sound planning.'

The war itself was peremptory in its execution, however, confounding many a pundit who had predicted Armageddon.* In only twenty-one days, a US force of nearly 250,000 forged its way to Baghdad, the Iraqi army of almost twice that size, including the famed Republican Guard with its capable Russian-made tanks, disintegrating before them. Meanwhile, the British contingent – three brigades, including 3rd (Commando) Brigade Royal Marines, under Headquarters 1st Armoured Division – took the key port of Basra and secured the oilfields of the south-east.

'High morale and self-confidence describe the mood of

---

* Simon Jenkins, for example, former editor of *The Times*, in an opinion piece of breathtaking unawareness – headlined 'Baghdad will be near impossible to conquer' – wrote: 'An astonishing event is about to happen. For the first time in modern history a city with the population of London is preparing to resist assault from a land army. The outcome of such a struggle is wholly imponderable.'

the 1st (UK) Armoured Division as it deployed for the second time in little over a decade to the head of the Gulf in 2003,' writes John Keegan in his concise but astute summary of operations, *The Iraq War*:

> The Gulf is one of the British army's historic campaigning areas. It had fought and won an eventually victorious, if difficult, campaign there in the First World War. It had put down a pro-Nazi rising in Iraq in 1941. It had fought again victoriously in 1991, beside its American comrades-in-arms, in whom it had confidence. It expected that the new campaign would have the same outcome.

British officers were indeed conscious of the historic dimension. In his now famous pre-invasion pep talk to his battalion, the Royal Irish, Lieutenant-Colonel Tim Collins invoked the words of fellow Irishman W. B. Yeats: 'Iraq is steeped in history. It is the site of the Garden of Eden, of the Great Flood and the birthplace of Abraham. Tread lightly there.'*

Nor was the divisional commander, Major-General Robin Brims, a deeply thoughtful bachelor Wykehamist, in any hurry. Basra is a large city; to have provoked an urban battle with Iraqi troops who would not yet have known the extent of the collapse of their comrades on the road to Baghdad would have been to risk increasing the body count of both his own troops and Iraqi civilians. Instead Brims adapted the techniques of intelligence-gathering and street control learned in Northern Ireland to play a clever and ultimately successful game of cat and mouse. 'Our snipers work in pairs,' explained Major Ben Farrell of the Irish Guards, 'infiltrating the enemy's territory to give us very good

---

* The reference is to Yeats's poem 'He Wishes for the Cloths of Heaven': 'Tread softly because you tread on my dreams'.

observation of what is going on inside Basra and to shoot the enemy as well when the opportunity arises . . . They don't kill large numbers but the psychological effect and the denial of freedom of movement to the enemy is vast.'

The Iraqi army in Basra was still active, however, and tried to provoke firefights in several sorties. But in this the advantage lay with the 1st Armoured Division, whose Challenger II tanks and Warrior armoured-infantry fighting vehicles were superior to anything the Iraqis could pit against them. On the night of 26–7 March, a squadron of Iraqi tanks fatally over-reached itself in one of these sorties, and was caught in open country at dawn by a squadron of the Royal Scots Dragoon Guards (who had also fought in the First Gulf War). Fourteen Soviet-made T55 tanks were destroyed in less than a minute.

By 8 April, the day before Baghdad fell to the Americans, Basra was in British hands, and the soldiers of 1st Armoured Division were being welcomed by the largely Shia Muslim inhabitants of the city, who rejoiced in the overthrow of Saddam's Ba'athist, largely Sunni, regime. The following day, the troops took off their helmets and put on regimental headdress instead: they would now patrol the city in bonnets, berets and caubeens as if it were Belfast. And if the wretched Basrawis were too short of food to give them tea and biscuits as they patrolled, as the residents of both Catholic and Protestant areas of Belfast had in the early days, then at least the soldiers could hand out sweets and chocolate to Iraqi children, and return the smiles of the shopkeepers. It had been a famous and clever victory.

So what went wrong? First, there was a collapse of law and order. The Iraqi army had melted away, but the US secretary of defense, Donald Rumsfeld, to whom President Bush had given the lead in the occupation strategy, had already decided on their complete disbandment. The Iraqi police, perhaps if only in fearful bewilderment, also quit their posts in large numbers. In the inevitable eruption of lawlessness neither the US nor the British – both of whom had rapidly drawn down their force numbers after the toppling of Saddam – had enough troops on the ground to restore order. Relations with Iraqi civilians turned sour as the coalition, which had become a very multinational force indeed for the 'humanitarian phase', struggled – at times heavy-handedly and at others provocatively weakly – to contain the disparate violence. Some of the excesses were more than just the re-action of the moment. One such incident, though nothing like as bad as some committed in the US sector, was the death in custody of a Basrawi (Baha Musa) in September which led to the unprecedented court martial of the commanding officer and several others of the Queen's Lancashire Regiment (QLR) – a severe blow to the army's self-esteem. Indeed, during this time in Basra there were breakdowns in the chain of command that were – if nothing like in scale and detail – akin to those in the aftermath of the storming of Badajoz in 1812, when the duke of Wellington's assault had not gone to plan and casualties among the officers had been high. In Basra, however, the crucial point was not that the officers were dead but that the troops were too thinly spread to contain the trouble. Sir Mike Jackson's response as CGS – a forceful and very public message to the chain of command – was the equivalent of Wellington's putting up the gallows in the main square.

From this eruption of lawlessness developed something more sinister, however: insurgency, and a war across the whole of Iraq between the militias, which the looted arsenals of the Iraqi army provided with weapons, ammunition and explosives. In the US sector, in the predominantly Sunni areas, the insurgency raged early and was met by an equally aggressive – indeed, violent – response. The 'downwash' into Basra and surrounding provinces took the already hard-pressed British-led Multi-National Division South-East (MND(SE)) by surprise, and they tried at first to contain the situation by the 'soft-hat' approach of Northern Ireland, rapidly losing control of the city (if control there had ever been) as casualties mounted.

Throughout 2004 the insurgency – or insurgencies – intensified across Iraq, not least in the British-led sector. Improvised explosive devices (IEDs) began taking their toll of men and vehicles; mortars pounded the bases, many of which had not yet been 'hardened'; and the insurgents pressed home attacks on patrols with machine guns and rocket-propelled grenades. It was in ambushes in Al Amarah, 100 miles north of Basra, that Private Johnson Beharry, a Grenadian-born soldier of the Princess of Wales's Royal Regiment, won the VC for valour in saving the lives of his comrades.

On 1 May 2004, Beharry's company had been ordered to replenish an isolated outpost in the centre of the city (Beharry himself was the driver of his platoon commander's Warrior). As they entered the city in the early hours they learned that a foot patrol was pinned down, and so drove at once to its relief. Beharry's Warrior was hit by multiple rocket-propelled grenades (RPGs) as they neared the patrol, the explosion knocking out both the platoon commander and the gunner in the turret, as well as the communications, and wounding several riflemen in the back of the vehicle. Beharry tried to drive on through the ambush, but as he came up to a barricade across the road the Warrior was hit again by RPG fire from insurgents in the adjacent alleyways and on the rooftops, and caught fire.

Beharry, now both driver and effectively commander, opened his hatch cover to let out the smoke and to work out what to do next. Without communications, and with five more Warriors hemmed in behind him, he decided to try to crash his way through the barricade, knowing it would probably be mined.

As he drove the 25 tonnes of aluminium armour at the obstruction another RPG hit them, further wounding the gunner and destroying Beharry's armoured periscope, forcing him to drive through the remainder of the ambushed route – a full mile – with his hatch open and his head exposed. With the rest of the Warriors following him in the bid to break clear, Beharry's own vehicle was struck repeatedly by RPGs and small-arms fire, one bullet penetrating his helmet and lodging inside.

Eventually they reached the outpost that they were meant to be replenishing, only to find that it too was under fire. Beharry's medal citation takes up the story:

Once he had brought his vehicle to a halt outside, without thought for his own personal safety, he climbed onto the turret of the still-burning vehicle and, seemingly oblivious to the incoming enemy small arms fire, manhandled his wounded platoon commander out of the turret, off the vehicle and to the safety of a nearby Warrior. He then returned once again to his vehicle and again mounted the exposed turret to lift out the gunner and move him to a position of safety. Exposing himself yet again to enemy fire he returned to the rear of the burning vehicle to lead the disorientated and shocked riflemen and casualties to safety. Remounting his burning vehicle for the third time, he drove it through a complex chicane and into the security of the defended perimeter of the outpost, thus denying it to the enemy. Only at this stage did Beharry pull the fire extinguisher handles, immobilising the engine of the vehicle, dismount and then move himself into the relative safety of the back of another Warrior. Once inside Beharry collapsed from the sheer physical and mental exhaustion of his efforts and was subsequently himself evacuated.

This was courage and devotion to duty of a high order. Just six weeks later, having returned to duty after medical treatment, Beharry was in another ambush in which he repeated his earlier valour and presence of mind, though on this occasion he was seriously wounded. This action, together with the first, gained him the VC – the British army's first in twenty years.

Beharry was not the only one in his battalion to be honoured for bravery in their six-month deployment. Conspicuous Gallantry Crosses were awarded to two serjeants; there were seven MCs* and no fewer than sixteen mentions in despatches, as well as DSOs for the commanding officer and one of the company commanders – a truly impressive list. And the Princess of Wales's Regiment are by no means élite or even 'fashionable' infantry (though their antecedent regiments can number fifty-seven VCs between them): during this time other battalions were having just as much of a fight and winning similar laurels. At that time in Iraq, uncommon gallantry was commonplace.

Despite the bravery and hard fighting, however, the situation across the country had become critical. The Americans, facing both the Sunni insurgency and Al-Qaeda fighters who had been pouring into the country for a year and more, were going increasingly 'kinetic' – shorthand for relying on firepower – and several nations contributing to the British-led MND(SE) were keen to pull out their troops. The war had never been popular in Britain, and as casualties mounted senior officers had real fears that irreparable damage would be done to morale. The army had in any case been operating at full stretch for many years – overstretch, in the opinion of many† – while

---

* The Military Medal having been discontinued, the MC is now open to all ranks.
† The army's so-called 'harmony guidelines' are that individual soldiers should not exceed 415 days of separated service in any period of 30 months, and regiments' tour intervals should be no less than 24 months. Statistics which suggested that most of the army was operating within these guidelines often obscured the reality of disruption at unit level, however, with regiments 'raided' for manpower to bring another up to strength for deployment. And since some parts of the army were not available for deployment – the training and support organizations, for example, and certain garrisons – repeat tours for key types of unit were becoming more frequent.

at the same time being progressively underfunded, and it was beginning to show. Retention rates were falling, with some of the best and most experienced officers and NCOs among those voting with their feet; regiments were simply going back on operations too quickly for their soldiers to be properly rested and retrained in between. And then towards the end of 2005 the government announced a significant increase in troop levels in Afghanistan, specifically Helmand province. The additional troops were meant to boost civil reconstruction, but within a year they were engaged in some of the heaviest fighting the army had seen since Korea.

By 2006, therefore, the British imperative was to reduce numbers in Iraq, where it was judged that there was only marginal operational advantage to be gained compared with Helmand. Efforts to train the new Iraqi army so that they could take the place of British troops were redoubled. The strategy was known as 'transition, with security'. But this was always going to be a difficult calculus. Judging the tactical capability of the Iraqi troops was relatively straightforward; assessing the conditions on the ground, however, the other element in whether the Iraqis were truly ready to take over operations, was a finer judgement, and one requiring the greatest integrity in dealing with Whitehall's pressure to keep up the momentum of withdrawal. In addition, the growing insurgency threatened the transition strategy: would the Iraqi army (and police) be able to cope with the accelerating violence?

The Americans faced a similar predicament in the north, but their response in 2007 was altogether different. At the instigation of three or four visionary generals, of whom David Petraeus and the retired former vice-chief of army staff Jack Keane are the best known, the US army wrote a new doctrine of counter-insurgency which drew deeply on the British experience in Malaya and elsewhere. Iraq-bound troops, and even those in theatre, were then retrained, and together with a huge increase in numbers (up to 20,000 at peak) the new doctrine was applied. The aim of the 'surge', as it became known, was to damp down violence in Baghdad and other centres of the insurgency to give the Iraqi army and police the chance they needed to take control of the streets. Some senior British officers argued for a similar surge in the south-east, not least because they perceived a demoralization in their troops, who were more and more on the 'back foot'. The SAS, for example, were reporting that British battalions had lost their offensive edge, and several senior officers were concerned that although there was no want of courage in the ranks, it was increasingly a reactive courage – which the medal citations seemed to bear out.

What had demoralized these troops was the residual 'Northern Ireland approach' – 'counting each round', and afterwards the seemingly pedantic after-action investigations by the military police and the lawyers – which had proved inadequate in the face of violent rebellion. And none of it was helped by the deep unpopularity of the war at home: at heart every soldier knew not only that the invasion had been of questionable legitimacy at the very least, but also, and more importantly, that the subsequent campaign had been botched.* The strategy of 'transition, with security' was beginning to look like 'cut and run'. Sir Richard Dannatt himself seemed to be saying as much when, in an interview shortly after becoming CGS, he hoped we would 'get ourselves out sometime soon because our presence exacerbates the security problems', a remark which dismayed many of those on the ground.

For their part, the Americans were critical of what they perceived as a British unwillingness to grip security in Basra.

---

* At the time of the final handover to the Iraqis, in April 2009, several soldiers were observed to be wearing tee-shirts saying 'Not the end of an era, the end of an error'.

To some extent this was a reaction to British criticism of their heavy-handedness in the early days by officers certain of their better understanding of counter-insurgency, but it came also from their exasperated conviction that the surge was having tangible effect in the north and that the continuing lawlessness in the south was threatening to undermine it. London was nevertheless wedded to its withdrawal programme, the more so because of the violence of the Taleban resurgence in Helmand, which threatened to draw in a vastly greater number of men than had been expected. And it suited the new prime minister, Gordon Brown, with Blair's Iraqi albatross hanging heavily around his neck, to announce progressive troop reductions – which he took every opportunity to do, sometimes counting the same troops twice.

Against this background, the least propitious since Suez, and embroiled in possibly the most unpopular war in the army's history, commanders had a hard job of it in Basra. In September 2006 the new commander of MND(SE), Major-General (now General Sir, Deputy SACEUR) Richard Shirreff, a forceful cavalryman who like General David Petraeus was deeply read in his operational art, launched Operation Sinbad, in large part to restore the offensive spirit at battalion level, but principally to root out ineptness and corruption in the Basrawi police and to get a greater number of Iraqi troops on to the streets. Sinbad, which lasted for six months, went a good way to restoring fighting morale, but in the end it failed to achieve what it might have done because of sheer lack of 'boots on the ground'. For unlike Petraeus, Shirreff did not have the ear of his commander-in-chief.*

---

* The command of operations is vested in the chief of the defence staff (at the time Air Chief Marshal Sir Graham 'Jock' Stirrup), but operations are run by the Permanent Joint (tri-service) Headquarters (PJHQ) at Stanmore, the chief of which is a three-star soldier, sailor or airman. Neither the CGS nor CinC (Land) has any direct operational responsibility: their influence is purely 'moral', though in practice a consensus

That the British public evidently had little stomach for the fight was also exercising the CGS, who was increasingly dismayed at the indifference – and sometimes even hostility – shown towards the troops in Iraq (and Afghanistan), and the consequent effect on morale. He spoke of 'the growing gulf between the Army and the nation', contrasting it with the support that US troops were shown both politically and by the nation as a whole, and called for homecoming parades for regiments returning to Britain from operations. This was to prove more successful than many in the MoD expected.

Garrison towns, not just the regiments' own recruiting areas, were soon turning out to applaud the marching ranks of desert combat dress. Even the least military of places were suddenly seized with a *Sun*-like feeling for 'our boys'. The Norfolk market town of East Dereham asked the Light Dragoons based at a lonely disused airfield 5 miles away – the only troops for 50 miles – to march through its streets when they returned from Afghanistan, though all its soldiers were from either Yorkshire or the North-East. Being cavalry, and perhaps because it was late November, the dragoons put on their best uniform rather than 'desert combats'. The local Iceni Brewery presented every soldier with a bottle of specially brewed beer named 'Breckland Hero', and two dozen public houses across the county sold it and made donations from the profits to the Army Benevolent Fund and the 'Help for Heroes' campaign.* Tough old serjeants who thought they had seen it all had lumps in their throats during the march. Similar events were taking place up and

must be reached among the service chiefs. PJHQ was established in the wake of the First Gulf War to professionalize the command arrangements for operations which hitherto had been ad hoc, based on what was perceived as the most appropriate of the three separate service HQs to take charge.

* Indeed, there has been a surge in contributions to service charities generally these past three years.

down the country as people tried to express support – sympathy – for the troops while distancing themselves from the war(s) itself. And in Wiltshire the little town of Wootton Bassett began what soon became the sad custom of lining the route for the hearses which take the repatriated bodies of servicemen from nearby RAF Lyneham on the first leg of their home journey to their final resting places. Dannatt had succeeded in decoupling esteem for the soldier from the political unpopularity of the war.

Not long afterwards, in December 2007, British troops pulled out of Basra altogether and consolidated at Basra airport in what was called an 'overwatch' role. This withdrawal remains hugely controversial. There was in it an element of 'tough love', judging that the Iraqis were as ready as they would ever be to take on operational responsibility for the city, with much reliance (including much hopeful thinking) placed on the respected Lieutenant-General Mohan al-Furayji, who commanded the two Iraqi army divisions in the south-east, though one of these was still in embryonic form. But in truth it was pragmatism: certain commanders had worked out that, in the absence of the resources, especially troop numbers, needed to do the job properly – a straight reflection of political–strategic will – the only sensible alternative left was in effect to 'retreat' into the firm base at the airport, having signed a deal with the devil: Muqtada al-Sadr, leader of the main Shia militia, the so-called Mahdi Army. The duke of Wellington used to say that the test of generalship was to know when to retreat and to have the courage to do it; but this was a retreat not of the soldiers' making, and the imputations of defeat were and remain deeply resented.

Despite the militias' ceasefire, which had allowed the British to quit the city without firing a shot, violence and general lawlessness once again flared up, in part because the Mahdi Army had been infiltrated by Iranian-trained saboteurs. General Mohan had planned to launch an operation with coalition forces to defeat the Mahdi Army and the criminal gangs working under its patronage once his second division was ready, and before the provincial elections in October. But in late March the prime minister Nouri al-Maliki, buoyed up by the reduction of violence in Baghdad and elsewhere brought about by the surge (among other things), and dismayed by the starkly contrasting situation in Basra, told Petraeus that he intended taking personal command in the city and launching an immediate offensive against the Mahdi Army with Iraqi troops brought from the north.

Petraeus urged caution, but Operation Charge of the Knights went ahead two days later, taking the British as much by surprise as it did the militias (indeed, MND(SE)'s commander was out of the country). Within days, however, British troops were able to get themselves back into the loop and lend support, although Whitehall spin suggesting that everything had gone according to plan fooled no one. The then defence secretary Des Browne made an airy statement in Parliament, announcing that 'We and our coalition partners are providing support to the Iraqis in line with our commitments under overwatch and in accordance with our usual rules of engagement. Requests for support are being made through the coalition, and I can confirm that UK forces have continued to meet all their obligations as part of the multinational corps.' He was at once chided by the opposition whip and former soldier Crispin Blunt for hiding behind senior officers' words, which to some extent have to be couched in such terms as to maintain morale.

Although the operation was at times chaotic – many Iraqi troops mutinied or deserted – it did succeed in clearing the Mahdi Army from significant areas of Basra, allowing the police and Iraqi army to consolidate control in the following months. By the end of the year, the city was as 'peaceful' as Baghdad was increasingly becoming. Even the *Guardian*, impaled uncomfortably on the dilemma of having supported the war while being viscerally unsympathetic to all things military, opined in December that 'The Basra which Britain will leave behind is a city that is rubbish-strewn, divided and impoverished. But it is safer – at least for now – than it has been throughout most of the army's five-year occupation.' British troops formally withdrew from operational responsibilities in Iraq on 30 April 2009, leaving only small specialist teams – trainers principally – behind. Significantly, the Iraqi government has asked the British to run their officer cadet training, which perhaps inevitably was dubbed 'Sandyhurst'.

The strategic tug-of-war between Iraq and Afghanistan had been the entirely unintended and unforeseen consequence of that second strategic miscalculation in as many years: the 'non-combat' mission in Helmand. NATO had taken over responsibility for the International Security Assistance Force (ISAF) in Southern Afghanistan, and it was unsurprising therefore that Britain would be asked to do more. Announcing the increase in troop numbers in January 2006, the defence secretary John Reid was at pains to explain that the 3,500 (a brigade's worth) extra men were being deployed to help the reconstruction effort, but he added not un-reasonably that

Although our mission to Afghanistan is primarily re-construction, it is a complex and dangerous mission because the terrorists will want to destroy the economy and the legitimate trade and the government that we are helping to build up. Of course, our mission is not counter-terrorism but one of the tasks that we may have to accomplish in order to achieve our strategic mission will be to defend our own troops and the people we are here to defend and to pre-empt, on occasion, terrorist attacks on us. If this didn't involve the necessity to use force we wouldn't send soldiers.

Predicting that security for the reconstruction of Helmand could be a three-year task, he distanced the operation from that of the Americans (Operation Enduring Freedom). It was, he said, 'fundamentally different to that of the US forces elsewhere in Afghanistan . . . We are in the south to help and protect the Afghan people construct their own democracy.' But then, in a sort of throwaway line which encapsulated a decade's woolliness in Westminster's military thinking, and having already implied that there was no unity of effort in the country, he added: 'We would be perfectly happy to leave in three years and without firing one shot because our job is to protect the reconstruction.'

While such hopes of peace were wholly worthy they would soon reveal a dangerous mis-appreciation of the situation in Afghanistan, and in Helmand province in particular, as well as quite remarkable hubris. Within months of the announcement forty servicemen had been killed, half of them in Helmand in one month alone, for 16 Air Assault Brigade (which includes two battalions of the Parachute Regiment and all the army's airborne combat support) at once began trying to pre-empt Taleban attacks, as Reid had intimated, as well as fighting them off. The brigadier, Ed Butler, a former commander of the SAS, seemed determined to take the fight to the Taleban rather than await the inevitable – as he saw it – counter-offensive, though in fact in terms of combat power he had not much more than an augmented 3 Para. In the late summer and autumn there were daily and heavy firefights with the Taleban, and platoon-size (thirty men) standing patrols dotted about Helmand in defended houses had to resort increasingly to calling in fire from artillery, RAF (and allied) ground-attack aircraft – the RAF, principally, with that old warhorse the

Harrier 'jump jet' – and the Apache attack-helicopter.* While this huge amount of ordnance killed Taleban in large numbers, the collateral damage was also heavy, and what little progress there was in 'nation-building' was soon either put on hold or even reversed.

This was the time of General Sir David Richards's 'conversations with Slim and Templer',† and the 'dozens of Rorke's Drifts every day'.‡ And although, unlike Iraq, Helmand was so far away as to be inaccessible to the casual news cameraman, the 'embedded' print journalists were able to paint pictures of the fighting which would have thrilled a Victorian audience. Under the headline 'Makeshift "Rorke's Drift" unit of medics and engineers hold out Taliban' (26 November 2006), the *Daily Mail*'s Matthew Hickley, for example, wrote a story that (with the odd cut) could have come straight from the pages of the *Boy's Own Paper*:

After a summer of intense fighting by British troops in Northern Helmand, attention was focussed on 16 Air Assault Brigade's epic defence of the besieged 'platoon house' garrisons in Sangin, Musa Qala and Nowzad. But hundreds of miles to the south and largely ignored, the frontier town of Garmisir was also under siege and had already fallen once to the Taliban – for whom it is a key transport hub for fighters crossing the nearby border from Pakistan. Helmand's provincial governor, an Afghan trusted by the British, was warning that if Garmisir fell again he would have to resign. On September 8 the town was overrun, presenting UK commanders with a crisis. Garmisir must be saved, but there were no British troops available.

Instead, three officers were given 24 hours to scrape together

---

* The Apache had turned the Army Air Corps into a combat arm, effectively, rather than one of combat support.
† See Introduction.
‡ See ch. 17. And, indeed, the VC was won, posthumously, by Corporal Bryan Budd of the Parachute Regiment.

what men and equipment they could, and ordered to lead around 200 Afghan National Army (ANA) and police on a desperate 100-mile dash across Taliban-held desert in open top Land Rovers and trucks, groaning with all the ammunition they could carry.

On the night of September 10 they paused outside Garmisir and at dawn – five years to the day after the Twin Towers fell – they advanced. Captain Doug Beattie of the Royal Irish Regiment was one of the three British officers, and recalls how things went disastrously wrong within minutes, when the ANA got lost and failed to secure a vital canal crossing.

'Our remit was to stay at the back and let the Afghans take the lead,' Doug said. 'But they took the wrong crossing and wouldn't move. We were under heavy fire and our attack had already stalled.'

Captain Paddy Williams, the Household Cavalry Regiment officer commanding the operation, realised decisive action was needed. Nine British soldiers in two Land Rovers raced forward to storm the correct bridge, braving mortar fire, RPGs and heavy machine-gun fire from the Taliban. The ANA soldiers quickly lost two soldiers killed and refused to go any further, leaving the tiny British force and the Afghan police to fight on. For 12 hours on the first day the fighting raged, with continuous airstrikes by UK and American aircraft guided in by tactical air controller Corporal Sam New of the Household Cavalry Regiment, who was to play a crucial role in the battle. By dusk, the British held the small town's main street, with Doug Beattie and Sam New established on a low hill outside – sheltering in the remains of an ancient fort built by Alexander the Great's armies.

At dawn on day two, they led the Afghan police further south, hoping to create a two mile buffer zone protecting the town. The Taliban had other ideas, and the British were soon pinned down under withering fire from three sides, sheltering in mud huts while allied jets screamed overhead, dropping precision bombs as close as they dared to the UK ground call sign 'Widow 77'.

Creeping forward in their vehicles, the UK troops pushed their luck too far. Doug Beattie, 42, recalled: 'We were shot up pretty badly at that point. We had no cover at all, and had to shelter behind the Land Rovers for four hours. At one point Sam New's radio went dead as he called in an airstrike. A round had severed the wire from the handset to the radio. Our antenna was shot away, and the water cooler and tyres were shot out.'

Again the attack had stalled, and again Paddy Williams decided on a bold attack. Patching up their vehicles as best they could, still under fire, the British soldiers made an astonishing 1.5 mile sprint south along the canal road to storm a Taliban stronghold, blasting a group of huts with heavy machine guns and clearing the buildings with grenades as the enemy fell back. As dusk fell the ANA finally advanced to help secure the position, and for the first time in 40 hours the British soldiers were able to break contact and pull back to sleep for a couple of hours.

At dawn they tried to advance again, but it was soon clear the small force had over-reached itself. The ANA commander was killed along with two of his men. Doug Beattie's driver Joe Cummings, a TA reservist, was wounded in the leg. Capt Beattie recalled: 'We were taking heavy fire from three sides – Taliban within 100 yards of us – and we had no cover. The ANA didn't move up to support us. That was when I really thought we were f*****. We got our arses kicked on that third day.'

Eventually Captain Tim Illingworth of the Light Infantry, the third British officer, stepped in and led the Afghan soldiers forward, and with allied jets dropping 500lb bombs on the Taliban firing points Doug's two vehicles were able to limp out of the killing zone. But there would be no more attempts to advance. The British troops began digging defences where they were.

At last a helicopter brought fresh ammunition supplies, but

no food. Now the battle settled into a bloody slogging match. Wave after wave of Taliban attacks were broken up by airstrikes and machine gun fire, while the British officers led occasional fighting patrols forward, trying to stiffen the ANA soldiers' wavering resolve.

After eight days a Danish reconnaissance squadron arrived, but their rules of engagement prevented them from actually fighting the Taliban. Time and again the embattled force came close to disaster. When an RAF Harrier jump jet aimed a 1,000lb precision bomb at a Taliban position its guidance fins failed to work, and it exploded just yards from British troops, blasting red-hot shrapnel over their heads. On his next run the pilot obliterated his target, killing the Taliban's regional commander and nine of his men. The Danish soldiers were soon interpreting their rules of engagement loosely, helping to clear enemy-held buildings with grenades and machine guns. Doug Beattie recalls an Afghan police officer, Major Showali, as 'the bravest man I ever met':

'He refused to take cover under fire. Every time he saw us in trouble he would run over and pick me up and throw me into cover, shouting "It's not your fight, Captain Doug, it's my fight!" Some of our guys didn't trust the Afghans, and I didn't always. But I trusted that man with my life. When he was shot dead on the last day, I was so sad.'

Finally on the fourteenth day the exhausted British troops were relieved by a force of Royal Marines. They had fired 50,000 rounds of 7.62mm machine gun ammunition, and thousands more from SA80 rifles. Some had even emptied their pistols – weapons of last resort – as they stormed buildings. Miraculously, when the dust settled, there were no UK fatalities. Dozens of Afghan soldiers and police were dead, along with an unknown but certainly large number of Taliban.

Within days the Taliban attacked again in force and the hard-won, narrow buffer zone south of Garmisir was lost. Today the frontline is back to where it was after day one of the battle, and

Garmisir remains under siege. Doug Beattie said: 'It's nobody's fault. The Taliban were too strong, with endless supplies of men and ammunition coming in from Pakistan.'

Beattie had been the RSM of the Royal Irish (the battalion commanded by Lieutenant-Colonel Tim Collins) during the invasion of Iraq, and had been commissioned subsequently. He was awarded the MC for his courage and leadership in Afghanistan, but soon after his next tour of duty was over he left the army: he had had enough.

Troop numbers were now increased once more, and the Taleban offensive was blunted. From this point the IED became as deadly as the assault rifle and RPG (and the cool courage of the bomb disposal officers of the Royal Engineers and the Royal Logistic Corps – formed in 1993 from the RCT, RAOC et al. – was daily tested, as it had been in Northern Ireland); but the civil reconstruction programme – the whole object of Reid's deployment – had seriously stalled.

There have been criticisms of the way Brigadier Ed Butler spread his forces in 'penny packets' around Helmand – and indeed Dannatt advised a redeployment in face of the mounting casualties – but Butler himself appeared to feel he had been given no choice in the absence of strategic clarity. In his post-operational report he wrote of 'the lack of early, formal political direction and a strictly enforced manning cap [upper limit of troop numbers], established upon apparently best case rather than most likely or worst case planning assumptions and taking little account of the enemy vote'. He complained that getting the right equipment and in the right numbers

was hampered because the MoD and Treasury were unwilling to commit funds to Urgent Operational Requirements (UOR) enhancements prior to any formal political announcements.

On-going UORs were halted during the 2-month delay period [before the political decision to commit was taken]. As a result, many key items of equipment arrived in theatre late and some even failed to meet the . . . deployment at all.

Butler resigned eighteen months after returning from Afghanistan, as did the commanding officer of 3 Para, Lieutenant-Colonel Stuart Tootal, who was dismayed at the inadequate treatment of his wounded soldiers on evacuation to Britain – not least the lack of a proper military nursing environment at Selly Oak hospital.*

Every foreign army has to withdraw from Kabul in the end – it is the lesson of history – and it does so with a heavy 'butcher's bill', for that is, in Kipling's words, the 'arithmetic on the frontier'. The new US administration of President Obama is now following the same strategy here as in Iraq, placing the main effort on training the Afghan forces to deal with the insurgency themselves, with a troop surge meanwhile to get the situation in Helmand in particular under control. Even if building the Afghan forces goes to plan, however, there will be a deal more fighting for the British army. For this reason, in his final six months at its head, Dannatt had determined on a review of the army's ability to go the whole fourteen rounds.†

But why had the army seemed so keen to quit the ring in Iraq, and why had it seemed to reel under the blows in Afghanistan?

* In a cost-cutting move, all the service hospitals had been closed. Servicemen were henceforward to be treated in a 'military-managed' wing of Selly Oak hospital, Birmingham. Predictably, this did not prove up to the sheer casualty load of Iraq and Afghanistan.
† General Sir David Richards, who had become CinC Land (Forces) a year before, and was to take over as CGS, had on his own initiative placed his command on a 'campaign footing', and was increasingly frustrated by the evident unwillingness of parts of the MoD to do likewise.

In a real sense the army was – *is* – still adjusting to the fall of the Berlin Wall. The Cold War was the only time a force of significant size (a BAOR of four divisions) had been kept in being to meet a specific threat. The cuts to the army's strength made in the Major government's determined pursuit of a 'peace dividend' were the origins of the permanent state of operational 'overstretch' in which the army has found itself in the first decade of the twenty-first century. Since the disbanding of BAOR there has also been, in the opinion of some former army chiefs, too little competition at the rank of major-general – except, that is, in the Ministry of Defence, where success can be too dependent on political acumen. For while policy is ultimately determined by ministers – quite evidently and properly – without unequivocal military advice it inevitably risks being flawed. Some senior officers had in fact argued at the time of 'Options for Change', as the programme of Tory cuts was euphemistically entitled, that the priorities were wrong, that there was too great a concern to preserve hardware across the three services at the expense of manpower. In a recent exchange with the author, the commander-in-chief of UK land forces at the time, General Sir John Waters, recounted how he had urged that the 'order of battle' – the regiments themselves – be preserved in preference to any of the equipment programmes then on the stocks, and how he still holds that such a priority would have served the nation better in the past decade and a half.* And yet the regiments were cut, with many ancient names disappearing in amalgamations and in truth plain takeovers.† Old names have no sacred right to continue, of course; but reputation is not built overnight, and the losses in both numbers and quality

* Nor was this wishful memory: Waters had made his position very clear at the time – hence the interview.
† The Royal Armoured Corps was hardest hit, being effectively halved in strength.

proved a high price to pay for cherished equipment programmes. History shows that it is easier to re-equip an army than to raise and train one.

But it did not stop with 'Options for Change'. When John Major's government fell in 1997, the new Labour administration's defence secretary George Robertson began a strategic defence review (SDR). Given that all reviews are undertaken in the expectation, one way or another, of cutting costs, most senior officers were pleased enough with the outcome, which involved a shift of resources towards 'expeditionary warfare' and away from what remained of BAOR (an armoured division of three brigades) and all the paraphernalia of Cold War in the other two services. SDR was, however, progressively underfunded to such an extent that the army – like the forces as a whole, indeed – was not able to do what SDR had envisaged, and it was soon trying to cope with demands well beyond those articulated in the review's original planning assumptions, even as restated in 2004 thus: 'As a norm, and without causing overstretch, the Armed Forces must be capable of conducting three simultaneous and enduring operations of small to medium-scale. Given time to prepare, the UK should be capable of undertaking a demanding large-scale intervention operation while still maintaining a commitment to a small-scale peace support operation.'* It did not prove possible, however, to prosecute the Iraq counter-insurgency with the utmost vigour while maintaining a brigade in Helmand.

And then, as if things were not bad enough, in late 2004 the situation was made even worse by a further 'peace dividend' drawn by the Treasury – a cut of 2,000 men following the Northern Ireland agreement. The Army Board decided that these cuts could best be borne by the infantry, but only if there was an end to the 'arms plot' as it was

* Defence White Paper 2004.

known – the rotation of regiments as formed bodies between stations and roles* every four or five years. Instead multi-battalion regiments were to be formed, with constituent battalions permanently based at specific stations, and with individual soldiers posted between battalions as the need arose. This in theory meant that there would be the same number of battalions available for operations because there would be none *hors de combat* during the months of transition from station to station and role to role. Few soldiers believed, however, that this was anything other than 'smoke and mirror' work, and it risked a very great deal of what had taken years – centuries in some cases – to build for the sake of a little gain in 'organizational flexibility'. In fact, in the way that all attempts at rationalizing the infantry's structure produce yet another obstinately untidy model, there are now more types of regiment, from those of a single battalion to those of five, than ever before.

So was it the army that faltered? Or was it the MoD, sliding into two wars without the right numbers, equipment or operational plans, encouraged by wishful thinking in No. 10 and the FCO – and, in the words of one former CDS, by the Treasury's 'unhelpfulness'?†

There is no guarantee that the British or any other army will always win its battles, or even its wars, or that its soldiers will always do their duty. Alanbrooke had after all confided to his diary in 1941: 'Cannot work out why troops are not fighting better' (and many of these regulars). And indeed the 'chattering classes' were soon firing ranging shots in the press once it was announced that combat troops would

---

* Some battalions equipped, for example, with the Warrior armoured-infantry fighting vehicle, and some helicopter-borne.

† General the Lord Guthrie, CDS 1997–2001, has said publicly that he found Gordon Brown 'unhelpful' as chancellor of the exchequer.

leave Iraq in July 2009;* but until the Iraq (Chilcot) inquiry reports, there will be no knowing whether any of these rounds have fallen near the mark. The evidence of the body bags suggests that shirking on the ground had not been widespread, however; and the evidence of medal citations such as Private Beharry's suggests there was no want of old-fashioned courage when the occasion demanded. Perhaps, old joke though it is, George Bernard Shaw was right when he wrote, 'The British soldier can stand up to anything except the British War Office.'†

---

* See e.g. Richard Beeston, foreign editor of *The Times*, 26 Feb 2009: 'The war went wrong not the build up'; and the former defence secretary Michael Portillo in *The Spectator*, 25 Mar 2009: 'Our departure from Iraq ends a dismal period in our military history'.
† From *The Devil's Disciple* (1897) – a story set, ominously, during the only war that the army has ever lost.

Epilogue

What *Will* the Army Be For?

2011

It is doubtless fitting that there should be some amongst us who propose to prepare men's minds for that happy time when war shall cease among men. It is also proper that there should be others who, regarding the world in its present state of hostility, seek to raise, as much as our nature will permit, the character of that necessary institution, an army.

... To be an efficient soldier, a man must be patient under suffering, forbearing, able to resist temptations, quick to comprehend commands, and ready of resource, so that he may effectively obey them ... A mere burst of valour, the daring recklessness which might lead a villain to rush into action, and perform therein great deeds of courage – this is not the sedate and steadfast habit which is necessary for the veteran soldier. Any bold, bad man may fight through one day of battle, but a well-trained soldier can alone, with honour to himself, and utility to his country, perform the arduous duties attendant on a long campaign. If, without this exciting hope ... the British soldier has performed those feats of valour here recorded, how great must be his spirit, how quick of impulse to good, how patient, how forbearing!

Although these words were written in 1828 in the *Edinburgh Review* on the publication of Volume I of General Sir William Napier's *History of the War on the Peninsula*, they have a decidedly contemporary ring. For as I have tried to show, the British army has been built brick by brick, to no architect's plan and in a number of styles, and the soldier today is conscious of his operational heritage as well as being formed by it. Except in passing, however, to explain a point better, I have not tried to demonstrate that the British army is different from other armies: I have taken it as self-evident that an army which is today smaller than that of Turkey, or of Greece, France, Germany or Italy, and not much bigger than that of Spain, and yet is still a major player on the world's military stage, must in some important ways be unique.

Britain was the last of the great powers to introduce conscription and the first to abandon it. For all but twenty-four years of the British army's continuous existence since 1660 it has relied on volunteers. Unlike that of any other major power during those three and a half centuries, however, the British army has never existed because of a clearly identified threat to the 'homeland': France, for example, with her long borders, was always vulnerable to attack from rival Austria or the Netherlands, as were they in turn from France; Prussia positively floated on the map of Europe during the period, her borders resting wherever the Prussian army could make a defensive position; and Russia, though always able to trade vast tracts of territory for time, relied ultimately on her army to settle matters in the marches of Eastern Europe, the Baltic and the Black Sea. For a century and a quarter after independence the US army, too, fought its frontier wars with native tribes and with Mexico. Britain, on the other hand, always felt secure enough behind its 'wooden walls': the enemy could never come by sea, as successive sea lords confidently asserted, and when the Germans once tried to come by air the retort was emphatic.

Because of this, the British army has always had to argue its rationale – and thus for its rations. And more often than not it has been half-starved, for trade and empire were activities of choice, while to politicians of all colours continental military entanglements were something to be avoided altogether. Indeed, as a national insurance policy the army has always been more 'third party' than 'fully comprehensive'. And even when from the middle of the nineteenth century its numbers grew larger for reasons of empire, these numbers were dispersed around the world, rarely combining in more than divisional strength, whereas other nations organized their forces on an altogether larger scale – into army corps and even discrete armies. The British army did not as a rule think big in this way, although it did think globally – or, at least, its soldiers were at ease globally. When for example the 25th Middlesex – a Kitchener battalion – was sent to Siberia via Vladivostok on its own in 1918 as part of the anti-Bolshevik intervention, the soldiers did not bat an eyelid. Afterwards the Middlesex's commanding officer merely reported that 'One and all behaved like Englishmen – the highest eulogy that can be passed upon the conduct of men.' For 150 years British army officers, often very junior ones, have had to relate what they were doing on the ground to the grand strategic object that London desired. Nowadays, young NCOs are doing the same.

From my time in uniform, and from reflecting on its past, I know that the army is a conservative organization: it mistrusts revolutionary change. Field Marshal Lord Carver, Britain's most intellectually able and battle-experienced soldier of the last century, perhaps explains why in his *Britain's Army in the 20th Century*: 'Using the experience of the past as a guide to the balance required to meet future demands has often proved unreliable; but imaginative visions of how to meet them have also been, if not false, at least premature.' But the army has more often than not proved exceptionally quick to change in wartime, when the requirement is crystal clear – if a little late.

The risk of false trails, of which Carver warns, is always present. Some senior officers have recently observed, for example, that as a result of working closely with the US army for the past ten years a certain 'intensity' is appearing in middle-ranking officers where before there was a 'breezier' style. This is more than just the old cricketing jibe 'Gentlemen out, Players in' when Montgomery arrived with his own men. It is rather that the easy pragmatism, 'amateurism' in the best sense of the word, that served the army so well when it was faced with 'impossible' situations may be giving way to a 'professionalism' which asserts that there is an absolute right way in any situation – a sort of military totalitarianism. But this sort of approach can work only when there are plentiful resources of men and materiel – which is not the usual situation a British officer finds himself in. Indeed, too doctrinaire an approach would rob the army of one of its true force multipliers: the original thinking of its officers. To what extent the Iraq bruising has dented self-confidence in British superiority in 'small wars' remains to be seen, but the dangers of an over-reaction, aping American methods when the British army has nothing like the US army's resources, are obvious. The Iraq bruising may indeed prove to be a significant break in the habit of victory, but the army's ability to pick itself up after a setback – as I have shown again and again – was a habit acquired much earlier.

Technology can also be beguiling, pointing down thoroughly *expensive* false trails. During the 2004 'stealth defence review', the then CDS, General Sir Michael (now Lord) Walker, wrote in *The Times*:

Our advantage will no longer be in numbers, but in effects . . . one Apache Longbow helicopter flying against a dispersed and well-hidden enemy can be more effective than a squadron of tanks . . . Imagine if targeting and intelligence information from that helicopter could be relayed simultaneously to commanders on ships via Awacs aircraft, and to individual soldiers among amphibious forces landing to search for that same enemy. All in real time. Imagine how quickly we could make decisions. That is what we call network enabled capability, and it is that technology which permits us to deliver an increased effect with fewer platforms.

'Imagine' indeed. As I have tried to show in the previous chapter, the reality of fighting the insurgency in Iraq and Afghanistan has been more 'Victoria's wars' than 'network enabled'. But even if technology were – had been – the answer in Iraq and Afghanistan, are the battles of Helmand and Basra pointers to what war in the future will look like? This is the permanent dilemma facing service chiefs in deciding where the money should go.

The parlous state of the public finances in the next decade will make the dilemma more acute than ever. No longer will the chiefs be able to 'advance on a broad front'. Instead they will have to choose between current operations and possible future operations. In this contest between wolves and sledges, as the former CDS Lord Guthrie calls it, the only sensible course is to shoot the nearest wolf: make sure you win the battles of today, for defeat today increases the chance of war tomorrow.* This means finding more men for the army and taking a risk that we can get away with fewer Eurofighters, say, and even aircraft carriers. But finding recruits and keeping trained manpower has proved extraordinarily difficult these past ten years. The infantry

* Interview with the author, April 2009.

is chronically some 1,500 to 2,000 men under-recruited – three battalions' worth – and many trained infantrymen are *hors de combat* through war wounds or injuries in training. In addition, an infantry battalion's establishment – its authorized strength – has been so pared down in the past twenty years that most battalions are scarcely able to deploy on operations without heavy reinforcement from another regiment, including the TA. In fact, for its true war footing the army is at least 10,000 men under-established. But could the under-strength army recruit more men?

Several former chiefs are convinced that they could, not least because the current economic climate with its rising unemployment is in the army's favour at last (and an expanding organization, as opposed to a diminishing one, has its attractions).* Just as important, a bigger army would mean a less stretched army, one in which the soldier felt there was the right balance of training, operations, rest and personal development – and fewer seeking discharge in consequence. Avoiding the loss of serving personnel is particularly important among middle-ranking field officers (captains and majors) and senior NCOs, the repositories of operational experience in the regiments. There are other possibilities. The Brigade of Gurkhas, which bore a good deal of the burden of Malaya and Borneo, and now consists of just two infantry battalions and supporting troops (engineers, signals, transport),† could easily be doubled in size. They are now virtually interchangeable with British

---

*The army remains, too, a 'smart' enough profession to attract quality officers: it is, for example, the largest single employer of Old Etonians. And the royal family's continuing close association must play a part in the army's prestige, with the two princes' training at Sandhurst a powerful endorsement. But active operations have always exercised a powerful draw too: 1968, the army's only year of peace, actually saw a fall in recruiting.

† With three independent companies at Sandhurst, the School of Infantry in Brecon, and at the infantry training centre in Catterick.

units, and indeed have other characteristics that make them especially effective in places such as Afghanistan – not least facility in local languages. The ratio of infantry to other arms is also manifestly too small: in 1918 the infantry accounted for over half the army's total manpower; today the proportion is well under a quarter, yet the nature of operations is once more becoming manpower-intensive. Imaginative schemes to integrate the TA more closely with the regular army have also long been discussed, although the sheer impracticalities and uncertainties of employing TA as 'formed units' as opposed to individual reinforcements except in times of national emergencies will continue to dog aspirations for greater integration.

Special Forces have never stood at a greater premium than now. The SAS are without question the world leader in clandestine operations against strategic targets, but their strength – at one regular regiment and two TA – is far less than generally supposed.* Their manpower is never 'capped', however: the standard is an absolute one, and only a few out of every hundred who begin the process for 'Selection' make it to Hereford. One of the 'critical mass' arguments in debates about the size of the army is indeed the need for a pool large enough to yield the required number of recruits to the SAS; and equally one of the arguments for retaining the Parachute Regiment – though parachute operations in any strength are now almost inconceivable – is to foster a *corps d'élite* which not only has its effect in the army as a whole but is a fertile seedbed for the SAS.† It is also striking just how many senior officers in the army today wear SAS

*The author was told by an impeccable source that on becoming prime minister, Tony Blair had believed the (regular) SAS to number around 40,000 – an exaggeration by two noughts.
† The Sierra Leone hostage rescue operation in 2000 – Operation Barras – was almost entirely an SAS/Parachute Regiment affair, brilliantly and boldly conceived and executed.

wings – in no small part due to Guthrie's championing of men who had served with 'the Regiment' when he (himself a former SAS man) was CGS.

To what future, then, do the interventions of the past decade and a half in the former Yugoslavia, East Timor, Sierra Leone, Iraq and Afghanistan point? What will the army be for,* and how will it be organized and equipped?

In *The Shield of Achilles*, published shortly after the attacks on the World Trade Center and the Pentagon, Philip Bobbitt – American law professor, strategist, and former adviser to the US National Security Council – describes the transformation of the traditional nation-state model of international relations into 'a new age of indeterminacy', pointing to the heightened danger of the old cliché that generals prepare to fight the last war rather than the next one – although he is sympathetic, because for centuries the past was all we knew about the future: 'Things were usually pretty much the way they have been' because war had always had three characteristics – one country fights another; war is waged by the government, not private parties; the victorious side defeats its adversary. But this would no longer be the case, suggests Professor Bobbitt. In reviewing *The Shield of Achilles* in *The Times* in 2002 I wrote:

> A year after the end of the 20th century's 'Long War', Saddam Hussein . . . (oblivious to the new model) invaded Kuwait in the old way. And while he and others with access to weapons of mass destruction fail to recognize the theory, military planners

* The question, on which the title of this Epilogue draws, echoes that of Haldane on becoming secretary of state for war in 1905 in the aftermath of the humiliations of South Africa. Haldane was an Hegelian philosopher, not content to accept the plans for army reform until the cabinet and the senior officers could answer the existential question: 'What is the army *for*?' (see ch. 20).

will still be obliged to maintain the capability to win in the old way. Persuading generals to give up cherished programmes designed to do just this will not be easy – and quite right too. How, therefore, are the additional resources for defence and security to be found?

Eight years on, the new CGS, Sir David Richards, suggested that in fact there may not need to be quite such a competition for resources to fight both kinds of war. Policy-makers need to do three things, he argued:

Firstly to decide whether they believe conflicts with dissatisfied and violent non-state actors are here for the long term or an historical aberration. Secondly do they believe that, despite globalisation and mutual inter-dependence, state-on-state warfare remains something for which they must prepare? And thirdly, but here I think there may be some comfort to be drawn, if it is decided our armies need to be capable of succeeding in both, would the two types of conflict in practice not look surprisingly similar, at least to those actually charged with conducting them at the tactical level? ... what would such [state-on-state] warfare actually look like? Would it really be a hot version of what people like me spent much of our lives training for [repelling a Soviet invasion of Germany]? I wonder; why would China or Russia, despite the predictable clamour after Georgia, risk everything they have achieved to confront us kinetically, or symmetrically as it is often termed? The costs of creating the scale of military might required and the risks of failure even then, are enormous. The presence of nuclear weapons reinforces a likely caution. If they really want to cause us major problems surely they will employ other levers of state power: economic and information effects, for example? Attacks are likely to be delivered semi-anonymously through cyberspace or the use of proxies and guerrillas. After all, it was Sun Tzu who again famously reminds us that 'the acme of military skill is to defeat one's enemy without firing a shot'. In other words, what I am suggesting, is that there is a good case for believing that even state-on-state warfare will be suspiciously similar to that we will be conducting against non-state groupings. My only caveat on this is to ensure that we do not rush headlong into scrapping all our tanks to the point that traditional mass armoured operations, for example, become an attractive asymmetric option to a potential enemy.[*]

Here, then, is a pointer to the road along which the army that Cromwell first envisaged in the aftermath of Edgehill 'as if through a glass darkly' – an army which has marched across five, arguably all six, continents over three centuries, has frequently been beaten but has very rarely been bowed, and under generals whose deeds are still studied – may find itself marching during the next military generation.[†] But history and the mental approach to change which that history has produced suggests too that whatever emerges will be recognizable in its essentials to the generations that have gone before. For war – soldiering – has an enduring face at the tactical level. I am myself convinced, from both experience and study, that the old adage remains good: it is the man who is the first weapon of war. Or, as the former Great War infantry officer and man of letters A. P. Herbert[‡] wrote in his 1944 encouragement to the 'P.B.I.' – 'the Poor Bloody Infantry':

---

[*] International Institute for Strategic Studies, 50th Anniversary Conference, Geneva, September 2008, and subsequently.
[†] See Postscript – *The Strategic Defence and Security Review* 2010.
[‡] He served in the Royal Naval Division at Gallipoli and on the Somme.

Hail, soldier, huddled in the rain,
Hail, soldier, squelching through the mud,
Hail, soldier, sick of dirt and pain,
The sight of death, the smell of blood.
New men, new weapons, bear the brunt;
New slogans gild the ancient game:
The infantry are still in front,
And mud and dust are much the same.
Hail, humble footman, poised to fly
Across the West, or any, Wall!
Proud, plodding, peerless P.B.I. –
The foulest, finest job of all.*

Perhaps, though, we may extend Herbert's sentiment to every soldier, for the rear area today has little security, or even meaning. Proud, plodding, peerless British Army – the foulest, finest job of all. An army long in the making, which yet remains very much a work in progress.

* 'Salute the Soldier'.

'Proud, plodding, peerless P.B.I.'

Postscript

The Strategic Defence and Security Review 2010

SINCE THE ORIGINAL HARDBACK PUBLICATION OF THIS BOOK IN September 2009, and then, revised, in trade paperback in May 2010, Gordon Brown's Labour government has been replaced by a Conservative–Liberal coalition led by the Tory David Cameron, with the Liberal Nick Clegg as deputy prime minister. There was an almost collective sigh of relief in the defence community at this change, for it was widely held that Brown had been ideologically hostile to the MoD, constitutionally unsympathetic to men in uniform, and so devoid of the usual human attributes as to be impossible to deal with. During his time in No 10, defence strategy had stagnated, the war in Afghanistan had stalled bloodily, and the black hole in the MoD's budget had grown to crisis proportions. There seemed to have been a revolving door on the Defence Secretary's office: between May 2005 and May 2010 there had been no fewer than five ministers. Nor were matters helped by the unwillingness or incapability of successive chiefs of the defence staff, permanent secretaries and their senior officers and officials to grasp that the armed forces were at war and not winning, and that this was due in no small measure to lack of funding because the spending programme was choked with cuckoos in the nest – expensive equipment projects of dubious operational value whose costs were rising by the day. There were, for example, enough fast jets on order to re-fight the Battle of Britain, while the troops in Afghanistan were being constrained – and killed – through want of helicopters. As W. H. Auden said of the 1930s, it had been a 'low dishonest decade'.

And so hopes were high, just as they had been in 1979 with the arrival of Margaret Thatcher, that a new Tory-led government would be one with a gut instinct for defence. If there were doubts about the Chancellor, George Osborne, there was confidence that Cameron's instincts – and not least his capacity to engage with the issues – were sound. And Nick Clegg had consistently asked pertinent defence questions in parliament. The secretary of state, Dr Liam Fox, had been the shadow minister for five years, as well as having held military rank (as a civilian army medical officer he had been an honorary major). The minister for international security strategy, Gerald Howarth, had been a pilot in the RAF reserve. The minister for personnel, welfare and veterans, Andrew Robathan, had been a regular officer in the Coldstream Guards and the SAS. Likewise Lord (John) Astor, under secretary of state, and defence spokesman in the Lords, had served in the Life Guards. Fox's deputy, the minister of state, Nick Harvey, had no military experience, but he had been the Liberal front bench defence spokesman for five years.

There could now follow the long-promised security and defence review (SDSR). The only problem was that the senior officers and officials in the MoD – in the central staff, where policy and the budget are shaped – were the same old people, in particular the CDS, Air Chief Marshal Sir Jock (now Lord) Stirrup. Stirrup had been extended in office for an unprecedented five years by Gordon Brown because, it was rumoured, his obvious successor, General Sir Richard Dannatt, had been so outspoken against Blair's and Brown's military funding cuts and policies.* David Cameron, while in opposition, invited Dannatt, who was then still in uniform, to be an adviser, to take a Tory peerage and to be a minister in the expected Tory government. This leaked out and Dannatt was widely reviled for his 'poor' judgement – not least, if more privately, by senior officers of all three services. In any event, he played no role in SDSR (although the PM ennobled him anyway in December 2010 – the first CGS to be thus honoured). So ended one of the most acrimonious periods in political–military relations in modern times.

Work on SDSR in the MoD went on apace throughout the summer of 2010. Despite the Tories' frequently stated intention in opposition that such a review would be policyled, the reality of the defence budget was imposing its own timetable: there had to be cuts, and quickly. Because the central staff was holding the cards frustratingly close to the chest, however, the single service chiefs fought an increasingly public campaign on the question of priorities. The RAF was viscerally protective of its fast jets, present and future, even though Stirrup had hinted there might be too many in the programme. The army believed that, to paraphrase General Curtis Le May, ('Bomber' Harris's American counterpart in the Second World War), 'Flying fighters is fun; flying helicopters is important' – indeed, not just helicopters but Hercules transport aircraft and all the other unglamorous workhorses of far-flung operations. The navy,

---

* Cameron short-toured Stirrup as soon as decently possible – at the end of SDSR, at which point the CGS, General Sir David Richards, succeeded him.

meanwhile, was curiously unable to articulate its future role with either moral authority or intellectual rigour. Their pathetic performance, from seaman to First Sea Lord, in the Iranian hostage and 'cash for sailors' affair had undermined their standing in a way that few admirals even now seem to understand; and the ineffectiveness in coming to grips with piracy off the Horn of Africa had done nothing to help their image. Retrieving holidaymakers stranded in Spain by Icelandic volcanic ash hardly recalled the glory days of rescuing Sir John Moore's army after Corunna, or the BEF at Dunkirk. Gordon Brown had promised them two huge through-deck aircraft carriers, however, to be built in ship-yards in Labour constituencies, and these totems of potency together with the hugely expensive specialist surface ships needed to defend them became the navy's priority. So much so that the Navy Board were prepared to emasculate the Royal Marines by cutting amphibious shipping, and to reduce the fleet's global endurance by cutting frigates, contrary to their warnings of the dangers of 'sea blindness'. They were certainly prepared to see a huge cut in army numbers, and briefed relentlessly that the army was over-manned, with the First Sea Lord saying infelicitously that 'Afghanistan isn't the only game in town'.

Against this background the CGS, General Sir David Richards, had to argue the army's case. He succeeded in doing so for two reasons. First, the army was at war: contro-versially he himself while CinC had placed it on a 'campaign footing'. Although many in Whitehall, in and out of uniform, seemed unable to grasp this or its implications, David Cameron did. The second reason was Richards' unrivalled operational experience and his vigour as a military thinker. He was the first CGS since FM Sir Roland Gibbs (1976–79) with a DSO, and his operational experience was both recent and at high level. He had been thinking deeply about the changing nature of war for several years. It was well known, for instance, that he wrote his own papers and speeches. He was not, therefore, arguing a brief when fighting his corner with Cameron and Fox: he was articulating his own thinking, and doing so with moral force.

Unsurprisingly, therefore, but to the surprise of most people nevertheless, his arguments carried the day: bucking the trend of British history, SDSR largely preserves the army (at least while operational commitments in Afghanistan remain at a substantial level) by taking money from the other two services. That said, the army's share of the savings is not insignificant. The principal 'headline' is a manpower cut of some 7,000. No capbadges will be lost, but neither will they be augmented to proper fighting strength, as many hoped, and, crucially, the army will not be able to field the six-brigade structure it had been developing to sustain future operations. The cuts will come in the main from yet another reorganization of the command structure and the training base, some of these by the withdrawal of the 1st Armoured Division and supporting infrastructure from Germany. The implication of cuts in the training organization are obvious, those in headquarters less so. Practical and moral support to subordinate units, supervision of their administration and training, guidance and example to subordinate commanders and staff, judgement of officers' potential and their development needs – these are difficult things to quantify and therefore to place a value on; but when something goes wrong (such as the civilian deaths in Basra), the audit trail goes all the way back to the command structure.

'Armies must, in the nature of things, change and develop very slowly during peace, since they have always to be ready for action, and can never discard a weapon, an organization, or a tactical doctrine until a new one has been proved by long and careful experiment'. So wrote one of the army's most dedicated thinkers, if ultimately a disappointing field commander, Brigadier Archibald (later FM Lord) Wavell in 1930, when futuristic ideas on warfare were being advanced by 'the prophets'. But in 1930 there was a common understanding of what constituted war – a clash of armed forces of nation states; and with only a dozen years since the 'war to end all wars', the prospect of such a clash was dim. 'Long and careful experiment' seemed a luxury the nation could afford; and, indeed, with the drastic reduction in defence spending, the army found it could do little else but experiment.

Whether the government will treat SDSR as an end- or a start-point remains to be seen (for example, the defence secretary has made his determination known to sort out procurement, which is notoriously profligate and inefficient), but the British army today continues to be very much a work in progress – as it always has been, even if at times more a work in *re*gress. One of the major challenges, as ever, will be to maintain the moral component of fighting power that comes from continuity – what might be called preserving 'operational heritage' – while making the necessary changes to fit the army for the future. There are no easy answers, but General Peter Chiarelli, the vice chief of (US) army staff, summed up the stakes when he told one of his generals recently, 'If you don't like change, you're going to hate being irrelevant.'

Appendix I

The Anatomy of a Regiment

Morning in an imaginary infantry battalion

IT IS A LITTLE BEFORE EIGHT O'CLOCK – 0800 HOURS – IN WATERLOO barracks, Colchester, Aldershot, Catterick, Tidworth or wherever the 2nd Battalion of the Duke of M—ster's Regiment are stationed. They are an infantry battalion in the 'light' role – they move on their feet, or in 'soft-skinned' vehicles, or by helicopter – but they are not historically 'light infantry', for the Duke of M—ster's Regiment was formed in 2006, an amalgamation of three 'county regiments' from the north of England. This 'regional regiment', as the new entities were gracelessly dubbed by the defence secretary at the time (and inaccurately, for the Royal Regiment of Scotland, the Royal Welsh Regiment and the Royal Irish Regiment – others in the great reorganization – are national not regional), has three battalions each some 650 strong, each stationed in a different place. Unlike in the past, when a battalion moved en bloc from one station to another every four or five years, individual soldiers will now be posted from battalion to battalion as the need arises. But these are early days, and no one quite knows how this 'trickle posting', as it is called, will work out; though other arms and services – artillery, engineers, logistics, etc. – have always done it this way.

The 2nd Battalion is predominantly the old Royal —shire Regiment. The 1st Battalion is in the armoured infantry role in Germany (equipped with Warrior), and the 3rd Battalion is in the light role in Cyprus (but currently on operations in Afghanistan). The 2nd battalion – 2DMR – returned from Afghanistan six months ago, and have been warned for a further tour in eighteen months' time. A few officers and other ranks have come and gone from one battalion to another, but by and large the three battalions are trying to 'keep things in the family' – and at the moment the 'close family' is the battalion; those soldiers in the other battalions are more the distant cousins.

Outside B Company's lines men are gathering for muster parade. The single soldiers have shaved and showered in their en-suite rooms (for Waterloo Barracks have been modernized: elsewhere there are still six-man 'barrack rooms' with communal 'ablutions') and breakfasted in the 'Wellington Restaurant', which most of them still call the cookhouse. The married soldiers have driven in by car, or walked in from their married quarters nearby.

One of the latter, Corporal Steele, thirty-two years old, with thirteen years' service, calls 'Fall-in first section.' Six private soldiers, or Dukesmen as they are known in this reg-iment,* and Lance-Corporal Takavesi, a 25-year-old Fijian, automatically form two ranks side-by-side with second and third sections of 7 Platoon (10 per cent of the battalion are from overseas, an increasing number of late from West Africa: a trend replicated throughout the army). The three

---

* In the armoured corps a private soldier is 'Trooper'; in the artillery 'Gunner'; in the RE 'Sapper'; in the REME 'Craftsman'; in the Guards 'Guardsman' – and so on. In fact, there are few places in the army where a private soldier is called 'Private' any longer.

section corporals take post in the front rank and report 'all present and correct' to the platoon serjeant (in the 2nd battalion 'serjeant' is spelled with a 'j', because the Royal —shires always spelled it that way. The other two battalions spell it with a 'g'. The concession took many hours of negotiating in the amalgamation process.)

Serjeant Acton is two years older than Corporal Steele, and next month he will be promoted to Staff-Serjeant (but called Colour-Serjeant), when he will take over as company quarter-master-serjeant (CQMS) responsible for all B Company's equipment and accommodation: on operations the CQMS is the focus of resupply for the company, except for ammunition, which is the business of the company serjeant-major.

Serjeant Acton gives the platoon a quick 'once over'. They're all decently turned out – unlike some of 6 Platoon, evidently, for Serjeant Prince is 'bollocking' a lance-corporal for letting Dukesman Dolan come on parade with a button unfastened: a button unfastened on parade reveals a sloppiness – barks Prince – which on the battlefield might be manifest in failing, say, to load his rifle correctly (though Serjeant Prince puts this to Dolan in different words, of course).

Meanwhile this same falling-in procedure is also taking place in 5 Platoon, the third 'rifle' platoon in B Company.

The company serjeant-major (CSM), who has been watching from the wings, marches to the front of the company, pace-stick wedged firmly under his arm. Warrant Officer (Class II) Barrow has been eighteen years in the army, winning the MC in Iraq as a staff-serjeant platoon commander (when there are not enough officers, a senior NCO will command a platoon), and CSM of B Company for two years. He will probably be the battalion's next RSM in a year's time.

From around the corner come four officers – Captain

Pattinson, B Company's second-in-command, and three subalterns (lieutenants and second lieutenants), the platoon commanders. They are all bachelors living in the officers' mess, and have walked the couple of hundred yards to B Company lines together, chatting about last night's news and the day's training ahead. They wait at the edge of the parade for the arrival of the company commander.

A minute or so later Major Farrell appears. He is thirty-five, married, and has been with the battalion off and on for thirteen years. He has been B Company commander for just over a year, having previously been chief of staff of 23rd (Armoured) Brigade in Germany. While with 23 Bde he was appointed MBE for the brigade's tour of duty in Iraq, his second stint in 'the sand pit' (the first was as the battalion's adjutant). He won the MC on the recent tour of duty in Afghanistan, and it is odds on that he will command the battalion, or one of the other battalions of the Duke of M—ster's, in about three years' time after an appointment as a lieutenant-colonel in the MoD.

CSM Barrow marches briskly up to him, halts and salutes, and reports the 'parade state' – two sick, one absent, ninety-seven on parade (2 DMR are more or less up to strength). Major Farrell calls 'Fall-in the officers,' and Captain Pattinson marches to the rear of the company and waits while the platoon commanders take post in front of their platoons. 'Platoon commanders, carry on,' orders Major Farrell.

Lieutenant Burgess turns about to Serjeant Acton, who informs him that *all* 7 Platoon are on parade – as usual, for no one in 7 Platoon *ever* goes sick, let alone AWOL. Together they inspect the platoon, the lieutenant chatting easily with the riflemen, for Burgess has had 7 Platoon for two years, and in that time they have been to both Iraq and Afghanistan together. He is an Old Etonian, twenty-six years old and a graduate of the university of St Andrews. At Sandhurst, where he was a junior under-officer, the Coldstream Guards and a cavalry regiment tried to 'poach' him, but his grandfather had commanded the Royal –shires in Korea and his father had been killed with the regiment in Northern Ireland (where he had earlier won the MC), and the young Burgess had been determined to wear the same cap badge as they. He drives his father's old Alvis, which he has restored with the help of a fitter from the battalion's REME detachment, and plays cricket for the army. He is unofficially engaged to a doctor (an admiral's daughter) whom he met in his second year at university.

5 Platoon is (unusually) commanded by a non-graduate: 21-year-old Second Lieutenant Ferrier, whose father had been the regiment's second-in-command in the Falklands, and who from an early age had wanted only to join the army. He went to Sandhurst straight from Wellington, and hopes for a trial for army rugby next season.

6 Platoon is commanded by 24-year-old Lieutenant Crosthwaite. His father had been regimental serjeant-major (RSM), then quartermaster, of a fusilier regiment. The Crosthwaites had managed to send their only son to public school, Pocklington, by supplementing the boarding school allowance (Mrs Crosthwaite ran a laundry business – took in washing – for the officers in Catterick garrison). Young Crosthwaite had won a scholarship to Brasenose, where he took a first in PPE, was elected secretary of the Union and gained a blue for fencing. His father had been anxious when he hadn't joined the OTC at Oxford, but the other honours were recompense – as was the Queen's Medal for academic studies at Sandhurst (second only in prestige to the Sword of Honour, which is awarded to the best cadet). Crosthwaite reads for an hour before first parade each day, and listens to Radio 3 every evening. The other officers think him rather intense.

The rest of the battalion call B Company 'Guards

Company' because all the officers were at public school and the CSM has been an instructor at Sandhurst. The other two 'rifle' companies, A and C, are a more mixed bunch. One of the company commanders is newly arrived from the 1st battalion, and the other is an acting major.

Two further companies complete the battalion. The 'Manœuvre Support' Company consists of the battalion's 'heavy weapons' (mortars and anti-tank missiles) and reconnaissance platoons – and the assault pioneers, the battalion's own jack-of-all-trade 'sappers'. The pioneers are traditionally commanded by a serjeant permitted to (in the Duke of M—ster's *required* to) grow a beard. The officers in Support Company have previously commanded rifle platoons, and the company commander is on his second tour as a major, having previously commanded C Company.

Headquarter Company comprises the battalion's command and administrative elements – clerks, signallers, drivers, quartermaster's storemen, the REME detachment, etc. – in which are found several other cap badges, notably the Adjutant General's Corps (whose initials are sometimes irreverently spelled out as the 'All-Girls Corps'), who are responsible for pay and documentation, and the RAMC (the medical officer and assistants). HQ Company Commander is an 'LE' ('Late Entry') officer – i.e. commissioned from the ranks after twenty-two years' service. The battalion has five LE officers.

While each company is holding muster parade (or fitness training, or however the company commander wishes to start the day), in battalion headquarters the routine business of running an enterprise of 700 or so military souls is getting under way. The adjutant, the commanding officer's executive, is reading the outgoing orderly officer's report. An orderly officer is a subaltern nominated each day. His job is to carry out a number of standing inspections and to oversee the guard each evening, and to be the point of contact for any action required in the 'silent hours'. The adjutant is the battalion's brightest and best captain, the commanding officer's confidant and the scourge of the subalterns. On operations he runs the battalion's headquarters and radio command net, the CO's right-hand man.

Down the corridor the RSM is receiving the orderly serjeant's report. The RSM runs the NCOs in the same way that the adjutant runs the junior officers. Warrant Officer Class I Banks has been in the battalion for nearly twenty-one years, and apart from two years as a corporal away training recruits, and two as a signals instructor at the School of Infantry, has served the entire time at 'RD' (regimental duty). It is expected that next year he will be commissioned and will take over as quartermaster. He has been in uniform since the age of fourteen, having first been an army cadet and then at sixteen going to the former Infantry Junior Leaders' Battalion where his leadership potential was developed. His eldest son is at the Army Apprentices' College, whence he will join the Royal Engineers. At nine o'clock RSM Banks and the adjutant will meet with the commanding officer to run through the company commanders' nomination of Dukesmen for promotion to lance-corporal.

The commanding officer and the RSM have known each other since they were both fresh faces in C Company of the Royal —shires. Lieutenant-Colonel Hills, forty, has been in command for almost two years, and wears the ribbon of the DSO from the battalion's recent tour in Afghanistan, as well as a Queen's Gallantry Medal from Bosnia. He had never intended joining the army – there wasn't even a CCF (Combined Cadet Force) at Bryanston – but in his last year at Edinburgh University he had watched the Gulf War daily on the news and found himself surprisingly drawn to the idea of army life, so decided to join for three years. Six years later he was a troop commander with 22SAS.

After the promotion meeting Colonel Hills will have a

long session with his second-in-command (2IC), Major Copeland, and the operations/training officer, Captain Hodges, to plan the training year ahead and to make sure they have enough suitably qualified instructors. Hills will not take the battalion to Afghanistan next time: he has been selected for promotion to (full) colonel and six months hence will go to the MoD for his first tour in the 'Main Building' (as something on the operations staff).

Major Copeland will not go either; his three years are almost up as 2IC, and although he is still hoping to 'pick up promotion' in due course he is beginning to look for a job which will allow him to see more of his family.

2nd Battalion the Duke of M—ster's are in no way remarkable. They have one or two 'stars', but are otherwise what the colonel of the regiment, Major-General Hardy, who commanded a battalion of the Royal —shires many years ago, calls a 'solid regiment of the line'. As colonel, his main concern is finding officers (as well as overseeing the regimental investments and charitable trust). He wishes the regiment were more fashionable at Sandhurst – like, say, The Rifles – and that there were a few more stars in the officers' mess, but on the whole he thinks the amalgamation has gone pretty well. These things you just have to get on with, as he says.

The colonelcy is an entirely honorary appointment, unpaid, demanding and time-consuming. General Hardy is fifty-nine and retired from the army three years ago; he now runs the Institute of Practitioners. He has to juggle his regimental responsibilities carefully with those of his job at the Institute (which he cannot afford to give up just yet because two children are still at university, and there are the school fees to pay off). His wife would like to see a little more of him now that he has left the colours, and he himself would like to play a bit more golf. However, one of the great hidden strengths of the British army, of the infantry and cavalry in particular, is the depth which the colonelcy and its network of former officers represent. So first he has to find someone to take over from him. The regiment has been his life for forty years, and for all that the regiment has changed in that time – as has the army – the more it has in essence remained the same; and he will not pass the baton lightly.

Appendix II

The Numbering and Naming of Infantry Regiments

THE REDESIGNATIONS, AMALGAMATIONS AND DISBANDMENTS OF the British infantry of the line would be a worthy specialist subject for *Mastermind*. A particular regimental number sometimes disappeared from the line altogether, during a period of retrenchment for example. At various times, therefore, a number had a different name attached to it, and the later name was not always the official successor of the earlier one; regiments were sometimes deemed to have been fully disbanded, and when a new regiment was raised during a period of expansion, or a regiment raised by the East India Company was transferred to the Crown, the 'spare' number was reassigned to the new unit. So, for example, the number 77 has at various times been attached to 'Montgomerie's Highlanders', 'the Atholl Highlanders', 'the East Middlesex Regiment of Foot' and 'the Hindoostan Regiment of Foot'.

A little learning is therefore a dangerous thing; but there is not room here to drink too deep of the spring. Instead I have tried to simplify the list by taking the name(s) most commonly associated with the number in the half-century before the final

'Cardwell–Childers' reforms of 1881 which produced the names of the infantry regiments in which perhaps the reader's grand-, great-grand- or great-great-grandfather fought during the First World War. We may perhaps call the list 'the line of battle at its fullest Victorian stretch':

1st Regiment of Foot (Royal Scots/Royal Regiment of Foot)

2nd (Queen's Royal Regiment of Foot)

3rd (East Kent Regiment of Foot) or The Buffs

4th (King's Own Regiment of Foot)

5th (The Northumberland Regiment of Foot); from 1836 Fusiliers

6th (1st Royal Warwickshire Regiment of Foot)

7th (Royal Fuziliers)

8th (The King's Regiment of Foot)

9th (The East Norfolk Regiment of Foot)

10th (The North Lincolnshire Regiment of Foot)

11th (The North Devonshire Regiment of Foot)

12th (The East Suffolk Regiment of Foot)

13th (1st Somersetshire Light Infantry Regiment of Foot)

14th (Bedfordshire Regiment of Foot/Buckinghamshire Regiment of Foot)

15th (The Yorkshire East Riding Regiment of Foot)

16th (Bedfordshire Regiment of Foot/The Buckinghamshire Regiment of Foot)

17th (The Leicestershire Regiment of Foot)

18th (The Royal Irish Regiment of Foot)

19th (North Yorkshire Riding Regiment of Foot); Green Howards

20th (East Devonshire Regiment of Foot)

21st (Royal North British Fusiliers/Royal Scots Fusilier Regiment of Foot)

22nd (The Cheshire Regiment of Foot)

23rd (Royal Welch Fusiliers)

24th (2nd Warwickshire Regiment of Foot)
25th (King's Own (Scottish) Borderers)
26th (The Cameronian Regiment of Foot)
27th (Inniskilling Regiment of Foot)
28th (North Gloucestershire Regiment of Foot)
29th (Worcestershire Regiment of Foot)
30th (Cambridgeshire Regiment of Foot)
31st (Huntingdonshire Regiment of Foot)
32nd (Cornwall Regiment of Foot)
33rd (1st Yorkshire West Riding Regiment of Foot)
34th (Cumberland Regiment of Foot)
35th (Sussex Regiment of Foot)
36th (Herefordshire Regiment of Foot)
37th (North Hampshire Regiment of Foot)
38th (1st Staffordshire Regiment of Foot)
39th (Dorsetshire Regiment of Foot)
40th (2nd Somersetshire Regiment of Foot)
41st (Welsh Regiment of Foot)
42nd (Royal Highland Regiment of Foot); Black Watch
43rd (Monmouthshire Regiment of Foot)
44th (East Essex Regiment of Foot)
45th (Nottinghamshire Regiment of Foot)
46th (South Devonshire Regiment of Foot)
47th (Lancashire Regiment of Foot)
48th (Northamptonshire Regiment of Foot)
49th (Hertfordshire Regiment of Foot)
50th (The Queen's Own Regiment of Foot)
51st (2nd Yorkshire West Riding Regiment of Foot)
52nd (Oxfordshire Regiment of Foot (Light Infantry))
53rd (Shropshire Regiment of Foot)
54th (West Norfolk Regiment of Foot)
55th (Westmoreland Regiment of Foot)
56th (West Essex Regiment of Foot)
57th (West Middlesex Regiment of Foot)
58th (Rutlandshire Regiment of Foot)

59th (2nd Nottingham Regiment of Foot)
60th (Royal American Regiment of Foot/The King's Royal
    Rifle Corps)
61st (South Gloucestershire Regiment of Foot)
62nd (Wiltshire Regiment of Foot)
63rd (West Suffolk Regiment of Foot)
64th (2nd Staffordshire Regiment of Foot)
65th (2nd Yorkshire North Riding Regiment of Foot)
66th (Berkshire Regiment of Foot)
67th (South Hampshire Regiment of Foot)
68th (Durham Regiment of Foot (Light Infantry))
69th (South Lincolnshire Regiment of Foot)
70th (Surrey Regiment of Foot)
71st (Glasgow Highland Regiment of Foot)
72nd (Duke of Albany's Own Regiment of Foot)
73rd (Perthshire Regiment of Foot)
74th (Argyllshire Highlanders)
75th (Prince of Wales' Regiment)
76th (Hindoostan Regiment of Foot)
77th (East Middlesex Regiment of Foot)
78th (Seaforth Highlanders)
79th (Cameron Highlanders)
80th (Staffordshire Volunteers)
81st (Loyal Lincoln Volunteers Regiment of Foot)
82nd (Prince of Wales' Volunteers)
83rd (County of Dublin Regiment of Foot)
84th (York and Lancaster Regiment of Foot)
85th (The King's Regiment of Light Infantry (Bucks
    Volunteers))
86th (Royal County Down Regiment of Foot)
87th (Royal Irish Fusiliers)
88th (The Connaught Rangers)
89th (The Princess Victoria's Regiment of Foot)
90th (Perthshire Volunteers (Light Infantry))
91st (Argyllshire Regiment of Foot)

92nd (Gordon Highlanders)
93rd (Sutherland Highlanders)
94th (Royal Welsh Volunteers)
95th (Derbyshire Regiment of Foot)
96th (Queen's Own Germans)
97th (Earl of Ulster's Regiment of Foot)
98th (Prince of Wales' Regiment of Foot)
99th (Lanarkshire Regiment of Foot)
100th (Prince of Wales' Royal Canadian Regiment of Foot)
101st (Royal Bengal Fusiliers)
102nd (Royal Madras Fusiliers)
103rd (Royal Bombay Fusiliers)
104th (Bengal Fusiliers)
105th (Madras Light Infantry)
106th (Bombay Light Infantry)
107th (Bengal Infantry)
108th (Madras Infantry)
109th (Bombay Infantry)
110th Regiment of Foot [without territorial designation]
111th Regiment of Foot
. . . and so on to the 135th, the highest number reached.

Note: In 1816 the 95th (Rifles) were 'taken out of the line' and redesignated The Rifle Brigade. A new regiment was raised in 1823 for service in Malta, given the number '95' and two years later the territorial affiliation 'Derbyshire'.

The Guards, by definition, have never been part of the line.

The chart on the following pages sets out what may be called the succession of regiments from the Haldane–Childers reforms to the present day. It will enable the reader whose ancestor served in the infantry during the First World War to see into which present-day regiment his ancestral regiment has been progressively incorporated.

The list is reproduced, with permission, from the website of the Army Museums Ogilby Trust. The trust was founded in 1954 by the late Colonel Robert Ogilby DSO, whose personal experiences in two world wars persuaded him that the fighting spirit of the British soldier stemmed from the *esprit de corps* engendered by the army's regimental structure. The trust has played, and continues to play, a significant part in the establishment and development of some 136 museums in the United Kingdom, in which this *esprit de corps* is enshrined.

There is scarcely a city or a county town which is not therefore the home of a regimental or a corps museum. And it is in these museums that the real story of the army – the simple devotion of regimental soldiering – is told. A list of these museums, together with information about location, opening times and so much more, can be found at the trust's website: www.armymuseums.org.uk.

Succession of infantry titles

1881   1959

1881		1959
	Grenadier Guards	Grenadier Guards
	Coldstream Guards	Coldstream Guards
	Scots Guards	Scots Guards
		Irish Guards
		Welsh Guards
1	Royal Scots (Lothian Regiment)	Royal Scots (Royal Regiment)
25	The King's Own Borderers	King's Own Scottish Borderers
21	Royal Scots Fusiliers }	Royal Highland Fusiliers (Princess Margaret's
71 & 74	Highland Light Infantry }	Own Glasgow and Ayrshire Regiment)
42 & 73	Black Watch (Royal Highlanders)	Black Watch (Royal Highland Regiment)
72 & 78	Seaforth Highlanders (Ross-Shire Buffs, } Duke of Albany's) }	Queen's Own Highlanders
79	Queen's Own Cameron Highlanders }	
75 & 92	Gordon Highlanders	Gordon Highlanders
91 & 93	Princess Louise's (Argyll and Sutherland Highlanders)	Argyll and Sutherland Highlanders (Princess Louise's)
2	Queen's (Royal West Surrey Regiment)	Queen's Royal Surrey Regiment
31 & 70	East Surrey Regiment }	
3	The Buffs (East Kent Regiment) }	Queen's Own Buffs
50 & 97	Queen's Own (Royal West Kent Regiment) }	
35 & 107	Royal Sussex Regiment	Royal Sussex Regiment
57 & 77	Duke of Cambridge's Own (Middlesex Regiment)	Middlesex Regiment (Duke of Cambridge's Own)
37 & 67	Hampshire Regiment	Royal Hampshire Regiment
4	King's Own (Royal Lancaster Regiment) }	King's Own Royal Border Regiment
34 & 55	Border Regiment }	
8	King's (Liverpool Regiment) }	King's Regiment (Manchester and Liverpool)
63 & 96	Manchester Regiment }	
30 & 59	East Lancashire Regiment }	Lancashire Regiment (Prince of Wales's
40 & 82	Prince of Wales's Volunteers (South Lancashire Regiment) }	Volunteers)
47 & 81	Loyal North Lancashire Regiment	Loyal Regiment (North Lancashire)
5	Northumberland Fusiliers	Royal Northumberland Fusiliers
6	Royal Warwickshire Regiment	Royal Warwickshire Regiment
7	Royal Fusiliers (City of London Regiment)	Royal Fusiliers (City of London Regiment)
20	Lancashire Fusiliers	Lancashire Fusiliers
9	Norfolk Regiment }	1st East Anglian Regiment (Royal Norfolk
12	Suffolk Regiment }	and Suffolk)
10	Lincolnshire Regiment	2nd East Anglian Regiment
48 & 58	Northamptonshire Regiment }	
16	Bedfordshire Regiment }	3rd East Anglian Regiment (16th/44th Foot)
44 & 56	Essex Regiment }	
17	Leicestershire Regiment	Royal Leicestershire Regiment

1968/70	1992/4	2006/7
Grenadier Guards	Grenadier Guards	Grenadier Guards
Coldstream Guards	Coldstream Guards	Coldstream Guards
Scots Guards	Scots Guards	Scots Guards
Irish Guards	Irish Guards	Irish Guards
Welsh Guards	Welsh Guards	Welsh Guards
Royal Scots (The Royal Regiment)	Royal Scots (The Royal Regiment) }	
King's Own Scottish Borderers	King's Own Scottish Borderers }	
Royal Highland Fusiliers (Princess Margaret's Own Glasgow & Ayrshire Regt)	Royal Highland Fusiliers (Princess Margaret's Own Glasgow & Ayrshire Regt) }	
Black Watch (Royal Highland Regiment)	Black Watch (Royal Highland Regiment) }	The Royal Regiment of Scotland
Queen's Own Highlanders }	The Highlanders (Seaforth, Gordons, Camerons) }	
Gordon Highlanders }	}	
Argyll and Sutherland Highlanders (Princess Louise's)	Argyll and Sutherland Highlanders (Princess Louise's) }	
}	}	
} Queen's Regiment	Princess of Wales's Royal Regiment	The Princess of Wales's Royal Regiment
}		
}		
}		
} Royal Hampshire Regiment }		
King's Own Royal Border Regiment	King's Own Royal Border Regiment }	
King's Regiment	King's Regiment	The Duke of Lancaster's Regiment
}		
} Queen's Lancashire Regiment	Queen's Lancashire Regiment }	
}		
}		
}		
} Royal Regiment of Fusiliers	Royal Regiment of Fusiliers	The Royal Regiment of Fusiliers
}		
}		
}		
}		
} Royal Anglian Regiment	Royal Anglian Regiment	The Royal Anglian Regiment
}		
}		

1881	1959

1881	1959
14 Prince of Wales's Own (West Yorkshire Regiment) }	Prince of Wales's Own Regiment of Yorkshire
15 East Yorkshire Regiment }	
19 Princess of Wales's Own (Yorkshire Regiment)	Green Howards (Alexandra, Princess of Wales's Own Yorkshire Regiment)
33 & 76 Duke of Wellington's Regiment	Duke of Wellington's Regiment (West Riding)
22 Cheshire Regiment	Cheshire Regiment
29 & 36 Worcestershire Regiment	Worcestershire Regiment
45 & 95 Sherwood Foresters (Derbyshire Regiment)	Sherwood Foresters Regiment (Nottinghamshire & Derbyshire Regiment)
38 & 80 South Staffordshire Regiment }	Staffordshire Regiment (The Prince of Wales's)
64 & 98 Prince of Wales's (North Staffordshire Regiment) }	
23 Royal Welsh Fusiliers	Royal Welch Fusiliers [Welch after 1922]
24 South Wales Borderers	South Wales Borderers
41 & 69 Welsh Regiment	Welch Regiment [Welch after 1922]
27 & 108 Royal Inniskilling Fusiliers	Royal Inniskilling Fusiliers
83 & 86 Royal Irish Rifles	Royal Ulster Rifles
87 & 89 Princess Victoria's (Royal Irish Fusiliers)	Royal Irish Fusiliers (Princess Victoria's)
	Parachute Regiment
2nd King Edward VII's Own Gurkha Rifles	2nd King Edward VII's Own Gurkha Rifles (The Sirmoor Rifles)
6th Gurkha Rifles	6th Queen Elizabeth's Own Gurkha Rifles
7th Gurkha Rifles	7th Duke of Edinburgh's Own Gurkha Rifles
10th Gurkha Rifles	10th Princess Mary's Own Gurkha Rifles
11 Devonshire Regiment }	Devonshire and Dorset Regiment
39 & 54 Dorset Regiment }	
28 & 61 Gloucestershire Regiment	Gloucestershire Regiment
49 & 66 Princess Charlotte of Wales's (Berkshire Regiment) }	Duke of Edinburgh's Royal Regiment
62 & 99 Duke of Edinburgh's (Wiltshire Regiment) }	
13 Prince Albert's (Somerset Light Infantry) }	Somerset and Cornwall Light Infantry
32 & 46 Duke of Cornwall's Light Infantry }	
53 & 85 King's Light Infantry (Shropshire Regiment)	King's Shropshire Light Infantry
51 & 105 King's Own Light Infantry (South Yorkshire Regiment)	King's Own Yorkshire Light Infantry
68 & 106 Durham Light Infantry	Durham Light Infantry
43 & 52 Oxfordshire Light Infantry	1st Green Jackets 43rd and 52nd
60 King's Royal Rifle Corps	2nd Green Jackets, King's Royal Rifle Corps
Rifle Brigade (Prince Consort's Own)	3rd Green Jackets, The Rifle Brigade
	Special Air Service Regiment
18 Royal Irish Regiment	Disbanded 1922
26 & 90 Cameronians (Scottish Rifles)	Cameronians (Scottish Rifles)
65 & 84 York and Lancaster Regiment	York and Lancaster Regiment
88 & 94 Connaught Rangers	Disbanded 1922
100 & 109 Prince of Wales's Leinster Regiment (Royal Canadians)	Disbanded 1922
101 & 104 Royal Munster Fusiliers	Disbanded 1922
102 & 103 Royal Dublin Fusiliers	Disbanded 1922

1968/70	1992/4	2006/7
Prince of Wales's Own Regiment of Yorkshire	Prince of Wales's Own Regiment of Yorkshire	} The Yorkshire Regiment
Green Howards (Alexandra, Princess of Wales's Own Yorkshire Regiment)	Green Howards (Alexandra, Princess of Wales's Own Yorkshire Regiment)	}
Duke of Wellington's Regiment	Duke of Wellington's Regiment	}
Cheshire Regiment	Cheshire Regiment	}
} Worcestershire and Sherwood Foresters Regiment (29th/45th Foot)	Worcestershire and Sherwood Foresters Regiment (29th/45th Foot)	} The Mercian Regiment
Staffordshire Regiment (The Prince of Wales's)	Staffordshire Regiment (The Prince of Wales's)	}
Royal Welch Fusiliers	Royal Welch Fusiliers	} The Royal Welsh
} Royal Regiment of Wales (24th/41st Foot)	Royal Regiment of Wales (24th/41st Foot)	}
} Royal Irish Rangers (27th (Inniskilling) 83rd and 87th)	Royal Irish Regiment (amalgamation of Royal Irish Rangers and Ulster Defence Regiment)	} The Royal Irish Regiment
Parachute Regiment	Parachute Regiment	The Parachute Regiment
2nd King Edward VII's Own Gurkha Rifles (The Sirmoor Rifles) }		
6th Queen Elizabeth's Own Gurkha Rifles }	Royal Gurkha Rifles	The Royal Gurkha Rifles
7th Duke of Edinburgh's Own Gurkha Rifles }		
10th Princess Mary's Own Gurkha Rifles }		
Devonshire and Dorset Regiment	Devonshire and Dorset Regiment	}
Gloucestershire Regiment }	} Royal Gloucestershire, Berkshire and Wiltshire Regiment	}
Duke of Edinburgh's Regiment (Berkshire and Wiltshire) }		}
}		} The Rifles
} Light Infantry	Light Infantry	}
}		}
} Royal Green Jackets	Royal Green Jackets	}
Special Air Service Regiment	Special Air Service Regiment	Special Air Service Regiment

The Cameronians – disbanded 1968
Y&L – suspended animation 1968
(subsequently formally disbanded)

Notes and Further Reading

Introduction

Professor John McManners, whose thoughts conclude my foreword, was a son of the parsonage. In my Hervey novels I write of officers in the days when commissions were bought and sold – the era of 'purchase'. In the duke of Wellington's army they were a reliable 'officer class', a branch of the gentry, usually the minor gentry, and the parson's son was characteristic. But the McManners parsonage was quite different from the characteristic gentry parsonage of my own novels and of Wellington's army; for Jack McManners' father, the vicar of West Pelton, had earlier been a Durham miner, a Labour activist and an atheist. In the extract I quote from *Fusilier* (2002), Professor McManners is actually reflecting on an early encounter during his time with the Royal Northumberland Fusiliers, at the siege of Tobruk, an encounter which left many Germans dead by his own hand. It caused him not only to conclude what he does of the ultimate reason of the social order, but also to re-examine his faith:

Since that day in Tobruk I had thought through and rationalized the impact of the sudden sight of those slaughtered Germans; what had happened was that Christian belief had become intensely personal. Up to then I had followed, in intellectual debates, the *via eminentia* of Christian apologetics: the order and beauty of creation, the working of laws in the universe, spiritual experience, show there must be a God. Then, look around for signs of God's activity in the moral sphere, and we find the story of a good man preaching and healing in Palestine; he is crucified, his followers worship him – but the sight of the dead bodies in the sand-bagged post at Tobruk ended that chain of argument. I had seen what men out of their God-given freedom do to each other. This was the face of evil, and I was part of the evil, being glad, even in revulsion, that they were dead . . . In face of this, you cannot believe in God, the God of the deists. But you can, almost in despair, turn to the God who suffers with his creation, accepting the burden of sin that arises from human freedom, and taking it on himself. Religious apologetics begin from Jesus on the cross: the Christian life is allegiance to him.

Not everyone will agree with Professor McManners's conclusions, but his experiences and reflections are nonetheless compelling.

As general references I recommend: the *Oxford* (beautifully) *Illustrated History of the British Army* (1994; revised 2003), edited by the late David Chandler and by Ian Beckett; the Marquess of Anglesey's multi-volume *History of the British Cavalry* (1973–95), a comprehensive account of the cavalry's post-Napoleonic experience and development, which tells much of the Victorian army besides; and Anthony Makepeace-Warne's 1995 *Companion to the British Army* (though now out of date on some regimental and technical detail).

For further detailed – if highly selective – reading I have

given recommendations below in relation to each chapter or group of chapters.

Chapter 1: Background, and the Civil War

In considering England's fortuitous geography (from the late thirteenth century the Welsh were no longer a threat), and her lack of a standing army other than a few garrisons in forts up and down the country, it is well to remember that not only was the border with Scotland relatively short (96 miles), it was difficult country to traverse, with few roads, and therefore requiring nothing like so great a frontier garrison as borders on the Continent (nor was Scotland like Austria to France, or Sweden to Prussia; rather more the 'nuisance neighbour'). The Scottish border was also a considerable distance from anywhere of strategic importance: there was ample time for the county militias to mobilize against an army marching south, and consequently little need of standing forces.

Nevertheless, Englishmen under arms did find themselves on the Continent from time to time: 100,000 (volunteers from the militia, as well as levies) saw service overseas during the Armada days – some 15 per cent of all able-bodied males out of a population of less than five million. And as a result the nation's finances were awry for decades. A century and half later the historian Edward Gibbon would write that 'It has been calculated by the ablest politicians that no State, without being soon exhausted, can maintain above a hundredth part of its members in arms and idleness.' After the Spanish threat had abated, England maintained scarcely a thousandth part in arms. These men were scattered about the country in little coastal forts, in the great border bastions at Carlisle and Berwick, in barracks in Ireland, and in the field with the Anglo-Dutch brigade to guarantee the frontiers of the newly independent United Provinces (and a Protestant Scheldt estuary, the feared springboard for

Spanish invasion). A few Scots units continued in French service too.

For the Civil War in detail there is no more literary a narrative than C. V. Wedgewood's classic *The King's War* (1959), which should ideally be read after her equally absorbing *The King's Peace* (1955). Trevor Royle's splendid *Civil War* (2004) is likely to remain the best account of the military operations for many a year. Charles (Earl) Spencer's *Prince Rupert* (2007) is more dashing than the equally authoritative 1996 biography by (General) Frank Kitson, and Lucy Worsley's *Cavalier* (2007) is both a scholarly and an engrossing account of one of the great families at war (the duke of Newcastle's). Antonia Fraser's *Cromwell, Our Chief of Men* (1973) is more engaging than its subject, while John Adamson's *The Noble Revolt* (2007), though contentious, is quite masterly.

Chapters 2 and 3: The Restoration and 'Glorious Revolution'

Parliament's fear of the King's using the army to coerce the country – and therefore of martial law – was not irrational, given the continuing and justifiable suspicion of Stuart inclinations to absolutism and Catholicism, and Charles's predisposition to secret negotiations abroad, especially with his mother's native France. When his father had declared martial law for the levies raised for the expedition to La Rochelle in 1627, and the countryside began to feel the depredations of billeting and arbitrary military justice, the Commons had petitioned the King, asserting in their Petition of Right that it was 'wholly and directly contrary to the said laws and statutes of your realm' to visit troops and their law on the civil population. Only if the King exercised his suspending power – a true red rag to the parliamentary bull – could martial law be declared, and the army properly discipline its soldiers. Charles II never dared to exercise those powers in England, although he came close to it in the large number of levies raised for the Dutch War in 1673, and the war with France in 1678.

How far Marlborough was influenced by Monck can only be conjecture. Richard Holmes, Marlborough's most recent biographer, believes that he read one or two classical texts on war, Vegetius almost certainly, but possibly nothing else, for his approach was always pragmatic and showed no sign of any theoretical study, unlike officers in the continental armies. That said, as General Sir Michael Rose has pointed out, in the small world in which the young John Churchill moved he could not have been unaware of Monck's methods and achievements. Indeed, it seems inconceivable that a young officer on the rise, moving in Court circles – sharing the King's mistress, indeed – would not hear, or want to hear, the thoughts of Monck, who was, after all, a kingmaker as well as a field soldier. Other than the classical texts and the various medieval treatises there was not yet a great deal to study: Monck's work was the most up-to-date interpretation of generalship in a modern context that Marlborough could have studied in the 1670s, even without regard to its actual merits (which are considerable). There is therefore every good reason to claim a continuity.

As for the various under-secretaries running the army, these arrangements applied to the general regulation of the army, not to its equipping or supply. The Board of Ordnance, originating in the fifteenth century, was responsible for providing the army (and, until 1830, the navy) with weapons, ammunition and warlike stores – generally referred to as 'ordnance' – and with the expertise to use them (the sappers, miners and gunners). Indeed, until 1855, when the board was incorporated with the War Office, engineers and artillery were the entire responsibility of the head of the board, the Master General of the Ordnance (MGO), rather than the commander-in-chief. In later years the MGO held cabinet rank. Between 1702 and 1722 the duke of Marlborough was both commander-in-chief (or captain-general, as it was more usually known) and MGO, the only time the two posts were held simultaneously by the same man. Non-warlike stores (principally food and forage) and transport were supplied under the direct orders of the Treasury by civilian agents known as commissariat officers. With the later supervision of the commissioners of public accounts and the Board of Audit, civil control over army expenditure became ever tighter.

Of the battle of the Boyne, which has of course passed into the legends – myths? – of Ulster, it is worth observing that the casualty figures were quite low for an engagement of such a scale. Of the 50,000 or so participants, about 2,000 died, three-quarters of whom were Jacobites. The reason for the low death toll was that at the Boyne there was no follow-up when the defeated army left the field – sometimes the phase of greatest slaughter – for James's cavalry very effectively screened the retreat. The Jacobites were badly demoralized by their defeat, however, and many of the Irish infantry deserted. William's men triumphantly marched into Dublin two days after the battle, the Jacobite army fleeing to Limerick beyond the river Shannon, where they were besieged. James, however, rode with a small escort to Duncannon and returned to exile in France, even though his army was still in the field. His speedy exit enraged his Irish supporters, who nevertheless fought on until the Treaty of Limerick in 1691. They deserved better – as did the deluded followers of his equally fickle son and grandson in 1715 and 1745 respectively. Indeed, James was derisively nick-named by the Irish *Seamus a' chaca* – a title that translates literally as 'James the Shit'.

For further reading, Charles FitzRoy's *Return of the King* (2007) is fascinating on the actual process of restoration, and on Monck's part in it. There is much in Barney

White-Spunner's *Horse Guards* (2007) on the Restoration forces, and Antonia Fraser is every bit as readable in *King Charles II* (1979) as she was in *Cromwell*, if only indirectly touching on purely military matters. There is a very fine chapter on the Restoration army by Professor John Childs in the *Oxford Illustrated History of the British Army*; and his *The Williamite War in Ireland* (2007) gives a fine picture of the army in the reigns of Charles, James and William. And then on Marlborough there is David Chandler's *Marlborough as Military Commander* (new edn 2003), Richard Holmes's *Marlborough, England's Fragile Genius* (2008) – and, of course, Churchill's monumental and by no means wholly hagiographical *Marlborough: His Life and Times.* And should anyone want to read more of the sad and wretched affair that was the battle of Sedgemoor, there is the irrepressible David Chandler again in *Sedgemoor 1685: from Monmouth's invasion to the Bloody Assizes* (new edn 2001).

Chapters 4 and 5: Blenheim and Afterwards

The foundation stone of Blenheim Palace was laid on 18 June 1705 on a site prepared by the royal gardener Henry Wise. Building continued at the Crown's expense until 1712, when, the Marlboroughs having fallen from favour, the Treasury ceased to provide funds. On the Queen's death in 1714 the Marlboroughs returned from voluntary exile, but little was done until debts to Blenheim workmen were partially settled in 1716. Building then continued at the Marlboroughs' expense, and the family took up residence in 1719. After the duke's death in 1722 Sarah, duchess of Marlborough, completed the chief features of Vanbrugh's house plan, together with outworks such as the Grand Bridge, the Triumphal Arch and the Column of Victory. Her work was substantially complete by the early 1730s. (These details are taken from *A History of the County of Oxford*, vol. 12.) However, Vanbrugh's baroque gardens were completely reworked by 'Capability' Brown in his characteristic English naturalist style before George III's visit in 1786.

For further reading, G. M. Trevelyan's *England under Queen Anne*, though published in the 1930s, is excellent on the general background. On Blenheim itself, besides Churchill and Holmes, there is David Chandler again: *Blenheim Preparation* (new edn 2004), and also the highly readable *Blenheim, Battle for Europe* (2004) by Charles (Earl) Spencer, one of Marlborough's descendants. For details of the supply arrangements for the 'scarlet caterpillar' there is Martin van Crefeld's *Supplying War: Logistics from Wallenstein to Patton* (1977), though not for the bedtime reader.

Chapters 6 and 7: Dettingen and Culloden

The battlefield at Culloden is the best preserved of any in Britain, with markers showing very clearly the positions of the two sides. There is an impressive visitor centre, though at times it gives the impression that the Forty-five was a war between England and Scotland (with eventual victory in the 1999 devolution). Nearby, Fort George, subsequently enlarged and strengthened, is open to visitors.

There is an excellent article on the findings of the court martial of Sir John Cope in the *History of Scotland* magazine, vol. 2, no. 3, May–June 2002: 'Unlucky or incompetent? History's verdict on General Sir John Cope, Part II: the battle of Prestonpans and the aftermath' by Martin B. Margulies, which gives a fascinating insight into the workings of the army at the time, its condition and its generalship. The monograph *Culloden Moor and Story of the Battle* (1867) by the Inverness antiquarian Peter Anderson, revised in 1920 by his son, a don of Aberdeen University, is a work of deep scholarship and local knowledge. There is an excellent Osprey volume on the battle – with all the superb illustration and mapping characteristic of that publisher –

as well as a colourful one on the Jacobite army itself.

The allied army in the War of Austrian Succession was and is sometimes referred to as 'the Pragmatic Army' after the 'Pragmatic Sanction' of 1713, a legal mechanism to ensure that the Austrian throne and Habsburg lands would be inherited by Emperor Charles VI's daughter, Maria Theresa. The Holy Roman Empire was famously neither holy, Roman, nor an empire, but by this time the residual *Reich* of German nations. It was elective, and as elector of Hanover George II was a member of the electoral college of eight. The empire had less and less real power but retained enormous prestige, being the remnant of the former Western Roman Empire (though its constitutional arrangements were more Byzantine than the Eastern Empire of Byzantium itself). It was as well that George II disappeared periodically to Hanover (spelt with a second 'n' by his German subjects) to attend to its affairs, for they would have taxed the brains of all his archbishops combined. Nevertheless, Reed Brown's *The War of Austrian Succession* (1993) manages to steer the reader through its complexities, and Michael Orr's *Dettingen 1743* (1972), though a little dated now, is a clear account of the battle. And although it is not directly germane to the story of the making of the British army, David Fraser's incomparable *Frederick the Great* (2000) describes the rise of Prussia and its army magnificently, while for the life of the man who wrote all those celebratory anthems, Jonathan Keates's *Handel: The Man and his Music*, first published in 1985 but revised and republished in 2008, is a delight.

Chapters 8 and 9: The Seven Years War

The outcome of the war was indeed a great amendment to the geography books. In India, France was emasculated. In Europe, Germany was shot of all French troops, Portugal was rid of the Spanish, and Minorca was restored to Britain. French Canada was now British. In the West Indies, three islands were returned to France – Martinique, Guadeloupe and St Lucia – but Britain kept Grenada, St Vincent, Dominica and Tobago, giving it the strategic hand in the Lesser Antilles, the guardians of the Caribbean and much of the Spanish Main. Florida was gained in exchange for Cuba, seized a year before the peace in a fine action by Admiral Sir George Pocock and the 38-year-old General George Keppel. The Spanish were pleased enough with the exchange – a troublesome and vulnerable peninsula with British and French land neighbours, for a rich spice island – but for Britain, gaining Florida was far more significant in strategic terms, if not nearly so profitable, for Spanish Florida had been a brooding threat to South Carolina. The fly in the ointment was still Louisiana (a vastly bigger territory than the present state), but it had at least changed hands – from French to Spanish.

This is a Cinderella of a period for the army as far as popular historians are concerned: between the Marlburian and the Wellingtonian feasts it must seem comparatively poor pasture. But there are some interesting collections of papers published by the Army Records Society (and Sutton Publishing), notably volumes VI, *Colonel Samuel Bagshawe and the Army of King George II, 1731–1762*; XVIII, *The Journal of Corporal William Todd, 1745–1762*; and XX, *Amherst and the Conquest of Canada.* Details of all the society's publications are available on its website: www.armyrecordssociety.org.uk.

Chapter 10: The American Revolutionary War

There was a certain paradox in Pitt's strategic aim in the Seven Years War, the destruction of French power in Canada, which some at the time recognized: 'I don't know whether the neighbourhood of the French to our North American colonies was not the greatest security of their dependence on the mother country, which I feel will be slighted by them when their apprehension of the French is removed,' as the duke of Bedford remarked to the duke of Newcastle. And so was fired in due course the 'shot heard round the world':

> By the rude bridge that arched the flood,
> Their flag to April's breeze unfurled,
> Here once the embattled farmers stood
> And fired the shot heard round the world.
> (opening stanza of Ralph Waldo Emerson's 'Concord Hymn', 1837)

The American Revolutionary War is altogether more popularly covered than the Seven Years War which gave the duke of Bedford such cause for concern. *Rebels and Redcoats: The American Revolutionary War* (2003) by Hugh Bicheno is one of the most recent works. *Fusiliers* (2007) by Mark Urban is an absorbing account of one battalion's experience of the shock of the new type of war, as its subtitle suggests: *How the British Army Lost America but Learned to Fight.* There is an excellent collection of documents and first-hand accounts by an officer of the Black Watch in the Army Records Society's volume XIII, *John Peebles' American War, 1776–1782.* The war is also fertile ground for comparison with later campaigns, of which (General Sir) Michael Rose's *Washington's War* (2007) is perhaps the most apt in its analysis of the war alongside that of the Iraq insurgency.

Chapters 11–14: The Napoleonic Wars

Prior to the formation of the experimental rifle corps in 1800, the Board of Ordnance held a trial at Woolwich to select a standard rifle pattern. The design by one Ezekiel Baker was chosen, and 800 rifles were produced. After several modifications, the third and final model had the barrel shortened from 32 to 30 inches and the calibre reduced to allow it to fire a .625-inch calibre carbine bullet with a greased patch to grip the seven rectangular grooves in the barrel. It took a 24-inch 'sword bayonet' (hence Rifle regiments always fixed – and still fix – swords not bayonets. The Baker was 45 inches from muzzle to butt, 12 inches shorter than the Brown Bess musket, and weighed 9 pounds. Black powder fouling in the grooves made the weapon much slower to fire and would affect the accuracy of the weapon, so a cleaning kit was issued to riflemen (the Brown Bess was not issued with a cleaning kit).

Vimeiro was a terrible baptism of British volley fire for Napoleon's army of Spain, but there had been an earlier example of devastating musketry – though it was largely unobserved, and its effects were not felt beyond the actual battlefield. On 27 June 1806 a British force of some 5,000 men commanded by Major-General John Stuart sailed from Messina in Sicily, landing in the Gulf of Sant'Eufemia, in Calabria in the Kingdom of Naples, three days later. A French force of about 6,000 moved to confront them, and on 4 July the two met on the plain of Maida, the British occupying a low ridge. As the French advanced, the British held their fire until at about 50 yards they began volleying. The French faltered, the British charged with the bayonet and the French were routed. The entire action lasted only some fifteen minutes. But Stuart, instead of following up (there were few other French in Calabria), marched south and after a series of minor skirmishes returned to Sicily. The opportunity thus lost to open a 'second front' would not come again for two years – when it presented itself in the Peninsula.

Books on the Peninsular War and Waterloo are legion. Professor Charles Esdaile's *The Peninsular War* (2002) is particularly good on the overall picture. Michael Glover's *The Peninsular War 1807–1814: A Concise Military History*, though first published in 1974, is still an excellent account. Christopher Hibbert's *Corunna* (1961) has never been bettered in its picture of the fighting and its cool assessment of Sir John Moore's generalship. Mark Urban's *Rifles: Six*

Years with Wellington's Legendary Sharpshooters (2004) gives the true background to Bernard Cornwell's legendary 'Sharpe' novels, and his *The Man Who Broke Napoleon's Codes* (2002) is an absorbing account of intelligence-gathering in the duke of Wellington's campaigns.

Waterloo is a spring that can never be drunk dry. But there are four 'essential' books. Andrew Roberts's brief *Waterloo: Napoleon's Last Gamble* (2005) is a brilliant overview which can be comfortably read during the time it takes to get to Brussels by Eurostar. Alessandro Barbero's *The Battle* (2006) is the book to walk the battlefield with, and David Howarth's *A Near Run Thing* (1968) the one to retire with. And then a week or so later, having thought things over, and with many questions in one's mind, David Chandler's *Waterloo: The Hundred Days* (1980) is the one to go to for the answers.

I have not written about the War of 1812 (with the United States) – an unedifying and inglorious affair on both sides – and nor has any-one else much. The Americans' attack on Fort York (Toronto), the retributive burning of the White House, the composition of the US national anthem in the wake of the siege of Washington Roads, and the attack on New Orleans make for 'colour', but it is the naval side of the war that is the most interesting in many ways: (President) Theodore Roosevelt wrote a bracing account of the war at sea, and Alfred Thayer Mahan developed his doctrine of sea power on an examination of the fighting – such as it was – on the Great Lakes and in the Atlantic. Jon Latimer's recent *1812: War with America* (2007) is probably the best place to begin if one must.

Chapters 15–18: The Crimea to Khartoum

The Gatling gun was not a machine gun in the modern sense – it did not feed rounds into a single breech. Instead it had multiple breeches connected to multiple rotating barrels, each of which fired a single shot as it reached a certain point in the cycle, ejecting the spent cartridge and loading a new round. The multi-barrel system was a far bulkier affair and less accurate than the later machine gun, but it allowed higher rates of fire without the problem of overheating with which single-barrel weapons had to contend. It was designed by the American Dr Richard J. Gatling in 1861 and saw much service in the American Civil War. The Maxim gun, invented in 1884, was the first true machine gun, with a rate of fire of 600 rounds per minute (the Gatling's was 200).

The Lee–Metford was the first .303-inch rifle, the calibre that was to see the army through the two world wars. A spring in the base of the magazine, which held ten rounds, forced up each round in turn to the level of the chamber, so that the firer merely had to operate the bolt to eject the spent round and feed another into the breech. In well-trained hands they were thus capable of extremely rapid and accurate fire, with a killing range of 2,000 yards.

The case of Major-General Sir Hector Macdonald – 'Fighting Mac' – needs a little further explaining, sad though the story is. When the allegations of homosexuality (and possibly paedophilia) were made, the governor-general in Ceylon instigated an inquiry which reported thus:

> In reference to the grave charges made against the late Sir Hector Macdonald, we, the appointed and undersigned Commissioners, individually and collectively declare *on oath* [author's italics] that, after the most careful, minute, and exhaustive inquiry and investigation of the whole circumstances and facts connected with the sudden and unexpected death of the late Sir Hector Macdonald, unanimously and unmistakably find absolutely no reason or crime whatsoever which would create feelings such as would determine suicide, in preference to conviction of any crime affecting the moral and irreproachable character of so brave, so fearless, so glorious

and unparalleled a hero: and we firmly believe the cause which gave rise to the inhuman and cruel suggestions of crime were prompted through vulgar feelings of spite and jealousy in his rising to such a high rank of distinction in the British Army: and, while we have taken the most reliable and trustworthy evidence from every accessible and conceivable source, have without hesitation come to the conclusion that there is not visible the slightest particle of truth in foundation of any crime, and we find the late Sir Hector Macdonald has been cruelly assassinated by vile and slandering tongues. While honourably acquitting the late Sir Hector Macdonald of any charge whatsoever, we cannot but deplore the sad circumstances of the case that have fallen so disastrously on One whom we have found innocent of any crime attributed to him.

There is a fine statue of him in his home town of Dingwall.

The Crimean War has a most devoted following, with its own research society: http://cwrs.russianwar.co.uk/. Trevor Royle's *The Great Crimean War* (2000) is excellent, as is W. Baring Pemberton's *Battles of the Crimean War* (1968), if a little dated. There are many collected letters and diaries, including William Howard Russell's despatches, and *A Bearskin's Crimea* (2005) by Algernon Percy, both published by Pen and Sword Books (whose Crimea list is a particularly good one). On the charge of the Light Brigade, besides Cecil Woodham Smith's *The Reason Why* (1953), the best account by far is *Hell Riders: The True Story of the Charge of the Light Brigade* (2004) by Terry Brighton. I am afraid that Mark Adkin's *The Charge* (1996), while beautifully written, is preposterously off-mark in suggesting that Nolan deliberately led the Light Brigade on to the Russian guns to prove that it could be done. David Murphy's *Ireland and the Crimean War* (2002) is quite superb, especially on the nursing work of the Sisters of Mercy and on the squabbling between the Jesuits and the secular clergy chaplains.

The Indian Mutiny is equally well served by chroniclers of both the military side and the civil. Michael Edwardes' *Red Year: The Indian Rebellion of 1857* (1973) is still a fine work some thirty-five years after publication; Julian Spilsbury's *The Indian Mutiny* (2008) is more recent. The wars of the expanding empire are the subject of Saul David's *Victoria's Wars* (2007). Mike Snook's *How Can Man Die Better: The Secrets of Isandhlwana Revealed* (2005) is a very measured account of the first disastrous encounter with the Zulus, as is his *Like Wolves on the Fold: The Defence of Rorke's Drift* (2006). The National Army Museum's *Book of the Zulu War* (2004) by Ian Robertson is comprehensive, and the late David Rattray's *Guidebook to the Anglo-Zulu War Battlefields* (2003) an authoritative handbook by a man who lived and breathed the country for most of his life. Michael Barthorp's *Afghan Wars and the North-West Frontier 1839–1947* (1982) is a good overview of the 'Great Game', and *The Savage Frontier: A History of the Anglo-Afghan Wars* (1990) by D. S. Richards gives a fine picture of the nature of the fighting. For a fictional account of the life of the soldier-spy on the frontier, John Masters's *The Lotus and the Wind* (1953) is a gripping read. Michael Asher's *Khartoum: The Ultimate Imperial Adventure* (2005) is superb. But above all, there is Kipling. No one seems to be able to get near him for capturing the essence of imperial soldiering; and of course he knows India. Consider the last two verses of 'Arithmetic on the Frontier' as a pointer to modern ideas of 'civil reconstruction':

One sword-knot stolen from the camp
Will pay for all the school expenses
Of any Kurrum Valley scamp
Who knows no word of moods and tenses,
But, being blessed with perfect sight,
Picks off our messmates left and right.

With home-bred hordes the hillsides teem.
The troopships bring us one by one,
At vast expense of time and steam,
To slay Afridis where they run.
The 'captives of our bow and spear'
Are cheap, alas! as we are dear.

Chapters 19–20: The Boer War and the Edwardian Reforms

Britain had approached the Orange Free State and the Transvaal Republic in 1875 to try to arrange a federation of the British and Boer territories (modelled on the federation of French and English provinces of Canada), but the Boers rejected the proposal, notwithstanding the economic benefits. Nor were the territorial disputes always fuelled by mineral wealth. Bechuanaland (modern Botswana, located north of the Orange River) was claimed by the Germans to the west and the Boers to the east. Although the area had almost no economic value, after the Germans annexed Damaraland and Namaqualand (modern Namibia) in 1884, the British annexed Bechuanaland in an intelligent and relatively bloodless operation.

The Boer War has long been a popular subject of both historical writing and also fiction. Stuart Cloete's *Rags of Glory* (1963) is arguably the best novel of the war with a fine portrait of the British army's dilemma in South Africa. For the history of the war, Thomas Pakenham's *The Boer War* (1974) is magisterial. Baring Pemberton is every bit as good in *Battles of the Boer War* (1964) as he is on those of the Crimea. Two volumes of papers in the Army Records Society's series by Professor André Wessels are rich in detail on the thinking behind the campaigns: volumes XVII, *Lord Roberts and the War in South Africa, 1899–1902*, and XXV, *Lord Kitchener and the War in South Africa*. On army reform in the decade after the war, the *Oxford Illustrated History of the Army* is succinct, the chapter being written by Edward Spiers whose earlier *Haldane: An Army Reformer* (1980) remains the definitive work. On the Curragh 'Mutiny' *The Army and the Curragh Incident, 1914* by Ian Beckett, volume II of the ARS's series, is fascinating.

Chapters 21–3: The First World War and Beyond

Sooner or later any discussion of the First World War gets to the question of Haig's generalship. Opinions among academics range from the 'butcher of the Somme' to the 'only man for the job'. Interestingly, most soldiers steer clear of the argument. This may well be out of respect for the sheer magnitude of the problem that faced Haig when he took over the reins in November 1915, and because he was still in the saddle three years later when the allies finally prevailed – Haig having been, indeed, the orchestrator of a greater part of the final victory in the 'Hundred Days'. And in fact if Haig were to be judged on his handling of that last campaign and also of First Ypres, his first offensive–defensive battle, he would be acclaimed as a great general. But there is of course the problem of the years in between. I myself believe it is not possible to come to any overall settled view of Haig: the pendulum must be allowed to swing. There was no one else who could have taken Haig's place with any assuredly greater insight (including Smith-Dorrien). If there had been, Lloyd George would have replaced Haig. Three books are worthy of study for the spectrum of opinion. At the 'anti-Haig' end of the spectrum are Trevor Wilson and Robin Prior with *Command on the Western Front* (1991). At the 'saintly' end is John Terraine's *Douglas Haig, the Educated Soldier* (1963); and closer to the middle, but nevertheless pro-Haig, is Gary Sheffield's *Forgotten Victory* (2002).

For the fighting itself, there is no single better account of any battle than the superb *Ypres: Death of an Army* (1967) by Anthony Farrar-Hockley, with all the fine turn of phrase and military judgement of that outstanding soldier-author. L. A. Carlyon's *Gallipoli* (2002) is first-rate. And for the war in the Middle East, about which I have otherwise written little, there is *The Last Crusade: The Palestine Campaign in the First World War* (2002) by Anthony Bruce.

Here it might be appropriate to address the business of the Guards. I wrote in Chapter 13 that they have never faltered. This is a remarkable claim, but the evidence bears it out. I asked several Guards generals if the Guards had ever failed. Their response was 'of course' – for guardsmen are only human, and history is long. But then none of them could give an example except an occasional bad tactical judgement. In the Falklands, for example, the Welsh Guards (who were the last regiment to be formed, in 1915) were horribly bruised by the airstrike on the troop-carrier *Sir Galahad*, with thirty-two killed and many more wounded, but they recovered quickly and the battalion went on to win an MC and three MMs. There was one occasion in Italy, in May 1944 at Monte Grande northwest of Cassino, when the 3rd Grenadiers were thrown back from a strongly held German position, shelled heavily, and although the regimental history is a little coy, it does not seem to have been an unreasonable withdrawal. Something of the Guards' 'majesty' in action is conveyed in Sir Arthur Conan Doyle's 1919 poem about the attack of the 2nd Guards Brigade at Loos in September 1915, 'The Guards Came Through':

Men of the Twenty-first [Division]
Up by the Chalk Pit Wood,
Weak with our wounds and our thirst,
Wanting our sleep and our food,
After a day and a night—
God, shall we ever forget!
Beaten and broke in the fight,
But sticking it – sticking it yet.
Trying to hold the line,

Fainting and spent and done,
Always the thud and the whine,
Always the yell of the Hun!
Northumberland, Lancaster, York,
Durham and Somerset,
Fighting alone, worn to the bone,
But sticking it – sticking it yet.

Never a message of hope!
Never a word of cheer!
Fronting Hill 70's shell-swept slope,
With the dull dead plain in our rear.
Always the whine of the shell,
Always the roar of its burst,
Always the tortures of hell,
As waiting and wincing we cursed
Our luck and the guns and the Boche,
When our Corporal shouted, 'Stand to!'
And I heard some one cry, 'Clear the front for the Guards!'
And the Guards came through.

Our throats they were parched and hot,
But Lord, if you'd heard the cheers!
Irish and Welsh and Scot,
Coldstream and Grenadiers.
Two brigades, if you please,
Dressing as straight as a hem,
We – we were down on our knees,
Praying for us and for them!
Lord, I could speak for a week,
But how could you understand!
How should your cheeks be wet,
Such feelin's don't come to you.
But when can me or my mates forget,
When the Guards came through?

'Five yards left extend!'

675

It passed from rank to rank.
Line after line with never a bend,
And a touch of the London swank.
A trifle of swank and dash,
Cool as a home parade,
Twinkle and glitter and flash,
Flinching never a shade,
With the shrapnel right in their face
Doing their Hyde Park stunt,
Keeping their swing at an easy pace,
Arms at the trail, eyes front!

Man, it was great to see!
Man, it was fine to do!
It's a cot and a hospital ward for me,
But I'll tell 'em in Blighty, whereever I be,
How the Guards came through.

The Guards have never formed Territorial battalions, nor did they raise Kitchener battalions. The few additional battalions raised in both world wars therefore had the benefit of a strong core of professional Guards NCOs and officers from the outset.

I wrote too that there were three particularly interesting developments – innovations – of the First World War as far as the making of the army is concerned. One was 'mission-specific' training and rehearsal. All individual reinforcements on the Western Front had to pass through the base training depot at Etaples to be put through a course in trench-craft, the first time there had been a training regime for a particular theatre of operations. 'The bull ring', as the training ground was known, was detested by all, the bayonet drill especially, but the instruction in grenade handling, sentry work and field hygiene was grudgingly acknowledged as useful. Training also took place before attacks, with briefings using large models constructed by the sappers, and over ground chosen for its similarity to the sector in which the attack was to take place. All this became standard procedure in the Second World War, and has continued ever since – with added impetus gained during the long Northern Ireland campaign. Indeed, mission-specific training based on thorough operational analysis is one of the things that gives the British army its current edge.

Another development was military chaplains. The Army Chaplains' Department had been formed in 1796 (hitherto chaplains had been a regimental responsibility), but it was not until the Great War, through the circumstances of static warfare and the huge expansion of the army, that chaplains became a prominent and significant factor in the maintenance of morale (and therefore in fighting power). They were not always well regarded, however. Robert Graves rails against them, except the Roman Catholic padres. But he probably never came across, for example, the Reverend Theodore Bayley Hardy, who won the DSO, MC and VC before dying of his wounds a few weeks before the end of the war, two days before his 55th birthday; or the Reverend Noel Mellish, VC, MC; or the Reverend W. R. F. Addison VC. The trouble was that many Anglican chaplains – who of course ministered to everyone except those who declared themselves RC, non-conformist, Jewish, etc. – went to the front on a year's contract full of missionary zeal and, finding the captive congregations unresponsive to evangelization, became disillusioned. But as the war went on, padres found their feet – aided by a certain sense that 'there are no atheists in an artillery bombardment'. And by 1939 the department* ('royalled' in 1919) had a very clear idea of how to prepare its chaplains, and how to deploy them.

---

* A 'department', not a corps, service or regiment, because, uniquely, the RAChD has no 'other ranks' (all chaplains have officer status) nor do they carry arms.

Interestingly, when the Berlin Wall came down and the former Eastern Bloc armies began 'democratizing', the RAChD's advice was much in demand by their new chaplains' departments. Every major unit (infantry battalion, artillery regiment, etc.) deploying on operations today takes a chaplain with it. The traditionally Catholic ones – the Irish Guards, for example – take an RC padre; but cover for every denomination is assiduously arranged. Every soldier likes to feel there is someone official to 'do the honours' if he or his comrades are killed in action.

The First World War, if it did not quite break the taboo on women in the army, dented it considerably. Besides the Imperial Nursing Service and various auxiliary nursing organizations, in February 1917 the Women's Auxiliary Army Corps was established, eventually employing 57,000 volunteers on lines-of-communication work (at home and in France): cooking, storekeeping, clerical work, telephony and administration, printing, vehicle maintenance. The earliest to wear khaki, however (albeit unofficially), were grooms attached to the Army remount depots in Britain. They wore (khaki) breeches and rode astride – to the dismay of the respectable side-saddle riders – but tied their hair up with gaily coloured scarves to emphasize their independence. There is a fine painting by Lucy Kemp-Welch of these 'gals' exercising remounts on permanent display at the Imperial War Museum. The WAAC was disbanded after the war, but was re-formed as the Auxiliary Territorial Service in 1938, after which women in khaki became commonplace except in the 'combat zone'. The Queen is the ATS's most illustrious former officer. Today, to a degree, women are making up the shortfall in the support arms, but the army's policy is firmly that women should not take part in *direct* combat: the consensus is that were they to do so this would be to 'cross a line' (as well as opening a can of practical worms for no appreciable gain). Not that this refers to the notion of *front* line, for in modern war and 'operations other than war' (or 'hybrid war', a mixture of the two) the term 'front line' is imprecise; rather, it refers to the idea that women and men are different. So far there are few equal-rights voices arguing that war and women have changed *that* much.

Conscription ended immediately with the Armistice. The problem of demobilization was enormous, as was the reshaping of the army that was to remain. Most new infantry battalions had been raised within existing regiments, the Northumberland Fusiliers being the most prolific, fielding fifty-one battalions. However, some new regiments had been formed, such as a fifth regiment of Foot Guards, the Welsh Guards, created in 1915 to honour the distinguished actions of the Welsh regiments in the opening battles of the war (the Royal Welch Fusiliers – the regiment of Sassoon, Graves and several other men of letters – were reportedly sounded out for conversion, but to their credit preferred to stick to their place in the line). The composition of the army had changed considerably, however. The cavalry of the BEF represented 9.28 per cent of the army, but by July 1918 was only 1.65 per cent. Infantry would also change from 64.64 per cent in 1914 to 51.25 per cent in 1918, while the Royal Engineers would increase from 5.91 per cent to 11.24 per cent in 1918.

Finally, the 'contemptible little army'. There is no documentary proof that the Kaiser ever said it. After the war he is said to have denied it, and that he would only have thought of saying 'contemptibly little'.

Chapters 24–6: The Second World War

The following statistics give an impression of the vast undertaking that D-Day was, and the vast repository of organizational experience the army possessed – possesses – as a result. On 6 June the allies landed around 156,000 troops in Normandy: the Americans 73,000, the British and

Canadians 83,115 (61,715 of them British) including 7,900 airborne troops. The landings were supported by 11,590 aircraft, flying 14,674 sorties; 127 of them were lost. In the airborne landings (on both flanks of the beaches), 2,395 aircraft and 867 gliders of the RAF and USAAF were used. The naval forces comprised 6,939 vessels: 1,213 warships, 4,126 landing ships and landing craft, 736 ancillary craft and 864 merchant vessels. By D+5 (11 June), 326,547 troops, 54,186 vehicles and 104,428 tons of supplies had been landed over the beaches. Total allied casualties on D-Day itself are estimated at 10,000, including 2,500 dead; British casualties were approximately 2,700. In addition, 100 glider pilots became casualties – either during landing or in subsequent fighting as infantrymen. Only one VC was awarded – 'for uncommon gallantry was commonplace' – to Company Serjeant-Major Stanley Hollis of the Green Howards.

British losses would undoubtedly have been greater without Hobart's 'funnies'. And his story is interesting as well as instructive. Percy Hobart, whose sister had been married to Montgomery (she died in 1937), had been commissioned into the Royal Engineers just after the Boer War, transferring to the Royal Tank Corps after the Great War. In 1934 he commanded the first permanent armoured brigade in Britain and was made Inspector Royal Tank Corps, and then in 1937 was made Deputy Director of Staff Duties (Armoured Fighting Vehicles) and later Director of Military Training as a major-general. Sent back from Egypt where in 1938 he had formed the 'Mobile Force (Egypt)', from which chrysalis 7th Armoured had emerged, he joined the Home Guard, in *Dad's Army* fashion taking the rank of lance-corporal and charge of the defences of his home village, Chipping Camden (Gloucestershire, not even a front-line county in the Home Guard). But at Churchill's instigation, after the success of the German *Blitzkrieg* had seemed to justify his unconventional ideas, he was reinstated to the

Active List and ultimately to command of the experimental 79th Armoured Division and development of the 'funnies' (by the end of the war, 79th Armoured had 7,000 vehicles farmed out to the fighting formations).

Of the many books about the several theatres of the war, some are classics; and there are more being written each year – of which Antony Beevor's *D Day: the Battle for Normandy* (2009) is the most recent of the 'big guns'. (General) David Fraser's *And We Shall Shock Them: The British Army in the Second World War* (1983) is truly essential reading. So is (Field Marshal) Lord Slim's *Defeat into Victory* (1956). *Quartered Safe Out Here* (1992) by Flashman's creator, George Macdonald Fraser, tells the same story from the infantry corporal's perspective. Correlli Barnett's *The Desert Generals* (1960) is infuriating but cannot be ignored. Jon Latimer's *Alamein* (2002) is excellent, as too is *Alamein: War Without Hate* (2002) by John Bierman and Colin Smith. (Field Marshal) Lord Carver's *The War in Italy, 1943–1945* (2002), John Keegan's and Max Hastings's books on June 1944 – *Six Armies in Normandy* (1982) and *Overlord: D Day and the Battle for Normandy* (1984) – are musts. Cornelius Ryan's *A Bridge Too Far* (1959) remains the most dramatic account of Arnhem, but that of the commander of 1st Airborne Division, Major-General Roy Urquhart, called simply *Arnhem* (1958), is also of course illuminating. So, too, is that of another who was there – *Men at Arnhem* (1986), a beautifully written study by (Colonel) Geoffrey Powell. And for a more recent and different perspective there is (Colonel) Robert Kershaw's *It Never Snows in September: The German View of Market Garden and the Battle of Arnhem* (1996). (Lieutenant-General Sir) Brian Horrocks was not to everyone's taste (a master publicist, he – played to a T in the film *A Bridge Too Far* by Edward Fox) but his *Corps Commander* (1977) is a spirited account of much of the fighting from Normandy to the Baltic. And Max Hastings's *Armageddon:*

The Battle for Germany 1944–45 (2004) from the failure to make a bridgehead on the Rhine at Arnhem to the final German capitulation is monumental.

Chapters 27–32: Post-war

An innovation which might have helped the Glosters at Imjin was the self-loading rifle (SLR) which entered service in the late 1950s. Sergeant-Instructor Snoxall's (1914) record of thirty-eight rounds in a 12-inch target at 300 yards in one minute with the bolt-action Lee–Enfield (see chapter 20) was dazzlingly greater than the national serviceman's average at the time of the Korean War. The Royal Ordnance Factory (Enfield) variant of the Belgian *Fabrique Nationale* semi-automatic rifle, known to soldiers simply as the 'SLR' (as the old rifle was known simply as the '303'), used the principle of the machine gun to gas-feed rounds into the chamber from a magazine of twenty: each time the trigger was squeezed a round was fired, the empty case was ejected and a new round was fed automatically into the chamber. The rifleman's aim was scarcely broken. The 7.62mm calibre round was very powerful, and although both weapon and ammunition were heavy the SLR was a superb 'defensive' rifle. In 1987 it was replaced by the SA80, a much lighter – both weapon and ammunition (5.56mm) – 'assault rifle', whose shorter length, reduced weight and thirty-round magazine makes it much handier in fire and movement. It is this weapon that is the infantry's mainstay in Afghanistan today.

The prolific Max Hastings's *The Korean War* (1987) is the best all-round account of its subject. Indeed the war is otherwise sparsely covered, although Anthony Farrar-Hockley's *The Edge of the Sword* (1954) is truly inspirational. *War of the Running Dogs: The Malayan Emergency, 1948–1960* (1971) by Noel Barber is the best in its field still, although now a little dated. *Gangs and Counter Gangs* (1960) by (General

Sir) Frank Kitson, a study in counter-insurgency, ranges over the British experience from Palestine to Kenya and beyond. Robert Jackson's *The Malayan Emergency and Indonesian Confrontation: The Commonwealth's Wars 1948–1966* (2008) is recommended, as is (with affection for my old teacher) *Jungle Warfare: Experiences and Encounters* (2008) by (Colonel) John Cross. And another book by one of my former superiors, (Major-General) Tony Jeapes's *SAS Operation Oman* (1980), is authoritative, gripping and wonderfully well written. *Brush Fire Wars* (1987) by (Colonel) Michael Dewar is a masterly and succinct overview of the British army's post-war experience east of Suez.

The Falklands War has inspired many books, but sadly few of them are what would conventionally be called 'good reads' except (again) Max Hastings's first-hand account *The Battle for the Falklands* (1983); *Above All, Courage: The Eyewitness History of the Falklands War* (1985), by Max Arthur; *Eyewitness Falklands* (1982) by Robert Fox; and *Forgotten Voices of the Falklands* (2007) by Hugh McManners, the Reverend Professor John McManners' son, who was a commando artillery officer during the war.

Of the most recent operations, I beg leave to omit any recommendation for Northern Ireland. The first Gulf War (Operation Desert Storm) is brought to life on the ground by the commander of 7th Armoured Brigade (Patrick Cordingley) *In the Eye of the Storm* (1996), and at the campaign level by (Lieutenant-General Sir) Peter de la Billière's *Storm Command* (1995). The Balkans – dull but important – are well covered in two books by British officers. First is (General Sir) Michael Rose's *Fighting for Peace* (1994). Although it received some excoriating reviews (perhaps the worst by Dr Noel Malcolm of Oxford, though Malcolm might be said to be *parti pris*) it is the view of the commander of UNPROFOR in 1994. *A Cold War: Front-line*

Operations in Bosnia 1995–1996 (2008) by (Brigadier) Ben Barry is a very professional account of operations at battalion level. The Iraq War is still a little close for definitive accounts, but John Keegan's short book by that title (2004) sets the scene. The insurgency in Basra is another story – or several stories. The same is true of Afghanistan, although Patrick Bishop's *3 PARA: Afghanistan Summer 2006* (2007) is a stirring account of what the Paras do best – fighting off attacks from all sides deep in 'enemy' territory – while Richard Doherty's *Helmand Mission* (2009) is a fine portrait of a line battalion (The Royal Irish) on campaign.

Acknowledgements

I have many former and serving officers and soldiers to thank for their help and influence, direct and indirect, in the writing of this book – a good deal of them on the basis of strict confidentiality – and in the subsequent correction of errors and 'infelicities'. Likewise I owe thanks to some distinguished names from the academic community, as well as curators, archivists and librarians of regimental museums and book depositories. I trust that those who have not expressly asked me *not* to mention them by name will in the interests of decorum accept my unspecific gratitude here, in addition to my separate personal thanks.

There are, however, some names that should be specified, for in a real sense their work reflects something of what the British army has become in the affection of the public. The former chief of the general staff, Sir Richard Dannatt, as I have explained in the book, called for shows of support for the troops returning from operations, and I believe that my publishers have in their own way responded to that call. There has been a very tangible sense of going the extra mile in the production of *The Making of the British Army*, and it is clear to me that this is out of respect for the institution which for all its faults has served the nation well over the centuries, but especially for the soldiers at duty today in Afghanistan – and until recently in Iraq – at constant risk of returning home on stretchers or in coffins.

This has been a demanding book to bring to production, especially against tight deadlines, so I should like to thank Katrina Whone for orchestrating things so tightly but sympathetically, Gillian Somerscales for the most assiduous copy editing, Elizabeth Dobson for her diligent proof-reading, Sheila Lee for inspired picture research, Phil Lord for the soldierly maps, Stephen Mulcahey for a rousing and most appropriate jacket, John Noble for his Herculean labours with the index; but especially Selina Walker, my editor, who first recognized that this was a story to be told, and who then patiently worked to ensure that it was told in a way that did justice to its subject and to its readers.

Picture Acknowledgements

Line drawings

p. 22: Detail from *Instructions and Demonstration of Postures for Musketeers and Pikemen*, engraving by Thomas Cockson, 1635: private collection/Bridgeman Art Library.

p. 392: Trench diagram from *British Trench Warfare 1917–1918: A Reference Manual*, General Staff, War Office.

Colour sections

First section

p. 1: *Oliver Cromwell*, unfinished miniature by Samuel Cooper, c.1657: private collection/Bridgeman Art Library; detail of *Plan of the Battle of Naseby, 14th June 1645*, vellum, 17th century: British Library, London /© British Library Board. All Rights Reserved/Bridgeman Art Library.

pp. 2–3: *Main image: The Battle of Blenheim on 13th August 1704*, by John Wootton, *c.*1743: National Army Museum, London/acquired with assistance of National Art Collections Fund/Bridgeman Art Library. *Top left: King George II at the Battle of Dettingen, with the Duke of Cumberland and Robert, 4th Earl of Holderness, 27th June 1743*, by John Wootton, *c.*1743: National Army Museum, London/acquired with assistance of National Art Collections Fund/Bridgeman Art Library. *Top right*: detail of *An Incident in the Rebellion of*

Second section

p. 1: 'Officers and Men of the 3rd (East Kent) Regiment of Foot (The Buffs)', from an album of fifty-two photographs associated with the Crimean War by Roger Fenton, 1855: National Army Museum, London/Bridgeman Art Library.

pp. 2–3: *Top left: Charge of the Queen's Bays, 1859*, by Harry Payne (1868–1940): Regimental Museum 1st The Queen's Dragoon Guards; *top right: Candahar: The 92nd Highlanders and the 2nd Gurkhas Storming Gaudi Mullah Sahibdad, 1880*, by Richard Caton Woodville, 1881: private collection/© Christie's Images/Bridgeman Art Library; *below left: The Defence of Rorke's Drift* by Alphonse Marie de Neuville, 1880: Art Gallery of New South Wales/Bridgeman Art Library; *below right:Charge of the 21st Lancers at Omdurman, 2nd September 1898*, after Richard Caton Woodville, 1898: private collection/The Bridgeman Art Library.

pp. 4–5: *Top left*: Albert Knowles being sworn in on enlisting with the 'Queen's Pals' at Leeds Town Hall on the outbreak of the First World War: Imperial War Museum Q 111825; *centre left*: trench at La Boutillerie, 1917: Imperial War Museum Q 49104; *below left*: British tank on the Western front, 1917: © akg-images/Alamy. *Right, main image*: British airborne troops of the 1st Airlanding Reconnaissance Squadron landing, Arnhem, 17 September 1944: Imperial War Museum BU 1163. *Right, from top*: British troops line up on the beach at Dunkirk, 26–9 May, 1940: Imperial War Museum NYP 68075; a Chindit column crossing a river in Burma, 1943: Imperial War Museum IND 2290; a German soldier surrenders at Alamein, 1942: © Lordprice Collection/Alamy; commandos of the 1st Special Service Brigade landing at La Brèche, 6 June 1944, with tanks of 13/18 Hussars: Imperial War Museum B 5103.

pp. 5–7: *Left, from top*: Revd 'Sam' Davies taking Sunday service for the 1st Battalion the Gloucestershire Regiment, Korea, April 1951: image courtesy of the Soldiers of Gloucestershire Museum, www.glosters.org.uk; Operation Motorman, Londonderry, 1972: Press Association Images; 5th Infantry Brigade lands at San Carlos, Falkland Islands, 2 June 1982: Imperial War Museum FKD 347; British Sultan command post vehicle, Gulf War, 1991: Sipa Press/Rex Features. *Right, top*: a Chinook helicopter comes in to the landing zone at Nowzad, Afghanistan, 31 July 2006: photograph by Cpl Mike Fletcher © Crown Copyright/MOD, image from www.photos.mod.uk, reproduced with the permission of the Controller of Her Majesty's Stationery Office; *inset*: riflemen from the Royal Green Jackets check for roadside bombs, Basra, Iraq, January 2007: Rex Features; *below (left)*: gunners from the 19th Royal Artillery Regiment fire 105mm light guns, Upper Gereshk Valley, Afghanistan, 26 July 2007: © Cpl Jon Bevan/epa/Corbis; *below (right)*: British soldier on patrol in Musa Qala, Helmand province, Afghanistan, 27 March 2009: © Omar Sobhani/Reuters/Corbis.

p. 8: *Main image*: coffins of Gurkha Corporal Kumar Pun, Royal Military Police Sergeant Ben Ross, Rifleman Adrian Sheldon and Corporal Sean Binnie are repatriated through the Wiltshire village of Wootton Bassett, 13 May 2009: SWNS; *inset*: 450 men from 29 Commando Regiment Royal Artillery march through Plymouth, Devon, 17 April 2009: SWNS.

Index

Allan Mallinson is a former infantry and cavalry officer of thirty-five years' service worldwide.

He is the author of the Matthew Hervey series, and of *Light Dragoons*, a history of four regiments of British Cavalry, one of which he commanded.

He writes on defence matters for *The Times* and the *Daily Telegraph*, is a BBC contributor, and is a regular reviewer for *The Times*, the *Spectator* and the *Literary Review*.

On His Majesty's Service

Allan Mallinson

**In the Eastern Balkans, Matthew Hervey faces
bloody war with the Turks.**

JANUARY 1829: GEORGE IV is on the throne, Wellington is England's prime minister, and snow is falling thickly on the London streets as Lieutenant-Colonel Matthew Hervey is summoned to the Horse Guards in the expectation of command of his regiment, the 6th Light Dragoons.

But the benefits of long-term peace at home mean cuts in the army, and Hervey is told that the Sixth are to be reduced to a single squadron. With his long-term plans in disarray, he undertakes instead a six-month assignment as an observer with the Russian army, an undertaking at the personal request of the commander-in-chief, Lord Hill.

Soon Hervey, his friend Edward Fairbrother and his faithful groom, Private Johnson, are sailing north to St Petersburg, and from there on to the Eastern Balkans, seat of the ferocious war between Russia and the Ottoman Empire.

Hervey is meant to be an impartial spectator in the campaign, but soon the circumstances – and his own nature – propel him into a more active role. In the climactic Battle of Kulewtscha, in which more troops were engaged than in any battle since Waterloo, Hervey and Fairbrother find themselves in the thick of the action.

For Hervey, the stakes have never been higher – or more personal.

9780593058169

The Ever-changing Face of War

Left: Padre Sam Davies of the Gloucestershire Regiment conducts a drumhead service before the battle of Imjin during the Korean War (1950–3) – fighting of a type familiar to any who had fought in the Second World War.

Right: Armoured cars and machine guns in the streets of Britain. The long war with the IRA in Northern Ireland – 'war among the people' – kept the army's blade sharp at a time of post-colonial peace and the 'Cold War' in Germany.

Above: War in its old familiar form came out of the blue in 1982 – in the Falklands, where, as in Burma, the battle was first with the terrain and the elements.

Right: War which would have been familiar to Montgomery's 8th Army – in the desert in the wake of Saddam Hussein's invasion of Kuwait at the end of 1990.

Left: After twenty years of post-war neglect, the British army was hopelessly wrong-footed in France and Belgium in 1940, to be saved only by a Corunna-like evacuation by the Royal Navy from, principally, Dunkirk.

Above: War in the Far East was a battle as much with climate and terrain as with the Japanese.

Above: The turning point in military fortunes and self-esteem came at Alamein in October 1942.

Right: In more ways than one Alamein paved the way for the opening of the 'Second Front': in June 1944 British, Canadian and US forces landed in irresistible strength in Normandy to begin the long slog to the Rhine and beyond.

Main picture: In trying to 'bounce' a Rhine crossing at Arnhem, British airborne troops wrote the definitive chapter of 'far, fast and without question' in the handbook of war – an example which underwrote the courage of the Falklands War, and underwrites that in Afghanistan today.

War on an Industrial Scale

Right: The patriotic surge of volunteers in 1914, and later conscription, brought vast numbers into the army. 'Men of good family' here answer the civic call to enlist, aided by Kitchener's famous recruiting poster, 'Your Country Needs You!'

Above: The war soon settled into a vast stalemate which consumed millions of artillery shells and the lives of men in hitherto unimagined numbers.

Left: The technological innovations of aeroplane and tank, unproven even by the Armistice, led afterwards to false promises of quick and cheap victory in any future war.

Above: the 21st Lancers at Omdurman during the suppression of the Mahdist revolt in the Sudan.

Left: the 24th (2nd Warwickshire) Foot at Rorke's Drift in the Anglo-Zulu War.

The Price of Empire

'*We must, with our Indian empire and large Colonies, be <u>Prepared for attacks and wars, somewhere or other, CONTINUALLY</u>*' wrote Queen Victoria to her prime minister, Gladstone, with emphatic capitals and underlinings. And Victoria's wars were a gift for artists.

Right: The 2nd Dragoon Guards at Lucknow (Indian Mutiny).

Far right: The 92nd (Gordon) Highlanders at the storming of Kandahar in the Second Afghan War.

It Beat the French!

The British soldier in the Crimea was virtually unchanged in appearance from Wellington's day. Some even carried the 'Brown Bess' musket with which their predecessors had fought at Waterloo forty years earlier (though the percussion cap had at least replaced the flint as the firing mechanism). The men of the 3rd (East Kent) Regiment of Foot (The Buffs) shown in this pioneering photograph by Roger Fenton, however, are equipped with the newly issued Minié rifled musket, with its much greater range and accuracy; but they will still fire volleys shoulder to shoulder as at Waterloo.

Homecoming

'The soldier's trade, verily and essentially, is not slaying, but being slain. And the reason the world honours the soldier is that he holds his life at the service of the State.' John Ruskin

Above: The 'unfinished business' of the First Gulf War: back to 'war among the people' again in Iraq, 2003–9.

Left and *below*: The curious hybrid that is the war in Afghanistan: helicopters and artillery, but finally a battle for the support of the civil population, where the infantryman (here a Gurkha) on his feet is the best resource – in the vernacular, 'boots on the ground'.

Left: At Dettingen in 1743, under George II's personal command (the last time a British monarch would take to the field), the army won a bruising victory. *Right:* At Culloden in 1746 George's younger son, the duke of Cumberland, defeated the French-backed 'Bonnie Prince Charlie' in an often-overlooked victory at the climax of a finely calculated campaign, and in doing so put paid to the Jacobite cause once and for all.

The Two Impostors: Triumph and Disaster

In 1759, with Wolfe's capture of Quebec from the French, a victory that cost him his life (*below*), the British grip on North America was complete. By 1777 the grip had been fatally loosened: the surrender of Sir John Burgoyne to American revolutionary troops at Saratoga (*top right*) was the beginning of the end for George III's colonies south of the Great Lakes. The legacy for the army would be two decades of neglect in which its stock would fall to an all-time low.

Throughout the eighteenth century, while every other power in Europe had conscription, the army continued to rely on volunteer recruits, who by some miraculous process were turned from bumpkin or petty felon (*left*) into fine figures of men such as Coldstreamers parading at the Horse Guards (*far left*) or light dragoons (*below*) – if often only in the artist's eye.

The Fightback

The army's first taste of victory against that of revolutionary France came in 1801 at the battle of Aboukir at the mouth of the Nile (*left*), in which Major-General John Moore emerged as a capable commander and a coming man.

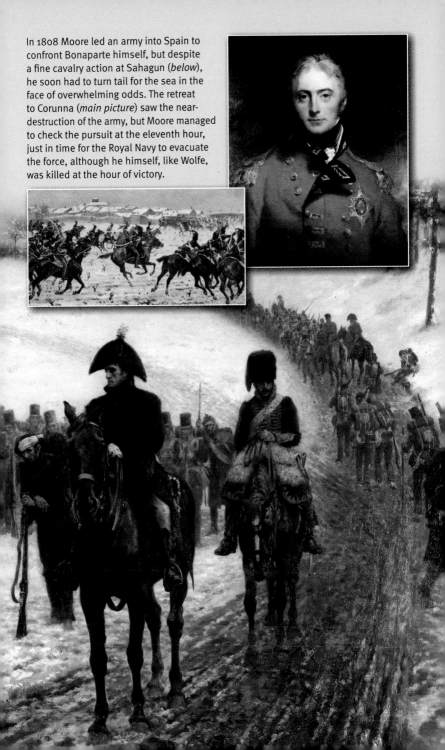

In 1808 Moore led an army into Spain to confront Bonaparte himself, but despite a fine cavalry action at Sahagun (*below*), he soon had to turn tail for the sea in the face of overwhelming odds. The retreat to Corunna (*main picture*) saw the near-destruction of the army, but Moore managed to check the pursuit at the eleventh hour, just in time for the Royal Navy to evacuate the force, although he himself, like Wolfe, was killed at the hour of victory.

The Fruits of Perseverance

'What is truly admirable in the battle of Waterloo is England, English firmness, English resolution, English blood. The superb thing which England had there – may it not displease her – is herself; it is not her captain, it is her army.' Victor Hugo's encomium in *Les Misérables* misses the point, however, that the duke of Wellington was the epitome of firmness and resolution. His example remains a powerful one in the army of today.

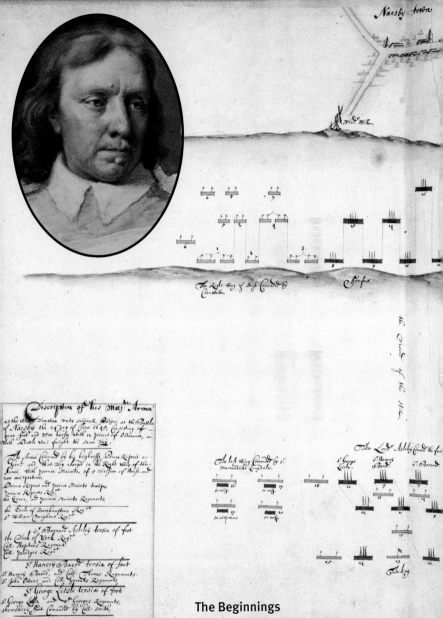

The Beginnings

Oliver Cromwell (*inset*), who saw in the débâcle of Edgehill, the first battle of the Civil War, the need for a proper military system. The New Model Army was eventually raised, whose right wing of cavalry he would command at its first and decisive battle – Naseby, in June 1645.

The Army Comes of Age

Marlborough's spectacular victory at Blenheim in 1704 would change the course of the eighteenth century: the redcoat was now a force to be reckoned with in Europe. But war with France would not end with 'Corporal John's' triumphs: the century which had opened with war against the 'Sun King' would close with war against Bonaparte. And in between there would be fighting on three continents, with resounding defeats as well as laudable victories.